MATHEMATICAL MODELS of INFORMATION and STOCHASTIC SYSTEMS

MATHEMATICAL MODELS of INFORMATION and STOCHASTIC SYSTEMS

PHILIPP KORNREICH

CRC Press
Taylor & Francis Group
Boca Raton London New York

CRC Press is an imprint of the
Taylor & Francis Group, an **informa** business

CRC Press
Taylor & Francis Group
6000 Broken Sound Parkway NW, Suite 300
Boca Raton, FL 33487-2742

Printed in the United States of America on acid-free paper
10 9 8 7 6 5 4 3 2 1

International Standard Book Number-13: 978-1-4200-5883-3 (Hardcover)

Library of Congress Cataloging-in-Publication Data

Kornreich, Philipp.
Mathematical models of information and stochastic systems / Philipp Kornreich.
p. cm.
Includes bibliographical references and index.
ISBN 978-1-4200-5883-3 (alk. paper)
1. Stochastic systems--Mathematical models. 2. System analysis--Mathematical models. I. Title.

QA274.2.K68 2008
003'.76015118--dc22 2007030349

Visit the Taylor & Francis Web site at
http://www.taylorandfrancis.com

and the CRC Press Web site at
http://www.crcpress.com

Contents

1 Introduction

1.1 HISTORICAL DEVELOPMENT AND ASPECTS OF PROBABILITY THEORY

By considering the past the future can be cast.

Probability theory was developed for the purpose of predicting the future on the basis of some **knowledge** of the present and the past. In this text the amount of information known about a system is used to derive its probabilistic properties. Every time one asserts that a particular event has a certain chance of occurring, one is trying to predict the future. Indeed, predicting the future goes way back in history to the earliest civilizations. Ancient rulers had soothsayers and other fortunetellers. National leaders continue the practice to this day. Of course, soothsayers are called pollsters and economists in the 21st century. As will be shown, how much is known about a system will play an important role in the mathematical models that will be used to foretell the future.

Technology is the application of the mathematical models of nature developed by humans. All these mathematical models of nature such as Newton's classical mechanics, Maxwell's electromagnetic theory, Einstein's general theory of relativity, chemistry, quantum mechanics, quantum electrodynamics, etc., are only valid for limited ranges of natural phenomena. To this day, no general field theory has been developed to completely explain all of nature. Perhaps this is not possible. Therefore, no natural phenomenon can be predicted nor can a machine designed with absolute certainty. Only love, beliefs, and, especially, hate can be expressed with absolute certainty.[1]

For example, a lever, as shown in Figure 1.1, consists of atoms that probably do not instantaneously follow its average motion. This stretches the more or less elastic bonds between atoms and sets the atoms vibrating. The vibration diverts some of the energy applied to tilting the lever. The atoms consist of electrons and nuclei, the nuclei consist of protons and neutrons, etc. Each particle does not instantaneously follow the average motion of the lever. Thus, the classical mechanics model of nature describing the motion of an idealized rigid lever is only an approximation. Because most of these effects are small, the classical mechanics model of nature is quite useful.

In the 14th century, William of Ockham* stated in Latin "*Pluralitas non est poneda sine neccesitate,*" which translates as "Entities should not be multiplied unnecessarily." This is known as *Occam's razor*. It has generally been interpreted as follows: if several theories are postulated to explain a certain physical phenomenon, the simpler one is probably correct. It is also used to eliminate unnecessary additions

* William of Ockham was born in 1280 in Ockham England, joined the Franciscans, taught philosophy in Paris, fought with Pope John XXII, and died April 10, 1349 in Munich Germany.

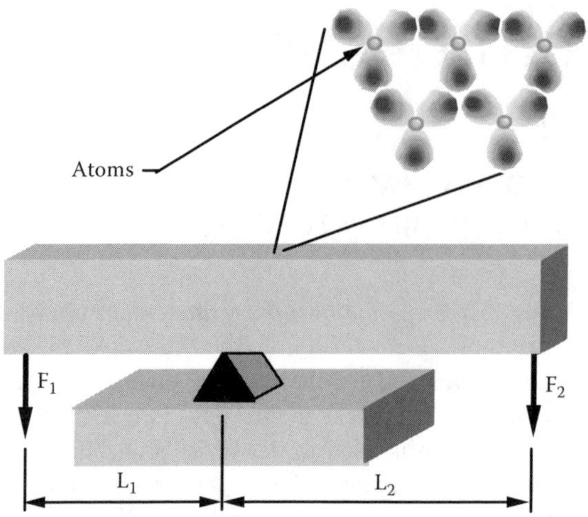

FIGURE 1.1 A lever consists of atoms that do not instantaneously follow its average motion. This stretches the more or less elastic bonds between atoms. It sets the atoms vibrating, which diverts some of the energy applied to tilting the lever. Thus, the classical mechanics model of nature describing the motion of an idealized rigid lever is only an approximation.

to scientific theories that have nothing to do with the physical phenomenon. In this text, too, this principle is used to select the simplest and most correct models.

Ancient soothsayers and astrologers tried empirically to deduce correlations between various natural phenomena such as the appearance of certain constellations in the sky, lines on people's palms, or patterns on tea leaves and various human events. They would notice that whenever a particular constellation appeared in the sky, the king would be victorious in battle. Most of these correlations were, of course, just accidental. Based on these observations, they would counsel the king when to go to war. Perhaps, if they were wrong the king would not survive to come back and chop their heads off. The problem with the ancient fortunetellers was that they used too few events to obtain their correlations. Suppose the king was successful in battle thrice under a certain constellation. The ancient soothsayers would have thought of this as a very good correlation on which future predictions could be based. Actually, the probability of the king being successful in battle thrice under the same heavenly sign is not that small even though the constellations probably did not have anything to do with the king's fortunes. Indeed, the probability of success is of the order of ⅛. This kind of foretelling survives to this day in the practice of astrology.

For example, a solid contains of the order of 10^{24} atoms per cm^3. The atoms execute small random motions about their equilibrium positions. The internal heat energy of the solid depends on these motions. Because of the very large number of atoms, one is able to predict the internal heat energy of the solid with a large degree of certainty. Thermodynamics, a science that has nothing whatsoever to do with dynamics, is based on the statistics of a very large number of particles. Thermodynamic parameters such as heat energy, pressure, entropy, etc., are average values of the motion of these particles. Other large stochastic systems such as words occurring in a text, the arrangements of nucleotides in DNA molecules, or the values of stocks

FIGURE 1.2 There is a small but finite probability that the thermal motion of most of the atoms in a chair will simultaneous be in the upward direction, causing a chair to jump into the air.

on the stock market can also be described by various macroscopic parameters. Indeed, a system is often described by various average "macroscopic" parameters rather than by the probabilities of individual events occurring.

There is an exceedingly small but finite probability that the majority of atoms in a chair may at some instant of time simultaneously move in an upward direction and the chair will spontaneously jump into the air, as shown in Figure 1.2. As discussed in Chapter 9, there is a practical problem in which such a small probability is important in current technology.

The probability of any given event occurring, such as a particular number of people voting for some candidate in an election, is one way of analyzing a set of events. Another way would be to compare the events of some random system that is to be analyzed to the events of a random set whose properties are very familiar. For example, the probability of, say, a candidate receiving exactly 32,768 votes is as random as, say, a set of 15 coins, or the probability of the same candidate receiving exactly 524,288 votes is as random as a set of 19 coins, etc.; that is, the candidate has a randomness of 15 coins to receive 32,768 votes and a randomness of 19 coins to receive 524,288 votes. The randomness values are just as good as the probabilities in describing the probabilistic properties of a system. One can calculate the average value of the randomness associated with each event. It is known as the *average randomness*, or *entropy*. This is a very useful quantity, as will be shown in the text.

There is an average randomness associated in every system with a large number of components, be they electrons, molecules, ants, reindeer, or people. For example, it is impossible to reduce the "waste" associated with the average randomness of the activities of people below a certain minimum value in a large organization such as a large corporation or a government. Politicians devote a considerable time to this

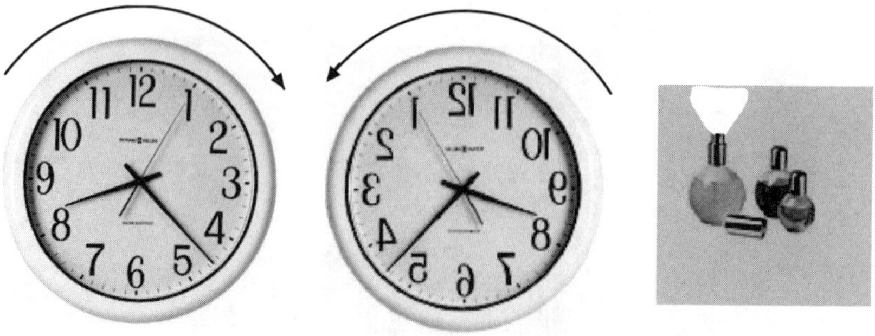

FIGURE 1.3 The mathematical model used to describe the motion of a mechanical clock can, without difficulty, be made to run either forward or backward. Indeed, clocks in barber shops that are meant to be viewed in a mirror run backward. However, the probability that the scent molecules escaping from a perfume bottle will simultaneously assemble back into the bottle is exceedingly small.

futile task. This is similar to the activity of alchemists in the 16th century trying to make gold out of lead.

It has been experimentally observed that the world is becoming more random as time progresses. The fact that a unique direction of time is observed itself might be a consequence of the increasing randomness of nature. Most mathematical models of physical systems work equally well for time that goes forward as well as for time that runs backward. Indeed, classical mechanics, a mathematical model that describes macroscopic phenomena that are readily observable by humans without the aid of any devices, works equally well for forward- or reverse-progressing time. For example, a mathematical model of a mechanical clock can, without difficulty, be made to run either forward or backward. Indeed, clocks in barber shops that are meant to be viewed in a mirror appear to us to run backward (see Figure 1.3). The real clock, its surroundings, and observers, which consist of a very large number of atomic-scale particles, progress monotonically in time.

However, the development in time of probabilistic systems seem to exhibit an arrow of time. A simple example of this is that perfume molecules escape from a perfume bottle, but are never observed to suddenly assemble and go back into the perfume bottle (see Figure 1.3). Time evolution of probabilistic systems will also be studied in this text. Time-dependent stochastic systems that at any instant of time can have any of a number of different values are known as *random processes*.

It is interesting to note that *information* and *randomness* (or entropy) are described by similar mathematical expressions. This mathematical model will be used, for example, to discuss the information transmission capabilities of languages and analyze the information content in nucleotide strings in genes.

At the time of this writing, the complete genetic information known as the genome of many animals, including humans, has been deciphered. In living organisms the genome information is stored in very large nucleotide strings. This information is now also stored in information storage media developed by humans. It can be stored in printed form, or as a recording on a compact disk, tape, or other storage medium. Perhaps in the distant future when the technology will be available to do this, a 21st century human can be reconstructed from this information.

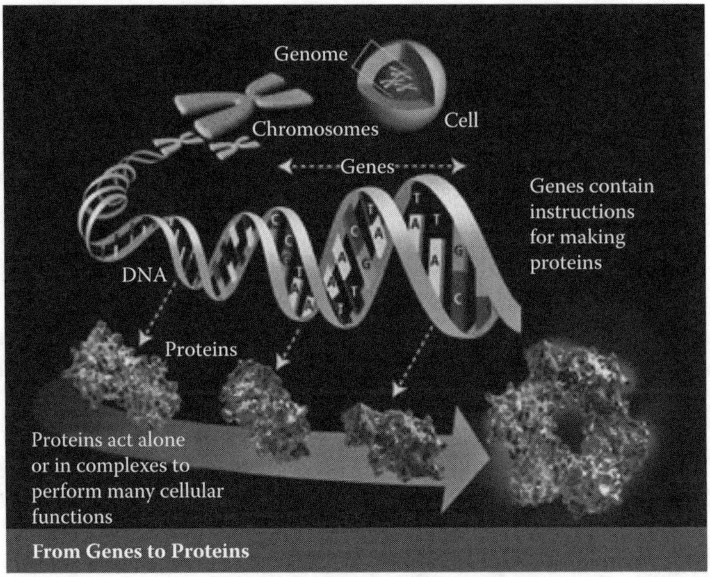

FIGURE 1.4 Nucleotide in a DNA (deoxyribonucleic acid) molecule.

The genetic code is encoded using just four nucleotide molecules: cytosine, guanine, adenine, and thymine. These are designated with the letters C, G, A, and T. The human genome is estimated to contain about 3 billion pairs of nucleotide molecules that are arranged in 20,000–25,000 genes. The nucleotide pairs are arranged end to end to form the DNA (deoxyribonucleic acid) molecule, which contains all of the genetic information. It is schematically shown in Figure 1.4. This order of the nucleotides spells out the exact instructions required to create a particular organism with its own unique traits. DNA from all organisms is made up of the same chemical and physical components.

Neither the atoms nor the nucleotide molecules in which the information to construct a living being is encoded is living matter. There probably is minimum information required for the encoded information to represent a living being. Thus, the difference between living and nonliving matter is information.

Information transmitted through a noisy channel is also a probabilistic system. The optimum rates of transmitting information through noisy channels will be calculated. A simple example of a transmission system with a noisy channel is illustrated in Figure 1.5.

1.2 DISCUSSION OF THE MATERIAL IN THIS TEXT

In Chapter 2 and the first part of Chapter 3, a small number of basic principles are developed. These are used in the rest of the text to develop various useful examples of probability theory. Two further basic concepts are introduced in Chapters 6 and 7: The concept of *macroscopic parameters* and *average values* is introduced in Chapter 6, and the concept of *randomness* is introduced in Chapter 7. These, too, are used in the subsequent text to develop various useful examples. The discussion of these examples takes up most of the pages of this text. The mathematical models discussed in these examples can be used to analyze analogous probabilistic problems. For example, the concept of systems with groups of events having the same probabilities and some prominent examples of this concept are discussed in Chapter 2.

Because discrete mathematical models are simpler, examples of discrete mathematical models are usually analyzed first. Often, the corresponding continuous probabilistic mathematical models are derived by analogy with the discrete models. In the discrete models it is assumed that the variables used have only discrete values with *no values* between them. The difference between a *zero value* and *no value* is that a zero value is a mathematical quantity subject to all the mathematical operations that other values are subject to, whereas a "no value" is not accessible to mathematical analysis. It is not part of the problem under analysis.

Average values, correlations, and covariances of functions of variables are discussed in Chapter 6. The correlations and covariance of variables are special cases of the correlations and covariances of functions of these variables. The probabilistic mathematical model of quantum mechanics is discussed in Chapter 6. This probabilistic mathematical model can also be used for other stochastic problems. No attempt is made to actually derive the physics of quantum mechanics, which follows deterministic principles.

The concept of *randomness* and the average value of the randomness or *entropy* for various systems are derived in Chapter 7.

The various discrete probabilities and continuous probability density functions are derived from what is known about a particular stochastic system in Chapter 8. Probabilities or probability densities of systems are derived for which the average randomness or entropy of the system is a maximum subject to what is known about a particular system. For example, if nothing is known about a system, all probabilities associated with it are equal; that is, all events, in this case, are equally likely to occur. Of course, if a measurement is performed on the system and more knowledge is gained about it, the probabilities or probability density functions will change. The probability functions derived include the Laplacian, Gaussian, and Poisson probabilities. A general method is given for deriving appropriate probabilities and probability density functions for any system subject to what is known about it.

It is worthwhile discussing the various tables in this text. Tables listing gamma function values and Gaussian distribution function values are given in this text. These tables date from a time before the arrival of small computers. Now the values in these tables can be obtained by software in most small computers and even in some pocket calculators. Thus, if gamma functions and Gaussian distribution functions were developed now, perhaps these tables would never have been compiled.

Chapter 9 is devoted to *information*. Both information of discrete and continuous systems are discussed. Transmission of information through noisy channels is discussed. The optimum transmission rates are derived by maximizing the information subject to what other facts are known about the system considered. This is similar to the derivation of the various probabilities and probability densities subject to what is known about a system, in Chapter 8. Effects of encryption and interesting examples are discussed. The bit error rate is discussed in Chapter 9.

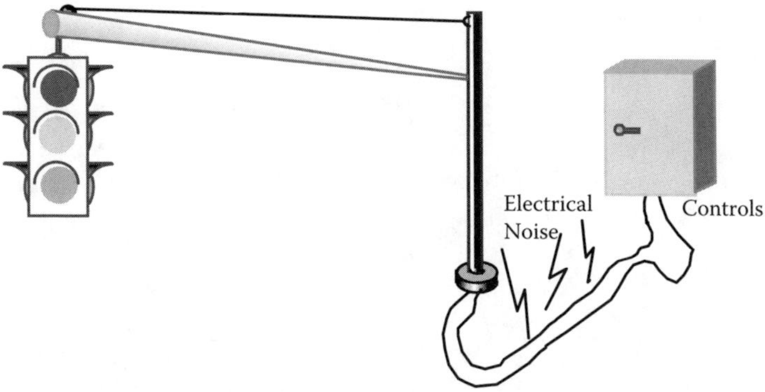

FIGURE 1.5 Information transmitted from a controller to a traffic light through a noisy channel. The messages transmitted here are "turn on the green light," "turn on the yellow light," or "turn on the red light." There is a probability that the signal to switch on the green light was transmitted and, because of the noise in the channel, a signal to switch on the yellow light was received.

Chapters 10 and 11 are devoted to the discussion of *random processes*, the discussion of time-dependent stochastic processes. Random walk problems are discussed in Chapter 10. The *correlation*, *autocorrelation*, *covariance*, and *autocovariance* functions are discussed. The degree of stationarity and time averages are discussed.

Basic data analysis is discussed in Chapter 12. A general discussion of *least square differences* is given.

Chapter 13 gives a brief discussion of chaotic systems and fractals. The Mandelbrot set is discussed in this chapter.

REFERENCE

1. Jacob Bronowski, *The Ascent of Man,* Little, Brown and Company, Boston. pp. 353–374.

2 Events and Densiy of Events

2.1 GENERAL PROBABILITY CONCEPTS

Probability theory can be derived from a few basic principles, which will be discussed in this chapter. Recall that in Chapter 1 some of the history of probability theory was discussed. It was stated that mathematical models will be used to develop forecasting techniques. To do this, the concept of a *set of events* will be introduced. A set of events S is a collection of occurrences. Consider the outcomes of a journey by airplane: the airplane could arrive early, the airplane could arrive on time, or the airplane could arrive late, as illustrated in Figure 2.1.

For example, the three outcomes of the journey by airplane are a set of events. A set can be represented mathematically by a letter such as $S_{Journey}$, where the subscript is the name of the set. This is followed by a list of the events in the set. The list of events is usually enclosed in braces, such as {early arrival, on time arrival, or late arrival}. Instead of verbally describing each event, symbols such as A_1, A_2, A_3, etc., are used to represent the events. Thus, a set of events can be expressed mathematically as $S_{Name}\{A_1, A_2, A_3, \ldots, A_n, \ldots, A_N\}$ where A_1, A_2, etc., stand for the individual events. A probability is associated with every event.

FIGURE 2.1 One current theory of the cosmos postulates that there are an infinite number of similar universes out there. Among those infinite number of universes, there must be some similar to ours. In one universe Bob arrives early by plane, in another one he arrives on time and is met by Alice, and in a third universe Alice is looking at her watch while waiting for Bob, who has yet to arrive.

$$A_1 \quad A_2 \quad A_3 \quad \ldots\ldots \quad A_n \quad \ldots\ldots \quad A_N$$

$$P_1 \quad P_2 \quad P_3 \quad \ldots\ldots \quad P_n \quad \ldots\ldots \quad P_N$$

Here, P_n is the probability that event A_n will occur. For the preceding example, the set of possible outcomes of the journey, the three P_n's, would be the probabilities that each outcome will occur; that is, P_1 is the probability that the airplane will arrive early, P_2 is the probability that the airplane will arrive on time, and P_3 is the probability that the airplane will arrive late.

The probabilities must be real and positive. This can mathematically be expressed as follows:

$$P_n = |\psi_n|^2 \tag{2.1}$$

where $|\psi_n|^2$ is the magnitude squared of the quantity ψ_n. The quantity ψ_n can be positive, negative, or complex. However, the magnitude squared of any quantity is positive and real.

One can choose various numbers to describe the probability of an event that is certain to occur. For example, one could say that there is a 100% probability that a particular event will occur. One can say that the number 1000 designates the probability of a completely certain event. However, most often, the number one (1) is chosen to describe the probability of an event that is certain to occur. Indeed, this convention will be used here. The probability of an event that is certain not to occur is equal to zero (0). Therefore, the probability of a discrete event occurring must be between zero and one, $0 < P_n \leq 1$.

The number of events discussed in this section is discrete. There is no event number 2.0001204. There are event 2 and event 3. However, the probability P_n that event n will occur can have any real value between zero and one.

Before proceeding further, it is very important that the following statements be made:

First, in this text, it is assumed that all events considered here are mutually exclusive unless otherwise stated; that is, it is assumed that if event A_3 occurred, none of the other events in the same set can occur. For example, if a coin lands on its "tail" side, it cannot simultaneously land on its "head" side. Thus, the events associated with tossing a coin, i.e., landing on its "head" and "tail," are mutually exclusive.

Second, it is assumed that at least one event in every set considered in this text will occur. Therefore, the probability that any event at all will occur is equal to one (1). For example, it is assumed that by throwing darts at a target, all darts will land somewhere on the target board but not necessarily at its center. Thus, the probability that a dart will land on the target is equal to one (1).

Next, the effect of making a measurement on a set of events on the probabilities of various events occurring is considered. The possible results of dropping a ball

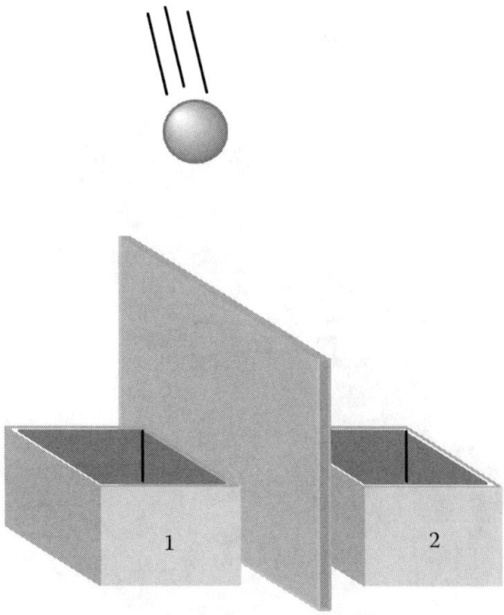

FIGURE 2.2 A ball can land in either box one or box two.

from some height into one of two boxes, as shown in Figure 2.2, are a set of two events, where it is certain that the ball will land in either box 1 or box 2. For example, these events could have the following probabilities of occurring before actually dropping the ball (Figure 2.2).

Event	**Probability**
Ball lands in box 1	$P_1 = 0.45$
Ball lands in box 2	$P_2 = 0.55$

The ball will land in one of the two boxes; say, it lands in box 1. After the ball has been observed to land in box 1, the probability P_1 that the ball will land in box 1 is equal to one, and the probability that it will land in box 2 is equal to zero.

Event	**Probability**
Ball lands in box 1	$P_1 = 1$
Ball lands in box 2	$P_2 = 0$

Thus, a person observing the ball lands in box 1, even though he cannot see box 2, will know that the ball has not landed in box 2, provided he knew that the ball could land in either box 1 or box 2. As shown later, probabilities are related to how much is known about a set of events.

The discussion now returns to a set $S\{A_1, A_2, A_3, A_4, \ldots,\}$ of N mutually exclusive events. The probability that either event A_2 or event A_4 will occur is equal to the sum of the probabilities of the two events.

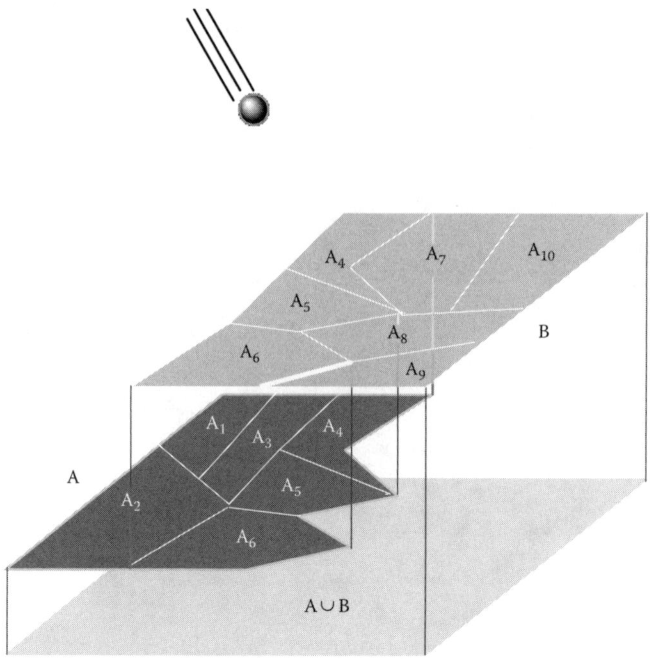

FIGURE 2.3 There is a probability P_A that the ball will strike area A. There is a probability P_B that the ball will strike area B. The probability P(A∪B) is the probability that the ball will strike either area A or B. The events of the ball striking anywhere in area A or area B are schematically illustrated here in a Venn diagram.

$$P(A_2 \cup A_4) = P_2 + P_4 \tag{2.2}$$

where the "union" sign ∪ is used to express the probability that either event A_2 or event A_4 will occur.

Two events A and B that are not mutually exclusive can be thought of as consisting of subsets of events $A\{a_1, a_2, a_3, a_4, a_5, a_6\}$ and $B\{a_4, a_5, a_6, a_7, a_8, a_9, a_{10}\}$. Note that here the events are not mutually exclusive. The union of events A and B would then be the set $(A \cup B)\{a_1, a_2, a_3, a_4, a_5, a_6, a_7, a_8, a_9, a_{10}\}$ as shown in the Venn diagram of Figure 2.3.

EXAMPLE 2.1

Consider three sets of events $S_A = \{A_1, A_2, A_3, A_4, A_5\}$, $S_B = \{A_4, A_5, A_6, A_7, A_8\}$, and $S_C = \{A_8, A_9, A_{10}, A_{11}, A_{12}\}$. There is a probability P_J associated with each of the 12 events.

A_1	A_2	A_3	A_4	A_5	A_6	A_7	A_8	A_9	A_{10}	A_{11}	A_{12}
P_1	P_2	P_3	P_4	P_5	P_6	P_7	P_8	P_9	P_{10}	P_{11}	P_{12}

Of course, the probability that any event at all of the 12 events will occur is equal to one.

$$\sum_{J=1}^{J=12} P_J = 1$$

a) The probability that either set S_A or S_B or both will occur is:

$$P(S_A \cup S_B) = P_1 + P_2 + P_3 + P_4 + P_5 + P_6 + P_7 + P_8$$

b) The probability that either set S_B or S_C or both will occur is:

$$P(S_B \cup S_C) = P_4 + P_5 + P_6 + P_7 + P_8 + P_9 + P_{10} + P_{11} + P_{12}$$

As was earlier assumed, it is certain that at least one event will occur. Also, the probability of any event at all occurring was taken to be equal to one.

$$\begin{aligned} P_{1\cup 2\cup 3\cup \ldots \cup N} &= P_1 + P_2 + P_3 + \ldots + P_N \\ 1 &= P_1 + P_2 + P_3 + \ldots + P_N \end{aligned} \tag{2.3}$$

This is actually only true for mutually exclusive events. Consider, for example, a set of events $S_A\{A_1[a_1, a_2, a_3, a_4], A_2[a_3, a_4, a_5, a_6, a_7], A_3[a_6, a_7, a_8, a_9]$, and $A_4[a_9, a_{10}, a_{11}]\}$, that are not mutually exclusive where each event consists of a number of subevents. Here, the subevents a_n are mutually exclusive. P_n is the probability that event A_n will occur, and p_n is the probability that the subevent a_n will occur. The probability that any event at all of the set of events A_n will occur is equal to one. This can be expressed mathematically in terms of the subevents as:

$$\begin{aligned} &P(S_A) = p_1 + p_2 + p_3 + p_4 + p_5 + p_6 + p_7 + p_8 + p_9 + p_{10} + p_{11} + p_{12} \\ &P(S_A) = 1 \end{aligned} \tag{2.4}$$

However, the probabilities that the events A_n will occur are:

$$\begin{aligned} &P_1 = p_1 + p_2 + p_3 + p_4, \quad P_2 = p_3 + p_4 + p_5 + p_6 + p_7 \\ &P_3 = p_6 + p_7 + p_8 + p_9, \quad P_4 = p_9 + p_{10} + p_{11} \end{aligned} \tag{2.5}$$

Therefore, in this case the sum of the probabilities P_n of the events A_n, which are not mutually exclusive, is not equal to one. Rather, the union of the probabilities of the events A_n is equal to one.

$$\bigcup_{n=1}^{n=N} P_n = 1 \tag{2.6}$$

where N is equal to 4 in this case. The four events A_1, A_2, A_3, and A_4, in this case, are not mutually exclusive.

Nevertheless, for mutually exclusive events, the sum of probabilities of all events is equal to one. Indeed, it was assumed that all events discussed in this text are mutually exclusive unless otherwise specified.

$$\sum_{\text{All Events}} P_n = 1 \tag{2.7}$$

Thus, the probabilities P_n for discrete mutually exclusive events A_n to occur must have the following properties:

1. P_n must be positive and real.
2. P_n must be between zero and one, $0 < P_n \leq 1$.
3. The probability of a completely certain event is equal to one.
4. The probability that any event at all of the N events will occur is equal to one; that is, the sum of the probabilities of all events is equal to one, $P_1 + P_1 + P_3 + , \ldots , + P_n + , \ldots , + P_N = 1$.

EXAMPLE 2.2

The probability of having a number n of consecutive sunny days in a week in December in some particular location is:

P (no sunny days)	=	0.200
P (one sunny day)	=	0.400
P (two consecutive sunny days)	=	0.200
P (three consecutive sunny days)	=	0.100
P (four consecutive sunny days)	=	0.050
P (five consecutive sunny days)	=	0.030
P (six consecutive sunny days)	=	0.012
P (seven consecutive sunny days)	=	0.008
P (any number of sunny days)	=	1.000

The probability that both events A and B will occur is equal to $P(A \cap B)$. This is known as the *joint probability*. The $\cap$ symbolizes the intersection of events A and B. The joint probability of two events A and B that are not mutually exclusive but are independent of each other is equal to the product of the probabilities of the individual events.

$$P(A \cap B) = P_A P_B \tag{2.8}$$

where P_A and P_B are the probabilities of the individual events occurring. By independent events, it is meant that, for example, event A can occur whether event B does or does not occur. For example, in tossing a coin and rolling a die, the coin can land

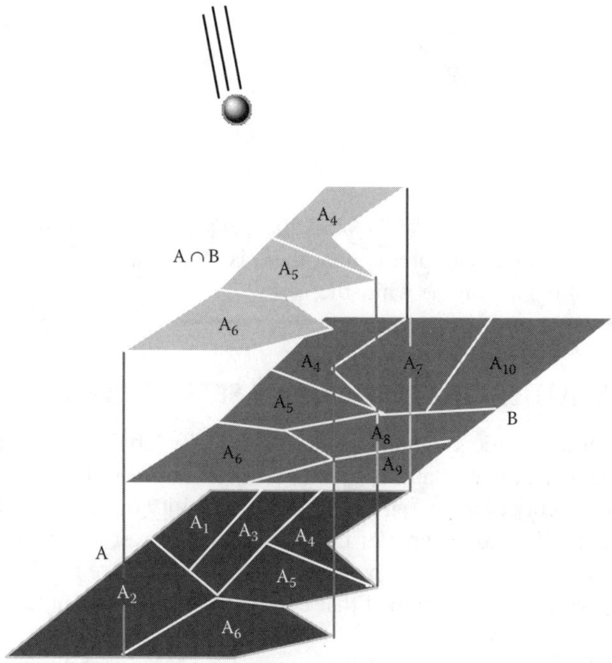

FIGURE 2.4 There is a probability P_A that the ball will strike area A. There is a probability P_B that the ball will strike area B. $P(A \cap B)$ is the probability that the ball will strike both areas A and B simultaneously. The events of the ball striking both areas A and B simultaneously are schematically illustrated here in a Venn diagram.

on either its head or tail independent of the number the die will show. The coin and die events are statistically independent. The concepts of dependent and independent probabilities of events will be developed in a later section.

Two events A and B not mutually exclusive can, again, be thought of as consisting of subsets of events $A\{a_1, a_2, a_3, a_4, a_5, a_6\}$ and $B\{a_4, a_5, a_6, a_7, a_8, a_9, a_{10}\}$. The intersection of events A and B would then be the set $(A \cap B)\{a_4, a_5, a_6\}$ as shown in the Venn diagram of Figure 2.4. Here, the upside-down U is used as the symbol of intersection.

EXAMPLE 2.3

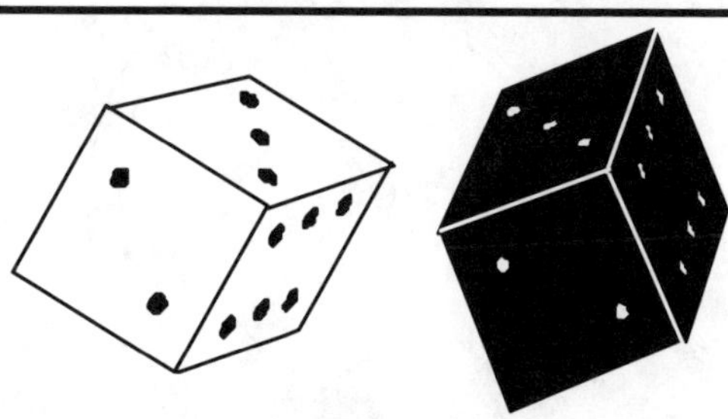

The probability of rolling any particular number on a die is equal to 1/N where N is the number of sides. Consider a white and a black die. The probability of rolling a 5 on the black die and a 2 on the white die is

$$P\left[(\text{black } 5) \cap (\text{white } 2)\right] = \frac{1}{6} \times \frac{1}{6}$$

$$= \frac{1}{36}$$

The probabilities of rolling a 5 on the black die and a 2 on the white die are not mutually exclusive. Thus, one can predict that there is a $\frac{1}{36}$ chance of rolling a 5 on the black die and rolling a 2 on the white die.

2.2 PROBABILITIES OF CONTINUOUS SETS OF EVENTS

The simplest probabilistic system is one of individually labeled events, as discussed earlier. For example, the probability that a particular number of birds will land on a wire is the probability of a *discrete event*. The probability that a shot when tossed in a shot-put event will land at some distance from the thrower is a *continuous event*; that is, the shot put can land at any point of a continuous infinite number of points on the ground. Likewise, it is assumed that continuous events are mutually exclusive, unless otherwise specified.

Consider the example of a wheel of fortune. There is a probability that the pointer of the wheel of fortune shown in Figure 2.5 can stop at any value of the angle θ. More accurately, there is a probability $p(\theta)d\theta$ that the pointer will stop at an angle between θ and $\theta + d\theta$. The angle θ can have values between zero and 2π, $0 < \theta \leq 2\pi$. Note that the probability is equal to the product of two quantities: the *probability density* $p(\theta)$ and the incremental angle $d\theta$. The quantity $p(\theta)$ can be interpreted as the probability density, the probability per unit angular distance.

Similar to the discrete case, the probability density $p(\theta)$ of a set of continuous events must also be positive and real.

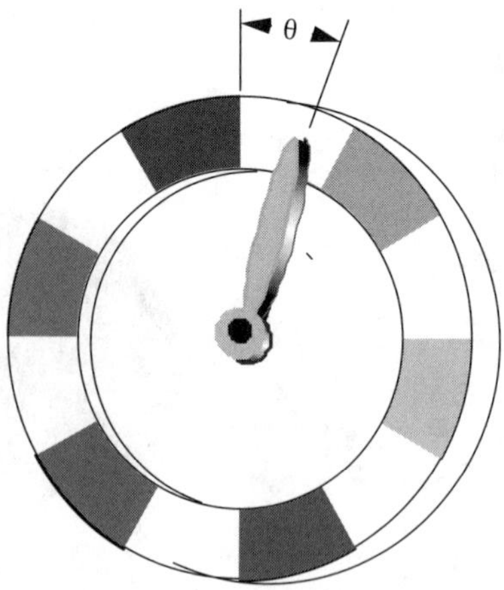

FIGURE 2.5 Wheel of fortune. The pointer can land at any angle θ.

$$p(\theta) = |\phi(\theta)|^2 \tag{2.9}$$

where $|\phi(\theta)|^2$ is the magnitude squared of the function $\phi(\theta)$. The quantity $\phi(\theta)$ can be positive, negative, or complex. However, the magnitude squared of any quantity is positive and real.

Recall, from the previous section, that the probability that either the discrete event A_J or event A_M will occur is equal to the sum of the probabilities P_J and P_M of events A_J and A_M. A similar concept exists in the continuous case. The probability $P\{\theta_A < \theta \le \theta_B\}$ that the pointer will stop in an angular region between angles θ_A and θ_B is equal to the sum of all the probabilities of the angles between θ_A and θ_B occurring. Therefore, the probability $P\{\theta_A < \theta \le \theta_B\}$ is given by the sum of the infinitesimal probabilities $p(\theta)d\theta$, the integral of the probability density $p(\theta)$ between the angles θ_A and θ_B.

$$P\{\theta_A < \theta \le \theta_B\} = \int_{\theta_A}^{\theta_B} p(\theta)\,d\theta \tag{2.10}$$

This concept will be discussed again in a later section. The probability that the pointer will land anywhere at all on the wheel of fortune is equal to one.

$$\int_0^{2\pi} p(\theta)\,d\theta = 1 \tag{2.11}$$

The integral of the probability density over the space of all events is equal to one. However, unlike the discrete probability, the probability density can be larger than one. The probability density $p(\theta)$ must have the following properties:

1. The probability density must be positive and real.
2. The probability that any event at all of the set of continuous events will occur is equal to one. In the previous example for $0 < \theta \le 2\pi$, $\int_0^{2\pi} p(\theta)\,d\theta = 1$.

In the previous example, the probability density per unit angle is considered. However, the probability density could be per unit length, per unit time, per unit area, per unit volume, or per unit energy, depending on the situation considered. Both the probabilities of a discrete set of events and the probability densities of continuous systems are used to predict the fortunes of various systems.

EXAMPLE 2.4

Consider a fair wheel of fortune where the probability that the pointer will stop between any angle θ and $\theta + d\theta$ is the same for all angles; that is, the probability density $p(\theta)$ is uniform:

$$p(\theta) = \frac{1}{2\pi}$$

The pointer of a biased wheel might be rigged to favor a particular region of the wheel. Indeed, the probability that the pointer will stop at any place at all on this wheel is equal to one:

$$\frac{1}{2\pi}\int_0^{2\pi} d\theta = 1$$

The probability that the pointer will stop between angles θ_1 and θ_2 is:

$$P\{\theta_1 \le \theta_2\} = \frac{1}{2\pi}\int_{\theta_1}^{\theta_2} d\theta$$

Thus, it is predicted that the probability of the pointer of a fair wheel stopping between angles θ_1 and θ_2 is:

$$P(\theta_1 \le \theta < \theta_2) = \frac{\theta_2 - \theta_1}{2\pi}$$

2.3 DISCRETE EVENTS HAVING THE SAME PROBABILITY

There are sets of events containing groups of events with the same probability, as for example:

Event	A_1	$A_2\ A_3$	$A_4\ A_5\ A_6$	$A_7\ A_8\ A_9$	$A_{10}\ A_{11}\ A_{12}\ A_{13}\ A_{14}$
Probability	P_1	P_2	P_3	P_4	P_5

Here, for example, events A_4, A_5, and A_6 have the same probability of occurring, P_3.

The sum of the probabilities of all the events in a set has to be equal to one, as described in the discussion of Equation 2.3. Thus, for the previous example:

$$P_1 + P_2 + P_2 + P_3 + P_3 + P_3 + P_4 + P_4 + P_4 + P_5 + P_5 + P_5 + P_5 + P_5 = 1 \tag{2.12}$$

This can be more simply written as:

$$G_1P_1 + G_2P_2 + G_3P_3 + G_4P_4 + G_5P_5 = 1 \tag{2.13}$$

where the number of events G_m is the number of events in each group of events. All events in each group have the same probability of occurrence. In this example, there is one event in group 1, there are two events in group 2, three events in group 3, three events in group 4, and five events in group 5. Thus, in this case $G_1 = 1$, $G_2 = 2$, $G_3 = 3$, $G_4 = 3$, and $G_5 = 5$. Equation 2.13 can be written more generally as:

$$\sum_{m=1}^{m=M} G_m P_m = 1 \tag{2.14}$$

where the sum is over the number of groups of events in which all the events in each group have the same probability. The total number N of events is equal to the sum of the G_m's, the number of events in each group.

$$\sum_{m=1}^{m=M} G_m = N \tag{2.15}$$

An example of particular interest where groups of events have the same probability is known as *Bernoulli trials*. Consider, first, a sphere having a small flat area on one side, as shown in Figure 2.6. Suppose the probability that it will land on this flat area is ⅕. Then, the probability that the sphere will not land on the flat area is 0.8. Next, consider four distinct spheres with flat sides: a red sphere, a green sphere, a blue sphere, and a yellow sphere, as shown in Figure 2.7.

FIGURE 2.6 Sphere with flat spots.

FIGURE 2.7 Four spheres with flat spots.

The probability that, say, the red sphere will land on its flat side and the remaining spheres will not land on their flat sides is $\frac{1}{5} \times \frac{4}{5} \times \frac{4}{5} \times \frac{4}{5} = \frac{64}{625} = 0.1024$. The probability that any one sphere will land on its flat side while the three remaining spheres do not land on their flat sides is the same as for the red, green, blue, or yellow spheres landing on their flat sides. Thus, the four events of one sphere landing on its flat side while the other spheres do not land on their flat sides have the same probability. The probability of two particular spheres landing on their flat sides while the remaining two spheres do not land on their flat sides is $\frac{1}{5} \times \frac{1}{5} \times \frac{4}{5} \times \frac{4}{5} = \frac{16}{625} = 0.0256$. There are six such events:

Blue	Green		
Blue		Yellow	
Blue			Red
	Green	Yellow	
	Green		Red
		Yellow	Red

where the colors of the spheres that land on their flat sides are listed. These six events have the same probability.

In the more general case, suppose there are M spheres with flat sides, and the probability of a sphere landing on its flat side is P. Then, the probability of the sphere landing on its round side is 1 – P. There is only one way for no sphere to land on its flat side. There are M ways for one out of the M spheres to land on its flat side while the other M – 1 spheres land on their round sides. There are $\left(M(M-1)\right)/\left(1\times 2\right)$ ways for two spheres out of the M spheres to land on their flat sides while the other M – 2 spheres land on their round sides; it was found, by counting, that there are six ways for two out of the four spheres to land on their flat sides and for the remaining two spheres to land on their round sides. Indeed, according to this formula, $\left(4(4-1)\right)/\left(1\times 2\right)=6$. There are $\left(4(4-1)(4-3)\right)/\left(1\times 2\times 3\right)$ ways for three spheres out of the M spheres to land on their flat sides while the remaining M – 4 spheres land on their round sides, etc.

Note that:

$$M(M-1)(M-2)\ldots\left[M-(K-1)\right]=\frac{M!}{(M-K)!} \tag{2.16}$$

Thus, there are $M!/\left(K!(M-K)!\right)\equiv (M/K)$ ways for K spheres out of the M spheres to land on their flat sides while the other (M – K) spheres do not land on their flat sides. The $M!/\left(K!(M-1)!\right)$ events have the same probability.

The probability that a *particular group* of K spheres out of M spheres will land on their flat sides while the remaining spheres land on their round sides is equal to $P^K(1 - P)^{M-K}$. To obtain the probability of *any* K spheres out of M spheres will land on their flat side while the remaining spheres land on their round sides, the probability that a *particular group* of K spheres out of M spheres will land on their flat sides is multiplied by the number of ways $M!/\left(K!(M-1)!\right)$ this can occur.

$$P(K \text{ out of } M) = \frac{M!}{k!(M-K)!} P^K (1-P)^{M-K} \tag{2.17}$$

Of course, the probability of any number at all of the M spheres landing on their flat sides while the remaining spheres do not land on their flat sides is equal to one.

$$1 = \sum_{K\sim0}^{K=M} \frac{M!}{K!(M-K)!} P^K (1-P)^{M-K} \tag{2.18}$$

This must be true for any values of M and P.

EXAMPLE 2.5

Consider a set of binary data consisting of 8 bits. Each bit can be a "one" or a "zero." In this case, the probability p(1) of a one arriving is equal to 0.6, whereas the probability p(0) of a zero arriving is equal to 1 – 0.6 = 0.4.

a) The probability of receiving 3 ones and 5 zeros is:

$$\frac{8!}{3!(8-3)!} 0.6^3 0.4^{(8-3)} = 0.12386304$$

b) The probability of either 3 ones and 5 zeros or 5 ones and 3 zeros arriving is:

$$\frac{8!}{3!(8-3)!} 0.6^3 0.4^{(8-3)} + \frac{8!}{5!(8-5)!} 0.6^5 0.4^{(8-5)} = 0.40255488$$

The probability of either 3 ones and 5 zeros or 5 ones and 3 zeros arriving is 0.40255488.

Equation 2.18 can also be written as follows according to Equation 2.14:

$$1 = \sum_{K=0}^{K=M} G_K P_K \tag{2.19}$$

where the probability P_K of the group K of events is:

$$P_K = P^K (1-P)^{M-K} \tag{2.20}$$

In this case, there is one event in group number 0, no balls landing on a flat side; N events in group number 1, one ball landing on a flat side; $(M(M-1))/(1\times2)$ events in group number 2, any two balls landing on their flat sides, etc. The total number of distinct events M from Equation 2.15 in this case is:

$$\sum_{K=0}^{KM} \frac{M!}{K!(M-K)!} = N \tag{2.21}$$

It is interesting to consider the case where the probability P of a sphere landing on its flat side is exactly ½; that is, the spheres have the same probabilistic properties as coins. The probability of a coin landing on its head side is ½. Note that

$$\left(\frac{1}{2}\right)^{K}\left(1-\frac{1}{2}\right)^{M-K} = \left(\frac{1}{2}\right)^{M} \tag{2.22}$$

By substituting Equation 2.22 into Equation 2.18, one obtains:

$$2^{M} = \sum_{K=0}^{K=M} \frac{M!}{K!(M-K)!} \tag{2.23}$$

This is the total *number of different events* N (as described in Equation 2.21) that can occur in this case. One distinct event of, say, rolling four spheres each with a flat spot is the following: spheres No. 1 and 3 land on their flat spots and spheres No. 2 and 4 land on their round surfaces. In this case, there are $2^4 = 16$ different such distinct events that could occur.

There is another interesting aspect to the term $M!/(K!(M-1)!)$; namely, it is equal to the polynomial coefficients resulting from multiplying out the binomial $(A + B)^M$:

$$(A+B)^{M} = \sum_{K=0}^{K=M} \frac{M!}{K!(M-K)!} A^{K} B^{M-K} \tag{2.24}$$

For example:

$$\begin{aligned}(A+B)^{3} &= \frac{3!}{0!(3-0)!}A^{0}B^{3-0} + \frac{3!}{1!(3-1)!}A^{1}B^{3-1} + \frac{3!}{2!(3-2)!}A^{2}B^{3-2} \\ &\quad + \frac{3!}{3!(3-3)!}A^{3}B^{3-3} \\ &= B^{3} + 3AB^{2} + 3A^{2}B + A^{3}\end{aligned} \tag{2.25}$$

Note that 0! is equal to one. The previous result is useful. Let us return to Equation 2.18. By using Equation 2.24, one can rewrite Equation 2.18 as follows:

$$\text{a)}\quad \sum_{K=0}^{K=M} \frac{M!}{K!(M-K)!} P^K (1-P)^{M-K} = \left[P + (1-P)\right]^M$$

$$\text{b)}\quad \sum_{K=0}^{K=M} \frac{M!}{K!(M-K)!} P^K (1-P)^{M-K} = 1 \tag{2.26}$$

Observe that the probability of any number at all of K spheres landing on their flat sides while the remaining spheres land on their round sides is equal to one, independent of the value of P.

A plot of the probability of any K out of 50 spheres, with a flat spot, landing on their flat surfaces is shown in Figure 2.8. The probability P of a single sphere landing on its flat surface is equal to 0.4. Note that this curve has a Gaussian bell–type shape. For large numbers M of spheres with flat spots or systems that behave like the spheres, one can approximate the probability of any K spheres out of M spheres landing on their flat spots by:

$$\frac{M!}{K!(M-K)!} P^K (1-P)^{M-K} \approx \frac{1}{\left[2\pi MP(1-P)\right]^{\frac{1}{2}}} - e^{-\frac{(x-MP)^2}{2MP(1-P)}} \tag{2.27}$$

Observe from Figure 2.8 that the probability of any K spheres out of M spheres landing on their flat spots has a shape approximated by the function of x shown in Equation 2.27. This is known as the *DeMoivre–Laplace theorem*. This approximation only works well for P close to 0.5. It is easier to integrate the preceding function of x than to calculate the ratio of the factorials for large values of N. A method of integrating the Gaussian exponential of Equation 2.27 is discussed in Chapter 8.

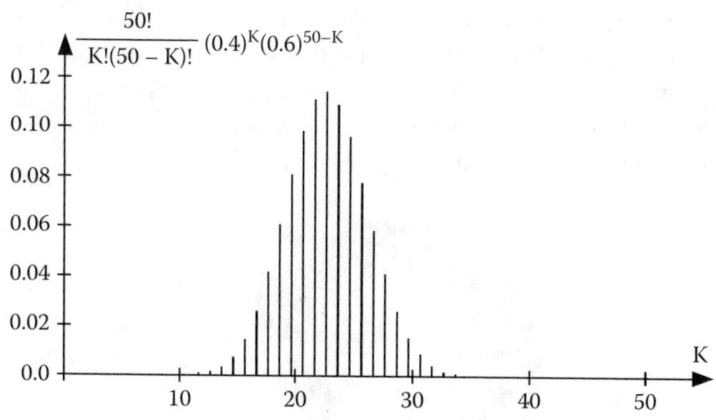

FIGURE 2.8 Plot of the probability of any K out of 50 spheres with a flat spot landing on their flat surfaces. The probability P of a single sphere landing on its flat surface, in this case, is equal to 0.4.

Two lists comparing exact values of the probability of M/2 spheres out of M spheres landing on their flat spots to the values calculated by using the approximation of Equation 2.27 is shown in the following text for two different values of the probability P of a single sphere landing on its flat spot.

For P = 0.5:

N	Exact Value	Approximation	Ratio
10	0.246094	0.252313	0.975350
20	0.176197	0.178412	0.987583
30	0.144465	0.145673	0.998170
40	0.125371	0.126157	0.993770
50	0.112275	0.112838	0.995013
60	0.102578	0.103007	0.995842
70	0.095254	0.095365	0.996434
80	0.088928	0.069206	0.996880
90	0.083871	0.084104	0.997226
100	0.079589	0.079788	0.997503

For P = 0.3:

N	Exact Value	Approximation	Ratio
10	0.102919	0.106215	0.9768970
20	0.030817	0.028977	1.0634910
30	0.010567	0.009129	1.1575820
40	0.003817	0.003050	1.2573800
50	0.001436	0.001525	1.3646440
60	0.000545	0.000371	1.4804411
70	0.000213	0.000132	1.6056834
80	0.000083	0.000048	1.7412632

Note that the approximation becomes better for larger values of the number of spheres M and for P = 0.5, as illustrated by the ratio that is approaching one. However, for P = 0.3 the approximation actually gets worse for large values of M.

Consider yet another special case, namely, the probabilities associated with spheres having more than one flat spot, as shown in Figure 2.9.

FIGURE 2.9 Sphere with multiple flat spots.

For example, consider N identical spheres, each having two flat spots. The probability that a sphere will land on its small flat spot is P_A, the probability that a sphere will land on its large flat spot is P_B, and $(1 - P_A - P_B)$ is the probability that a sphere will land on its round surface. The probability that any K spheres will land on their small flat spots, while L spheres will land on their large flat spots, and the $(N - K - L)$ remaining spheres land on their round surface is:

$$P\{K \text{ small flat, L large flat, and } (N-K-L) \text{ round}\} = G_{KL}P_A^K P_B^L (1-P_A-P_B)^{N-K-L} \tag{2.28}$$

Here, G_{KL} is the number of arrangements that will result in K spheres landing on their small flat spots, while L spheres land on their large flat spots, and $(N - K - L)$ spheres land on their round surfaces. It is the number of events with the same probability; that is, G_{KL} is the number of events in group KL. All events in group KL have the same probability, where the *trinomial coefficients* G_{KL} are

$$G_{KL} = \frac{N!}{K!L!(N-K-L)!} \tag{2.29}$$

The probability that any number at all of spheres will land on their small flat spots, while any number at all of spheres land on their large flat spots and the remaining spheres land on their round surfaces must be equal to one.

$$1 = \sum_{K=0}^{K=N} \sum_{L=0}^{L=N-K} G_{KL}P_A^K P_B^L (1-P_A-P_B)^{N-K-L} \tag{2.30}$$

The G_{KL}s are the coefficients of a trinomial.

$$(x+y+z)^N = \sum_{K=0}^{K=N} \sum_{L=0}^{L=N-K} G_{KL} x^K y^L z^{N-(K+L)} \tag{2.31}$$

Because in the stochastic case, $x = P_A$, $y = P_B$, and $z = 1 - P_A - P_B$, the trinomial of Equation 2.31 is equal to one.

$$(P_A + P_B + 1 - P_A - P_B)^N = 1 \tag{2.32}$$

The trinomial expansion in this case has another interesting property. Let x = ⅓, y = ⅓, and z = ⅓.

a) $$1 = \frac{1}{3^N} \sum_{K=0}^{K=N} \sum_{L=0}^{L=N-K} \frac{N!}{K!L!(N-K-L)!}$$

or (2.33)

b) $$3^N = \sum_{K=0}^{K=N} \sum_{L=0}^{L=N-K} \frac{N!}{K!L!(N-K-L)!}$$

This is the total *number of different events* that can occur in this case. Thus, say, by rolling six spheres each with two flat spot spheres, No. 1 and 3 land on their small flat spots and spheres, No. 2 and 4 land on their large flat spots and spheres, No. 5 and 6 land on their round surfaces. This is one distinct event. In this case there are $3^6 = 729$ different such distinct events that could occur.

This can be extended to a *multinomial* $(u + w + x + y + z + \ldots)^N$ of any size. Indeed, the Bernoulli trials are a special case of multinomials and its associate stochastic systems.

2.4 DIGRESSION OF FACTORIALS AND THE Γ FUNCTION

In the previous section, the fact that 0! (zero factorial) is equal to one was accepted without trying to explain its origin. Consider, for example, the following integral:

$$\int_0^{\infty} x^3 e^{-x} dx \tag{2.34}$$

This integral can be integrated by parts:

$$\int_0^{\infty} x^3 e^{-x} dx = -\left[x^3 e^{-x}\right]_0^{\infty} + 3\int_0^{\infty} x^2 e^{-x} dx \tag{2.35}$$

Although x^3 is equal to zero at $x = 0$ and e^{-x} is equal to zero at $x = \infty$, the term in the square bracket is equal to zero.

$$\int_0^{\infty} x^3 e^{-x} dx = 3\int_0^{\infty} x^2 e^{-x} dx \tag{2.36}$$

Integrating the integral on the left by parts again:

$$3\int_0^{\infty} x^2 e^{-x} dx = 3 \cdot 2\int_0^{\infty} x e^{-x} dx \tag{2.37}$$

Once more integrating the integral on the left by parts, for the original integral one obtains:

$$\text{a)}\quad \int_0^{\infty} x^3 e^{-x} dx = 3 \cdot 2 \cdot 1 \quad \text{or} \quad \text{b)}\quad \int_0^{\infty} x^3 e^{-x} dx = 3! \tag{2.38}$$

Equation 2.38b can be generalized:

$$\int_0^{\infty} x^K e^{-x} dx = K! \tag{2.39}$$

where K is a positive integer. Indeed, the integral of Equation 2.39 can be taken as the definition of a factorial of a positive integer K. Returning to 0!:

$$\int_0^{\infty} x^0 e^{-x} dx = 1 \tag{2.40}$$

Indeed, the integral for K equal to zero is equal to one.

Suppose, K is positive but not an integer, say, $K = \alpha$.

$$\Gamma(\alpha + 1) = \int_0^{\infty} x^{\alpha} e^{-x} dx \tag{2.41}$$

where $\Gamma(\alpha + 1)$ is the *gamma function* of $\alpha + 1$. This integral will converge for, …, $-4 < \alpha < -3$, $-3 < \alpha < -2$, $-2 < \alpha -1$, and $\alpha > -1$. It is conventional to define:

$$u \equiv \alpha + 1 \tag{2.42}$$

In this case, the gamma function integral takes the form:

$$\Gamma(u) = \int_0^{\infty} x^{u-1} e^{-x} dx \tag{2.43}$$

Next, consider $\Gamma(u + 1)$.

$$\Gamma(u+1) = \int_0^{\infty} x^{u} e^{-x} dx \tag{2.44}$$

Integrating by parts:

$$\text{a)}\quad \Gamma(u+1) = -\left[x^u e^{-x}\right]_0^{\infty} + u\int_0^{\infty} x^{u-1} e^{-x} dx \quad \text{or} \quad \text{b)}\quad \Gamma(u+1) = u\Gamma(u) \tag{2.45}$$

Equation 2.45 is a recurrence relation of the gamma function. A useful value of the gamma function is

$$\Gamma\left(\frac{1}{2}\right) = \pi^{\frac{1}{2}} \tag{2.46}$$

The recurrence relation, Equation 2.45b, and Equation 2.46 can be used to calculate $\Gamma(3\pi/2), \Gamma(5\pi/2), \Gamma(7\pi/2)$, etc. A list of gamma functions is listed in the following table. This table was compiled before the advent of personal computers. Now, the values of gamma functions can readily be obtained from most small computers and even from some hand-held calculators.

Table of Some Gamma Functions

u	Γ(u)	u	Γ(u)	u	Γ(u)	u	Γ(u)
1.00	1.00000	1.25	0.90640	1.50	0.88623	1.75	0.91906
1.01	0.99433	1.26	0.90440	1.51	0.88659	1.76	0.92137
1.02	0.98884	1.27	0.90440	1.52	0.88704	1.77	0.92376
1.03	0.98355	1.28	0.90072	1.53	0.88757	1.78	0.92623
1.04	0.97844	1.29	0.89904	1.54	0.88818	1.79	0.92877
1.05	0.97350	1.30	0.89747	1.55	0.88887	1.80	0.93138
1.06	0.96874	1.31	0.89600	1.56	0.88964	1.81	0.93408
1.07	0.96415	1.32	0.89464	1.57	0.89049	1.82	0.93685
1.08	0.95973	1.33	0.89338	1.58	0.89142	1.83	0.93969
1.09	0.95546	1.34	0.89222	1.59	0.89243	1.84	0.94261
1.10	0.95135	1.35	0.89115	1.60	0.89352	1.85	0.94561
1.11	0.94740	1.36	0.89018	1.61	0.89468	1.86	0.94869
1.12	0.94359	1.37	0.88931	1.62	0.89724	1.87	0.95184
1.13	0.93993	1.38	0.88854	1.63	0.89592	1.88	0.95507
1.14	0.93642	1.39	0.88785	1.64	0.89864	1.89	0.95838
1.15	0.93304	1.40	0.88726	1.65	0.90012	1.90	0.96177
1.16	0.92980	1.41	0.88676	1.66	0.90167	1.91	0.96523
1.17	0.92670	1.42	0.88636	1.67	0.90330	1.92	0.96877
1.18	0.92373	1.43	0.88604	1.68	0.90500	1.93	0.97240
1.19	0.92089	1.44	0.88581	1.69	0.90678	1.94	0.97610
1.20	0.91817	1.45	0.88566	1.70	0.90864	1.95	0.97988
1.21	0.91558	1.46	0.88563	1.71	0.91258	1.96	0.98374
1.22	0.91311	1.47	0.88560	1.72	0.91057	1.97	0.98768
1.23	0.91075	1.48	0.88575	1.73	0.91466	1.98	0.99171
1.24	0.90852	1.49	0.88595	1.74	0.91683	1.99	0.99581
						2.00	1.00000

By the use of the recursion relation of Equation 2.45b, other gamma functions can be calculated.

2.5 CONTINUOUS SETS OF EVENTS HAVING THE SAME PROBABILITY, DENSITY OF STATES

Similar to the case in which a number of events have the same probability, a set of continuous events can have regions with the same probability. Similar to Equation 2.14 for the discrete case, a normalization condition for the continuous case where regions g(x) of events have the same probability can be derived.

$$1 = \int_{-\frac{a}{2}}^{\frac{a}{2}} g(x)p(x)dx \tag{2.47}$$

where g(x) is the region of space with the same probability or density of states, and p(x) is the probability density of this region. The region can be a line, an area, or a three- or higher-dimensional volume. The density of events having the same probability of a continuous set of events is often called the *density of states* with the same probability. As before, the total region A of the continuous events is:

$$A = \int_{-\frac{a}{2}}^{\frac{a}{2}} g(x)dx \tag{2.48}$$

This is also best illustrated by an example. Consider the archery target shown in Figure 2.10. Assume that the probability of an arrow hitting the target between radii r and r + dr depends only on the radius, and is independent of the angle from a line through the center of the target; that is, all points on the target in an area $2\pi rdr$ of a circle of incremental width dr have the same probability p(r) per unit area of being hit by an arrow. The quantity $2\pi r$, in this case, is interpreted as the density of states with the same probability. In a real physical case, an archer can miss the target altogether. However, for the sake of this problem, it is assumed that the target being hit by an arrow at some point is a certainty; that is, the probability of an arrow hitting the target at some point is equal to unity.

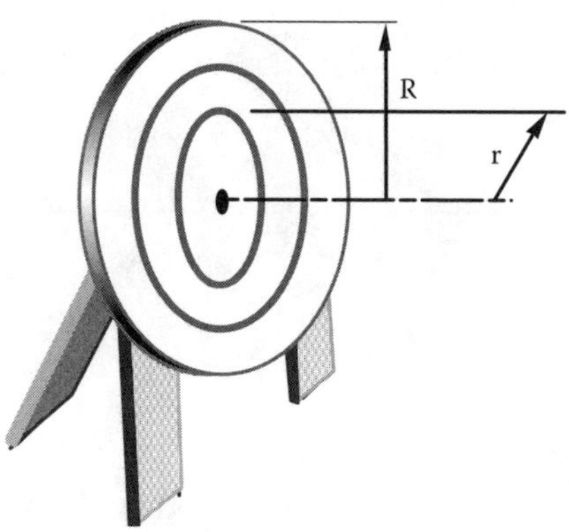

FIGURE 2.10 Archery target.

$$1 = 2\pi \int_0^R rp(r)dr \tag{2.49}$$

where the density of states g(r) is the region of a set of continuous events that has the same probability per unit region. Here:

$$g(r) = 2\pi r \tag{2.50}$$

The total area A of the target, from an equation similar to Equation 2.48, is

$$\text{a)} \quad A = 2\pi \int_0^R rdr \qquad \text{or} \qquad \text{b)} \quad A = \pi R^2 \tag{2.51}$$

EXAMPLE 2.6

A particle that can be described by classical mechanics has a probability density p(E) per unit energy that it will have a certain energy E is proportional to $\exp\left[-E/kT\right]$, where $k = 1.380662 \times 10^{-23}$ Joules per °K is Boltzmann's constant, and T is the absolute temperature. It is assumed that the particle under consideration has only a kinetic energy $E = ½\ mv^2$, where m is the mass of the particle and v is its velocity. Because the velocity v is a vector,

$$v^2 = v_x^2 + v_y^2 + v_z^2$$

All particles with velocities having the same magnitude will have the same energy, $E = ½\ mv^2$ and all particles with the same energy have the same probability. Therefore, all velocities in a spherical shell of radius v in velocity space correspond to the same probability of the particle having a certain energy.

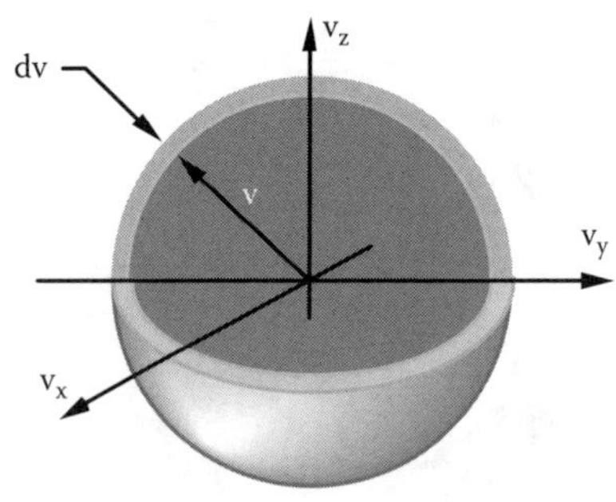

The density of states g(E) will be derived first:

The volume of the spherical shell of thickness dv is equal to $4\pi v^2 dv$. However, the probability density is given in terms of energy $E = ½\ mv^2$. Therefore, one must express the velocity in terms of the energy.

$$V = \sqrt{\frac{2}{m}}\sqrt{E} \quad dV = \frac{1}{2}\sqrt{\frac{2}{m}}\frac{dE}{\sqrt{E}}$$

By substituting into the incremental volume in velocity space, one obtains:

$$4\pi V^2 dV = 2\pi\left[\frac{2}{m}\right]^{\frac{3}{2}}\sqrt{E}\,dE$$

Thus, the density of states in energy, that is, the number of states g(E) having the same probability per unit energy, is equal to:

$$g(E) = 2\pi\left[\frac{2}{m}\right]^{\frac{3}{2}}\sqrt{E}$$

Indeed, the preceding expression is the result that one has been seeking here. However, it is interesting to extend this problem further to obtain a more interesting result. As stated earlier, the probability density p(E) is proportional to $[-E/kT]$. The probability, of course, has to be normalized.

$$p(E) = A\exp\left[-\frac{E}{kT}\right]$$

where A is the normalization constant. Multiplying the density of states g(E) by the probability and integrating:

$$A2\pi\left[\frac{2}{m}\right]^{\frac{3}{2}}\int_0^{\infty} E^{\frac{1}{2}}\exp\left[-\frac{E}{kT}\right]dE = 1$$

To integrate this equation, let:

$$x \equiv \frac{E}{kT} \quad \text{or} \quad E = kTx \quad \text{and} \quad dE = kTdx$$

Substituting this into the integral:

$$A2\pi\left[\frac{2kT}{m}\right]^{\frac{3}{2}}\int_0^{\infty} x^{\frac{1}{2}}\exp[-x]dx = 1$$

This is a Γ function integral. Note that the integral

$$\int_0^\infty x^{\frac{1}{2}} e^{-x} dx = \frac{\sqrt{\pi}}{2}$$

Substituting this for the integral:

$$A\left[\frac{2\pi kT}{m}\right]^{\frac{3}{2}} = 1 \quad \text{or} \quad A = \left[\frac{m}{2\pi kT}\right]^{\frac{3}{2}}$$

Next, the normalization constant A is substituted into the expression for the probability density p(E).

$$p(E) = \left[\frac{m}{2\pi kT}\right]^{\frac{3}{2}} e^{-\frac{E}{kT}}$$

As will be shown in Chapter 6, the average value ⟨f(E)⟩ of some function f(E) is equal to the integral of the product of the function f(E), the density of states g(E), and the probability density p(E).

$$\langle f(E)\rangle \equiv \int_0^\infty f(E)g(E)p(E)dE$$

In this case, f(E) is simply equal to E. By using the previous formula and the values of g(E) and p(E), one obtains:

$$\langle E\rangle = \frac{2kT}{\sqrt{\pi}}\int_0^\infty \left(\frac{E}{kT}\right)^{\frac{3}{2}} e^{-\frac{E}{kT}} d\left(\frac{E}{kT}\right)$$

By making the same transformation of variables as done previously, where $x = E/kT$, one obtains:

$$\langle E\rangle = \frac{2kT}{\sqrt{\pi}}\int_0^\infty x^{\frac{3}{2}} e^{-x} dx$$

This is another Γ function integral. As before, note that the integral

$$\int_0^\infty x^{\frac{3}{2}} e^{-x} dx = \frac{3\sqrt{\pi}}{4}$$

By substituting for the integral, one obtains for the average energy ⟨E⟩ of a free classical particle moving in three dimensions:

$$\langle E \rangle = \frac{3}{2} kT$$

This is known as the law of *equal partition of the energy* among the degrees of freedom of the motion of the particle. The particle has an average energy of kT/2 per degree of freedom of the motion.

PROBLEMS

PROBLEM 2.1

Given:

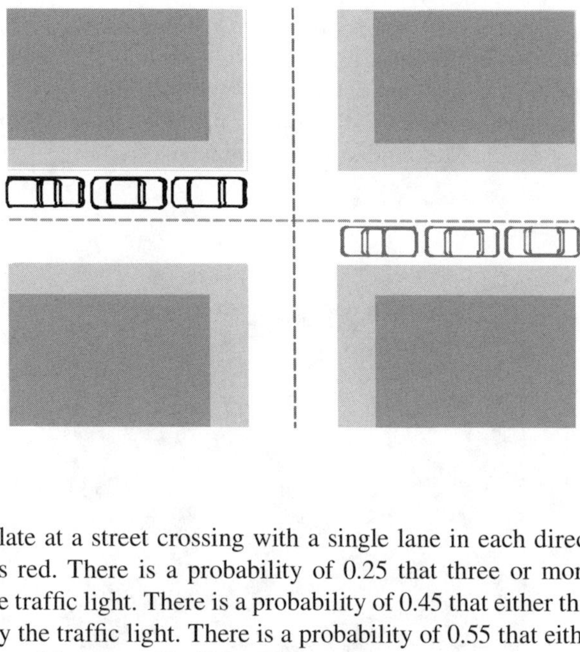

Cars accumulate at a street crossing with a single lane in each direction when the traffic light is red. There is a probability of 0.25 that three or more cars will be trapped by the traffic light. There is a probability of 0.45 that either three or two cars are trapped by the traffic light. There is a probability of 0.55 that either two cars, or one car is trapped by the traffic light. There is a probability of 0.30 that either more than three cars or no cars are trapped by the traffic light.

Problem:

a) Calculate the probability P_{more} that more than three cars, the probability P_3 that three cars, the probability P_2 that two cars, the probability P_1 that one car, and the probability P_0 that no cars are trapped by the traffic light.
b) What is the probability that more than three and only two cars are simultaneously trapped by the traffic light?

Problem 2.2

Given:

During World War II, the Germans buried five fuel depots between Morocco and Egypt in northern Africa. The German Field Marshal, Rommel, led an army to conquer North Africa. He started from Morocco and went east. These "Five Graves to Cairo" would serve to supply fuel for his tanks and other vehicles. Before the Africa campaign started, Rommel estimated that some of these fuel depots would be detected by the British. The probability of the British finding the first or second fuel depots is 0.5. The probability of the British finding the first or third fuel depots is 0.4. The probability of the British finding the third or fourth fuel depots is 0.45. The probability of the British finding the first or fifth fuel grave is 0.15.

Problem:

a) Calculate the probabilities P_1, P_2, P_3, P_4, and P_5 of the first, second, third, fourth, or fifth fuel depots being found.
b) The only fuel depot the British found was the last one near El Alamein in Egypt. This denied Rommel fuel for the conquest of Egypt and led to his defeat by the British. What values do the probabilities P_1, P_2, P_3, P_4, and P_5—that the first, second, third, fourth, or fifth fuel depots will be found—take after the end of Rommel's Africa campaign?

Problem 2.3

Given:

A star in the vicinity of a nova, an exploding star, has a probability density p(r) per unit volume of being hit with some debris. Here, the probability density p(r) per unit volume of being hit by debris is

$$p(r) = \frac{\alpha^4}{24\pi} \exp[-\alpha r] \quad \text{for} \quad 0 \le r < \infty$$

where α is 6.61×10^{-18} per meter.

The star has a probability 0.1 of being a size 1 star. It has a probability 0.25 of being a size 2 star, a probability 0.35 of being a size 3 star, and a probability 0.3 of being a size 4 star.

Note: One light-year is equal to $9.454254955 \times 10^{15}$ m.

Problem:

a) Calculate the probability that objects located between 14 and 18 light-years from the center of the nova will be hit by debris.
b) Calculate the probability that the star is either a class 1 or class 4 star and will be hit by some debris from the nova if it is located between 14 and 18 light-years from the nova.

Problem 2.4

Given:

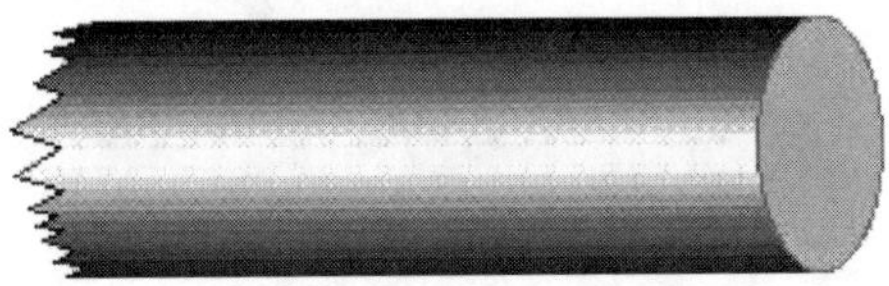

A shaft for a pipe-threading machine has a length x. The probability density p(x) per unit shaft length is

$$p(x) = A\frac{1}{(x-L)^2 + a^2} \qquad \text{where } L - b < x \le L + b$$

where the nominal length of the shaft L = 500 mm, the tolerance b = 0.050 mm, and a = 0.025 mm.

Note:

$$\int \frac{du}{u^2+1} = \tan^{-1} u$$

Problem:

a) Calculate the normalization constant A.
b) Calculate the probability that the shaft will have a length x between 500.025 mm and 499.975 mm.

Problem 2.5

Given:

A neon sign has 11 letters. Each letter dissipates about 120 W. Thus, when all letters are lit the sign will dissipate 1320 W. The probability that any letter will fail is equal to 0.05.

Problem:

a) Calculate the probability that the sign will dissipate only 960 W of light.
b) Calculate the probability that all letters except the "E" in HOTEL and the "I" in CASINO will light up.
c) Calculate the total number of different events of various letters failing to light.

Problem 2.6

Given:

By pouring sugar into a cup, some sugar crumbs will always spill outside the cup. Let us assume that, the probability density per unit area of finding a sugar crumb on the table at a distance r from the center of the bottom of the cup is:

$$p(r) = \frac{\beta^2}{2\pi(1+\beta a)} \exp\left[-\beta(r-a)\right] \quad \text{for} \quad a < r \leq \infty$$

where a = 40 mm is the radius of the cup and where $\beta = 0.02$ per mm. Although the probability density depends only on the radius from the center of the cup, there is an area equal to $g(r) = 2\pi r dr$ where the probability density is the same.

Problem:

a) Calculate the probability of finding a sugar crumb between the cup and a radius of 140 mm.
b) Calculate the probability of finding a sugar crumb outside a radius of 140 mm centered at the cup.

PROBLEM 2.7

Given:

A standard soccer ball consists of 12 pentagons and 20 hexagons. Usually, the pentagons are black and the hexagons are white. The probability P_P that a soccer ball will land on a pentagon is proportional to the total area of the 12 pentagons, and the probability P_H that a soccer ball will land on a hexagon is proportional to the total area of the 20 hexagons. About 250 years ago, Leonhard Euler devised the following formula for polyhedrons:

$$\text{Number of Vertices} - \text{Number of Edges} + \text{Number of Faces} = 2.$$

If the pentagons and hexagons forming the soccer ball were flat surfaces, it would be a polyhedron. The edges of the pentagons must be equal to the edges of the hexagons. Suppose a particular soccer game is a draw, 0 to 0. There are also no goals scored in the extra time periods. Thus, the game has to be decided by each team taking five penalty kicks. Therefore, there are a total of 10 penalty kicks at the end of the game.

Problem:

a) Calculate the probability P_P that a particular ball will land on a pentagon, and calculate the probability P_H that a particular ball will land on a hexagon.
b) Calculate and plot on a bar graph the probabilities of the various numbers $n = 0, 1, 2, \ldots, 9, 10$ of times the ball will land on a pentagon.

c) Calculate the number of individual possible events of the ball landing on either a pentagon or a hexagon during the 10 penalty kicks.

Problem 2.8

Given:

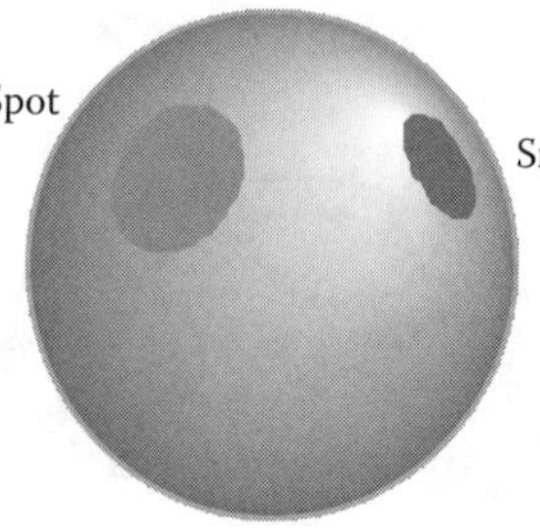

A game consists of rolling eight identical balls each with two flat spots as shown. The probability P_{SF} of any one of the balls landing on its small flat spot is equal to 0.12. The probability P_{LF} of any one of the balls landing on its large flat spot is equal to 0.28.

Problem:

a) Calculate the probability that either two balls will land on their small flat spots and no balls will land on their large flat spots, or three balls will land on their small flat spots and three balls will land on their large flat spots.
b) Calculate the number of individual possible events in rolling the eight balls with two flat spots.

Problem 2.9

Given:

The Dow Jones stock averages are computed each day, summing the values of some 30 selected stocks that are traded on the New York Stock Exchange and multiplying the sum by a number A. Currently, the number A = 0.27247214. There are better stock indices that take into account the number G_n of stocks traded. They use the number G_n of stocks traded as a weighting. However, the Dow Jones averages are the most widely quoted stock indices. There is a probability P_{UP} of 0.5 that on any particular day the Dow Jones averages will increase, and a probability P_{DOWN} of 0.48 that the Dow Jones averages will decrease. There is a probability of 0.02 that the stock averages will not change. There are six trading days in a week.

Problem:

a) Calculate the probability that on two out of the six trading days the stocks will increase, and on three days the stock will decrease, and on the remaining day the stock values will not change.
b) Calculate the probability that on one out of the six trading days the stocks will increase, and on two days the stock will decrease, and on the remaining three days the stock values will not change.

3 Joint, Conditional, and Total Probabilities

3.1 CONDITIONAL PROBABILITIES

Previously, probabilities of individual sets of events were discussed. Here, probabilities of several sets of events will be discussed. Consider two sets of discrete events, set $S_A\{A_1, A_2, \ldots, A_n, \ldots, A_N\}$ and set $S_B\{B_1, B_2, \ldots, B_m, \ldots, B_M\}$. For example, spinning a roulette wheel (Figure 3.1) and rolling a die can be considered as two sets of events. The set of events S_A are the 36 numbers on which the ball of the roulette wheel can land, and events S_B are the six events corresponding to the different number of dots of the top face. The events in each set might be mutually exclusive; that is, if event A_{28} occurs, none of the other events A_n can occur, and likewise for the events in set B. However, any event in set A can occur together with any event in set B. This was discussed in Chapter 2. The probability that event A_n from set A *and* event B_m from set B will occur is known as the *joint probability* $P(A_n \cap B_m)$. Here, the symbol $\cap$ stands for the "intersection" of A_n and B_m, as was shown in the Venn diagram of Figure 2.3 of Chapter 2.

For two *statistically independent events* A_n and B_m, the joint probability is equal to the product of the probabilities of the individual events. For example, the events of the ball landing on number 22 on the roulette wheel and a die landing so that the surface with 5 dots is on top are statistically independent events. The ball landing on any number does not depend on how the die will land and the die showing any number of dots does not depend on which number the roulette ball will land. As mentioned before, by two independent events is meant that one event can occur whether the second event occurred or not.

$$P(A_n \cap B_m) = P_n Q_m \tag{3.1}$$

In general, events A_n and B_m are not necessarily mutually exclusive. From the Venn diagram of Figure 2.4 of Chapter 2, one observes that:

$$P(A_n \cap B_m) = P(A_n) + Q(B_m) - P(A_n \cup B_m) \tag{3.2}$$

In the more general case, the event A_n might not be independent of event B_m. The probability of event A_n occurring might depend on event B_m having occurred. The probability that both events A_n and B_m will occur will have the following form in this case:

$$\text{a)}\ \ P(A_n \cap B_m) = P(A_n|B_m)Q(B_m) \quad \text{or} \quad \text{b)}\ \ P(A_n \cap B_m) = P(A_n|B_m)P(A_n) \tag{3.3}$$

FIGURE 3.1 Roulette wheel.

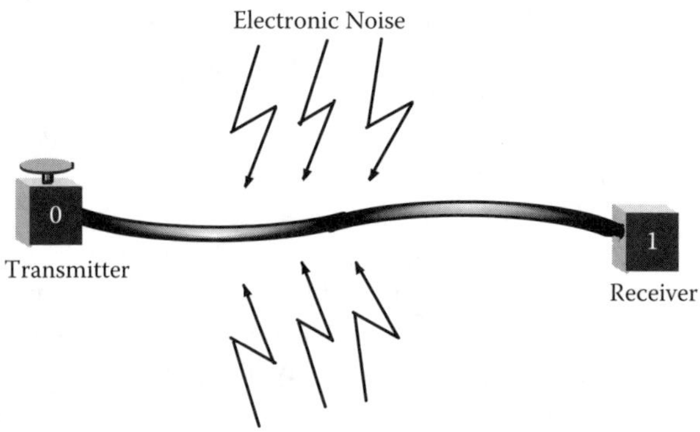

FIGURE 3.2 Binary transmission system subject to electronic noise. A ZERO has been transmitted but because of noise a ONE was received.

where $P(A_J|A_M)$ is the *conditional probability* that event A_n will occur provided that event B_m has occurred. For example, the events of receiving a particular signal from a binary system that sends either a ONE or ZERO through a noisy channel are statistically dependent events; that is, there is a conditional probability p(ONE received|ZERO transmitted) that a ONE was received provided a ZERO was sent through the noisy channel, as shown in Figure 3.2.

Because the conditional probability is a probability, it has to be normalized; that is, the sum of all conditional probabilities depending on the same condition has to be equal to one.

$$\sum_{n=1}^{n=N} P\left(A_n \middle| B_m\right) = 1 \quad \text{for constant m} \tag{3.4}$$

In early use of Morse code transmission, the messages were often received by listening to the clicking of the receiver solenoid. It was easy to make mistakes, especially

FIGURE 3.3 Original telegraph key.

when sending letters that are not part of a word. In that case, an example of a conditional probability P(- -|• •) is the probability a - -, an M is received provided a • • , an I is transmitted. Figure 3.3 features an approximately 150-year-old Morse key.

EXAMPLE 3.1

Suppose the joint probability $P(A_n \cap B_m)$ that events A_n and B_m will occur is equal to 0.025, and the probability that event B_m will occur independent of event A_n occurring is equal to 0.4.

The conditional probability that event A_n will occur provided that event B_m has occurred, in this case, is:

$$P(A_n|B_m) = 0.025/0.4$$

or

$$P(A_n|B_m) = 0.0625$$

The joint probability that events A_n, B_m, and C_k will occur is

$$P(A_n \cap B_m \cap C_k) = P(A_n|B_m)P(B_m \cap C_k) \tag{3.5}$$

where, as before, $P(A_n|B_m)$ is the conditional probability that event A_n will occur provided that event B_m has occurred. The joint probability $P(B_m \cap C_k)$ that both events B_m and C_k will occur is given by an expression similar to Equation 3.3a. Therefore, Equation 3.5 can be written as:

$$P(A_n \cap B_m \cap C_k) = P(A_n|B_m)P(B_m|C_k)P(C_k) \tag{3.6}$$

This concept can be extended to a large chain of events:

$$P(A_i \cap B_j \cap C_k \cap D_m \cap F_n \cap \ldots) = P(A_i|B_j)P(B_j|C_k) \\ P(C_k|D_m)P(D_m|F_n)\ldots Q(X_z) \tag{3.7}$$

3.2 DEPENDENT, INDEPENDENT, AND EXCLUSIVE EVENTS

Consider two events A_n and B_m, each having a nonzero probability of occurring. The events are *statistically independent* if event A_n can occur whether event B_m has occurred or not, and similarly, if event B_m can occur whether event A_n has occurred or not. This has been discussed before; that is, the conditional probabilities are equal to the independent probabilities.

$$P(A_n|B_m) = P(A_n) \quad \text{and} \quad P(B_m|A_n) = Q(B_m) \tag{3.8}$$

As has been shown before, for statistically independent events the joint probability is equal to the product of the individual independent probabilities.

$$P(A_n \cap B_m) = P(A_n)Q(B_m) \tag{3.9}$$

Events A_n and B_m are not necessarily mutually exclusive. Obviously, events that are not statistically independent are *statistically dependent* events.

Two events A_n and A_k are *mutually exclusive* if the occurrence of event A_n precludes the occurrence of event A_k, and vice versa. For example, if one rolls a 4 with a particular die one cannot simultaneously roll a 2 with the same die. Thus, the events of rolling a 4 and rolling a 2 with the same die are mutually exclusive. In this case, the joint probability for the two such events occurring is equal to zero.

$$P(A_n \cap A_k) = 0 \tag{3.10}$$

Another example is the events of a coin landing on its head or tail. These two events are also mutually exclusive. If the coin landed on its head, it cannot also land on its tail.

Without confusing the reader, the concept of exclusive events is only true in a physical world that can be described by classical models. For example a coin cannot simultaneously land on its tail and head side. For physical systems described by quantum mechanics, this is not true. For example, in a quantum mechanical description an electron has a spin angular momentum. The classical equivalent would be the electron spinning about its own axis. However, in the quantum mechanical description, it can only spin with a single magnitude of the angular momentum, or angular velocity. It can spin either clockwise or counterclockwise with an angular momentum equal to Plank's constant divided by 2π. Unlike a classical particle, an electron can have, say, a probability of 0.2 of spinning clockwise and a probability of 0.8 of spinning counterclockwise at the same time. This is Schrödinger's cat paradox, which will be discussed in a later section.

3.3 TOTAL PROBABILITY AND BAYES' THEOREM OF DISCRETE EVENTS

Consider two sets $S_A\{A_1, A_2, A_3, \ldots, A_n, \ldots, A_N\}$ and $S_B\{B_1, B_2, B_3, \ldots, B_m, \ldots, B_M\}$ of mutually exclusive events; that is, if, say, event A_k occurs, no other event in the same set S_A can occur. Similarly, if, say, event B_j occurs, no other event in the same set S_B

can occur. One can sum the joint probability $P(A_n \cap B_m)$ that event A_n of the set of events S_A and event B_m of the set of events S_B will occur over just one set of events.

$$\sum_{n=1}^{n=N} P(A_n \cap B_m) = \sum_{n=1}^{n=N} P(A_n|B_m)Q(B_m) \tag{3.11}$$

where Equation 3.3a was used in the right side of Equation 3.11. Note from Equation 3.4 that the sum of the conditional probabilities $P(A_n|B_m)$ that event A_n will occur provided event B_m has occurred over the events of the set S_A is equal to one.

$$\sum_{n=1}^{n=N} P(A_n \cap B_m) = Q(B_m) \tag{3.12}$$

where $Q(B_m)$ is the *total probability* of event B_m occurring. There is a similar expression for the total probability $P(A_n)$ of event A_n occurring, derived by the use of Equation 3.3b.

EXAMPLE 3.2

Consider a roulette wheel with 32 indents, half of them painted red and half painted black. The events of a ball landing in indents are mutually exclusive. Suppose indents 1 to 16 are painted red and indents 17 to 32 are painted black. Therefore, the conditional probability P(Red|Indent J) of a ball landing in the red area and in a particular indent of indents 1 to 16 is equal to $\frac{1}{32}$. The conditional probability P(Red|Indent J) of a ball landing in the red area and in a particular indent of indents 17 to 32 is equal to zero.

The total probability of a ball landing in the red area is

$$P(\text{Red}) = 16 \times \frac{1}{32} + 16 \times 0$$

$$= \frac{1}{2}$$

Indeed, as half of the wheel is painted red and half black, the probability of the ball landing in the red area is ½ ⅓2.

Consider again the two sets $S_A\{A_1, A_2, A_3, \ldots, A_n, \ldots, A_N\}$ and $S_B\{B_1, B_2, B_3, \ldots, B_m, \ldots, B_M\}$ of mutually exclusive events. Recall that mutually exclusive events are discussed in this text unless otherwise specified. From Equation 3.3b, the conditional probability $P(B_m|A_n)$ that event B_m will occur provided event A_n has occurred is:

$$P(B_m|A_n) = \frac{P(A_n \cap B_m)}{P(A_n)} \tag{3.13}$$

Suppose the values of the conditional probabilities $P(B_m|A_n)$ that event B_m will occur provided any one of the events A_n had occurred and the probabilities $P(A_n)$ that an event A_n will occur are known. However, suppose one would like to find the value of the conditional probabilities $P(A_n|B_m)$ that event A_n will occur provided the event B_m had occurred. Equation 3.3a can be rearranged as follows:

$$P\left(A_n|B_m\right)=\frac{P\left(A_n \cap B_m\right)}{Q\left(B_m\right)} \tag{3.14}$$

Equation 3.12 can be used for the total probability $Q(B_m)$ in Equation 3.14 to obtain

$$P\left(A_n|B_m\right)=\frac{P\left(B_m|A_n\right)\left(A_n\right)}{\sum_{k=1}^{k=N} P\left(B_m|A_k\right)P\left(A_k\right)} \tag{3.15}$$

where Equation 3.13 was used for the joint probability. This process is known as *Bayes' theorem*. Bayes' theorem is named after the Reverend Thomas Bayes, an English vicar and mathematician who lived from 1702 to 1761.

EXAMPLE 3.3

Direct current differential electronic amplifiers require that one obtain pairs of amplifying devices such as transistors that are almost identical. Previous to integrated circuit (IC) technology, one had to individually select pairs of devices that were more or less identical. Because pairs of transistors can be fabricated in close proximity on a single die, almost identical pairs of amplifying devices can be fabricated. A large number of devices can be fabricated on a single semiconductor wafer.

The conditional probability that a pair of devices come from wafer No. 04051244 provided they are identical within a certain tolerance T is given by:

P(No. 04051244|T = –0.01%) = 0.76
P(No. 04051244|T = –0.005%) = 0.78
P(No. 04051244|T = 0.01%) = 0.8
P(No. 04051244|T = 0.015%) = 0.82
P(No. 04051244|T = 0.011%) = 0.84

These conditional probabilities do not have to add to one, because each probability expresses a different tolerance condition. The probability that a pair of devices is within a certain tolerance T is

P(T = 0.001%) = 0.025
P(T = 0.005%) = 0.05
P(T = 0.01%) = 0.85
P(T = 0.015%) = 0.05
P(T = 0.011%) = 0.025

These probabilities, of course, have to add to one. The conditional probability that a device is within 0.0105% of specifications provided it came from wafer No. 04051244 is

$$P\left(T=0.015 \middle| \text{No. } 04051244\right)=\frac{P\left(\text{No. } 04051244 \middle| T=0.015\right)P\left(T=0.015\right)}{P\left(\text{No. } 04051244\right)}$$

where

$$\begin{aligned} P(\text{No. } 04051244) = {} & P(\text{No. } 04051244|T = 0.001\%)P(T = 0.001\%) + \\ & P(\text{No. } 04051244|T = 0.001\%)P(T = 0.001\%) + \\ & P(\text{No. } 04051244|T = 0.005\%)P(T = 0.005\%) + \\ & P(\text{No. } 04051244|T = 0.010\%)P(T = 0.010\%) + \\ & P(\text{No. } 04051244|T = 0.015\%)P(T = 0.015\%) \end{aligned}$$

or

$$P(\text{No. } 04051244) = 0.8$$

Thus, the conditional probability that a device that is within 0.0105% of specifications provided and it came from wafer No. 04051244 is:

$$P\left(T=0.015 \middle| \text{No. } 04051244\right)=\frac{0.84\times 0.025}{0.8}$$

or

$$P(T = 0.015\ \%|\text{No. } 04051244) = 0.02625$$

3.4 MARKOV PROCESSES

Another important example of a stochastically dependent system is the repeated transmission of data through some transmission system. As usual, this is best illustrated by an example. Consider three messages: message 1, message 2, and message 3, which are transmitted through some medium. A_1, A_2, and A_3 are the events that messages 1, 2, and 3 were received and B_1, B_2, and B_3 are the events that messages 1, 2, and 3 were transmitted. There are probabilities Q_m that each message is transmitted. The probabilities P_n that a particular message is received are

$$\begin{pmatrix} P_1 \\ P_2 \\ P_3 \end{pmatrix} = \begin{pmatrix} P(A_1|B_1) & P(A_1|B_2) & P(A_1|B_3) \\ P(A_2|B_1) & P(A_2|B_2) & P(A_2|B_3) \\ P(A_3|B_1) & P(A_3|B_2) & P(A_3|B_3) \end{pmatrix} \begin{pmatrix} Q_1 \\ Q_2 \\ Q_3 \end{pmatrix} \tag{3.16}$$

where the $P(A_n|B_m)$ are the conditional probabilities that message n is received provided message m has been sent. The conditional probability matrix does not necessarily have an inverse. The aforementioned message transmission is a *Markov process.*

This process could be repeated by transmitting the received messages again through the same transmission medium. The messages could be received and repeatedly be retransmitted through the transmission medium. This process is known as a *Markov chain.*

For example, consider a Markov chain with conditional probabilities $P(A_n|B_m)$ that characterize the transmission medium and probabilities Q_m that each of three messages are transmitted as given by Equation 3.16. Passing the information k times through the medium can be expressed mathematically as follows:

$$\begin{pmatrix} P_{1,k} \\ P_{2,k} \\ P_{3,k} \end{pmatrix} = \begin{pmatrix} 0.98 & 0.01 & 0.01 \\ 0.01 & 0.98 & 0.01 \\ 0.01 & 0.01 & 0.98 \end{pmatrix}^k \begin{pmatrix} 0.99990 \\ 0.00005 \\ 0.00005 \end{pmatrix} \tag{3.17}$$

The probabilities $P_{n,k}$ of receiving the messages after they have been passed successively k times through the transmission medium are shown in Figure 3.4

After the process is repeated a sufficient number of times, the conditional probability matrix describing multiple transmissions will not change anymore. It is interesting to investigate the form of the conditional probability matrix describing an infinite number of repeated transmissions.

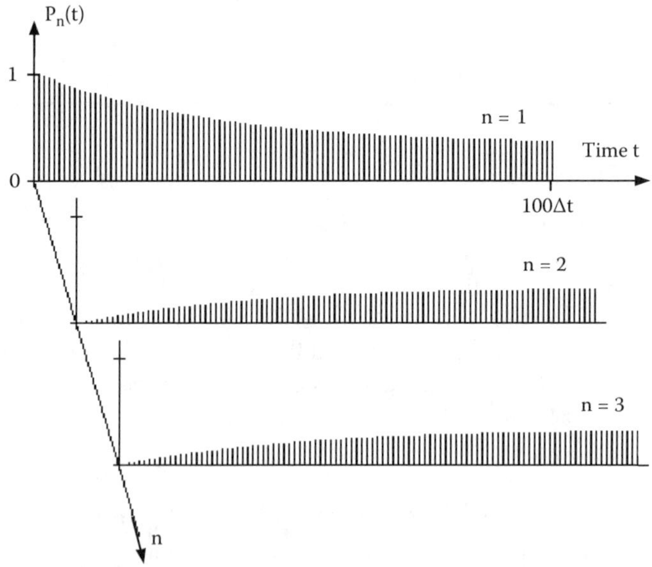

FIGURE 3.4 The result of passing messages up to 100 times through the transmission system described by Equation 3.17.

$$\lim_{N \to \infty} \begin{pmatrix} P(A_1|B_1) & P(A_1|B_2) & P(A_1|B_3) \\ P(A_2|B_1) & P(A_2|B_2) & P(A_2|B_3) \\ P(A_3|B_1) & P(A_3|B_3) & P(A_3|B_3) \end{pmatrix}^N = \begin{pmatrix} a & c & e \\ b & d & g \\ 1-a-b & 1-c-d & 1-e-g \end{pmatrix} \tag{3.18}$$

where the matrix on the right is equal to the conditional probability matrix describing an infinite number of repeated transmissions.

Three-dimensional conditional probability matrices were used in the foregoing examples. However, much larger matrices could be used in practice. After passing the information an infinite number of times through the transmission system, the transmission matrix should not change if the message is passed through the transmission system yet one more time.

$$\begin{pmatrix} P(A_1 | B_1) & P(A_1 | B_2) & P(A_1 | B_3) \\ P(A_2 | B_1) & P(A_2 | B_2) & P(A_2 | B_3) \\ P(A_3 | B_1) & P(A_3 | B_3) & P(A_3 | B_3) \end{pmatrix} \begin{pmatrix} a & c & e \\ b & d & g \\ 1-a-b & 1-c-d & 1-e-g \end{pmatrix} = \begin{pmatrix} a & c & e \\ b & d & g \\ 1-a-b & 1-c-d & 1-e-g \end{pmatrix} \tag{3.19}$$

Note that the multiplication of each column of the matrix representing the conditional probability matrix raised to the infinite power, the second and third matrices of Equation 3.19, results in an identical set of linear equations. Thus, the columns of the second and third matrices of Equation 3.19 are identical.

$$\lim_{N \to \infty} \begin{pmatrix} P(A_1|B_1) & P(A_1|B_2) & P(A_1|B_3) \\ P(A_2|B_1) & P(A_2|B_2) & P(A_2|B_3) \\ P(A_3|B_1) & P(A_3|B_3) & P(A_3|B_3) \end{pmatrix}^N = \begin{pmatrix} a & a & a \\ b & b & b \\ 1-a-b & 1-a-b & 1-a-b \end{pmatrix} \tag{3.20}$$

This can be illustrated with a two-dimensional example. After passing the information an infinite number of times through the transmission system, the transmission matrix should not change if the message is passed through the transmission system yet one more time.

$$\begin{pmatrix} a & b \\ 1-a & 1-b \end{pmatrix}\begin{pmatrix} p & p \\ 1-p & 1-p \end{pmatrix} = \begin{pmatrix} p & p \\ 1-p & 1-p \end{pmatrix} \tag{3.21}$$

where $a = P(A_1|B_1)$ is the conditional probability that message A_1 was received provided message B_1 was transmitted, $1 - a = P(A_2|B_1)$ is the conditional probability that message A_2 was received provided message B_1 was transmitted, b is the conditional probability $P(A_1|B_2)$ that message A_1 was received provided message B_2 was transmitted, and $1 - b$ is the conditional probability $P(A_2|B_2)$ that message A_2 was received provided message B_2 was transmitted. It is usually the case that the various conditional probabilities $P(A_n|B_m)$ of the transmission matrix are known. Note that in agreement with Equation 3.20, the columns of the second and third matrix of Equation 3.21 are equal. The element p of the matrix that describes the passing of the information an infinite number of times through the transmission system can be calculated by multiplying out the matrices of Equation 3.21:

$$\begin{aligned} &\text{a)} \quad ap + b - bp = p \\ &\text{b)} \quad p - ap + 1 - b - p + bp = 1 - p \end{aligned} \tag{3.22}$$

where Equations 3.22a and 3.22b are, of course, identical. By solving for p, one obtains:

$$\text{a)} \quad p = \frac{b}{1-a+b} \quad \text{and} \quad \text{b)} \quad 1-p = \frac{1-a}{1-a+b} \tag{3.23}$$

An interesting example is the error correction process in programs with a very large number of lines of digital code. Programs containing a very large number of lines of digital code contain a number of errors. Attempts are made by successive scans through the program to correct these errors. The error correction process can be characterized by conditional probabilities. Here:

P(Program is correct|Provided program was correct) = a
P(Program is correct|Provided program had an error) = b
P(Program has error|Provided program was correct) = 1 – a
P(Program has an error|Provided program had an error) = 1 – b

Suppose in this example the probability Q(correct) that the initial code had no errors is equal to 0.2 and the probability Q(not correct) that the code had some errors is 0.8.

First, it will be investigated what one would obtain by passing the code an infinite number of times through the correction process. For this one needs to calculate the matrix with the p's and q's of Equation 3.21. Assume that in this case $a = 0.9999$ and $b = 0.99$. By passing the program an infinite number of times through an error correction process, one obtains from Equation 3.23 that p is equal to 0.9802950787. Thus, after the program is passed an infinite number of times through the correction process, one obtains:

$$\begin{pmatrix} 0.9998990001 & 0.9998990001 \\ 0.000100999899 & 0.000100999899 \end{pmatrix}\begin{pmatrix} 0.2 \\ 0.8 \end{pmatrix} = \begin{pmatrix} 0.9998990001 \\ 0.000100999899 \end{pmatrix} \quad (3.24)$$

That is, the result after passing the program an infinite number of times through the error correction process is that one still does not obtain an error-free program. There is a 0.0100999899% chance that the program still contains an error. Perhaps it is not necessary to pass the program through an infinite number of error correction processes. Passing the program once through the error correction process:

$$\begin{pmatrix} 0.9999 & 0.99 \\ 0.0001 & 0.01 \end{pmatrix}\begin{pmatrix} 0.2 \\ 0.8 \end{pmatrix} = \begin{pmatrix} 0.99198 \\ 0.00802 \end{pmatrix} \quad (3.25)$$

It is found that all but 0.802% of the errors have been corrected. This is probably not good enough. Passing the program a second time through the error correction process:

$$\begin{pmatrix} 0.9999 & 0.99 \\ 0.0001 & 0.01 \end{pmatrix}\begin{pmatrix} 0.99198 \\ 0.00802 \end{pmatrix} = \begin{pmatrix} 0.999820602 \\ 0.000179398 \end{pmatrix} \quad (3.26)$$

It is found that all but 0.0179398% of the errors have been corrected. By comparing this result to the one obtained from Equation 3.24, we note that this is within 99.9921594% of what can ultimately be achieved by passing the program an infinite number of times through the error correction process. Thus, it might only be necessary to pass the program twice through the error correction process.

The other extreme is a transmission matrix that completely scrambles the information; that is, a = ½, 1 – a = ½, b = ½, and 1 – b = ½. For this case one still obtains p = ½; that is, if passing the information once through the transmission system completely scrambles it, passing it an infinite number of times through the transmission system will surely not unscramble it.

3.5 JOINT, CONDITIONAL, AND TOTAL PROBABILITIES AND BAYES' THEOREM OF CONTINUOUS EVENTS

Consider two sets of continuous events $S_A\{-a/2 < x \le a/2\}$ and $S_B\{-b/2 < y \le b/2\}$. The joint probability that the continuous event u will have values between x and x + dx and the continuous event v will have values between y and y + dy is

$$P([x < u \le x + dx] \cap [y < v \le y + dy]) = p(x \cap y)dxdy \quad (3.27)$$

where p(x ∩ y) is the *joint probability density* of the continuous events x and y (Figure 3.5). The conditional probability that the continuous event u will have values between x and x + dx provided that the continuous event v has values between y and y + dy is

$$P([x < u \le x + dx] | [y < v \le y + dy]) = p(x|y)dx \quad (3.28)$$

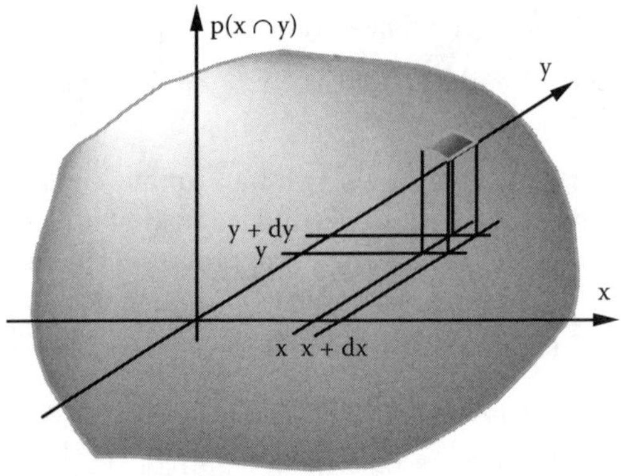

FIGURE 3.5 The region where the continuous event u has values between x and x + dx and the continuous event v has values between y and y + dy.

where p(x|y) is the *conditional probability density* that the continuous event u has values between x and x + dx provided the continuous event v has values between y and y + dy. Similarly, the conditional probability that the continuous event v will have values between y and y + dy provided that the continuous event u has values between x and x + dx is

$$P\left(\left[y < v \le y + dy\right] \middle| \left[x < u \le x + dx\right]\right) = p\left(y \middle| x\right) dy \tag{3.29}$$

where p(y|x) is the conditional probability density that the continuous event v has values between y and y + dy provided the continuous event u has values between x and x + dx.

The probability that the continuous event u will have values between x and x + dx is

$$P\left(x < u \le x + dx\right) = p\left(x\right) dx \tag{3.30}$$

where p(x) is the probability density of the continuous event x. Similarly, the probability that the continuous event v will have values between y and y + dy is

$$P\left(y < v \le y + dy\right) = p\left(y\right) dy \tag{3.31}$$

where p(y) is the probability density of the continuous event x.

Similar to the discussion leading to Equation 3.3, the joint probability that the continuous event u will have values between x and x + dx and the continuous event v will have values between y and y + dy can be expressed in terms of the conditional probabilities and the total probabilities.

$$P\left(\left[y<v\le y+dy\right]\cap\left[x<u\le x+dx\right]\right)=$$
$$P\left(\left[y<v\le y+dy\right]\middle|\left[x<u\le x+dx\right]\right)P\left(x<u\le x+dx\right) \tag{3.32}$$

or

$$P\left(\left[y<v\le y+dy\right]\cap\left[x<u\le x+dx\right]\right)=$$
$$P\left(\left[x<u\le x+dx\right]\middle|\left[y<v\le y+dy\right]\right)P\left(y<v\le y+dy\right) \tag{3.33}$$

By substituting Equations 3.27, 3.29, and 3.30 into Equation 3.32 and canceling dxdy on both sides of the resulting equation, one obtains an expression for the joint probability density in terms of the conditional probability density and the probability density per unit length y. This result is similar to the one obtained in the discrete case.

$$p(x\cap y)=p(y|x)p(x) \tag{3.34}$$

Similarly, substituting Equations 3.27, 3.28, and 3.31 into Equation 3.33 and canceling dxdy on both sides of the resulting equation:

$$p(x\cap y)=p(x|y)p(y) \tag{3.35}$$

Returning to the properties of a joint probability of two variables x and y, it is required that the probability of both x and y having any values at all be equal to one. For example, for $-(a/2) < x \le (a/2)$ and $-(b/2) < y \le (b/2)$, it is required that the double integral of the joint probability density equal one.

$$\int_{-\frac{a}{2}}^{\frac{a}{2}} dx \int_{-\frac{b}{2}}^{\frac{b}{2}} dy\, p(x\cap y)=1 \tag{3.36}$$

It is required that the probability of obtaining any value of x at all be equal to one, and of obtaining any value of y at all be equal to one.

$$\text{a)}\ \int_{-\frac{a}{2}}^{\frac{a}{2}} dx\, p(x)=1 \quad \text{and} \quad \text{b)}\ \int_{-\frac{b}{2}}^{\frac{b}{2}} dy\, p(y)=1 \tag{3.37}$$

It is also required that the conditional probability of obtaining any value at all of x provided the continuous event v has values between y and y + dy be equal to one. Similarly, it is required that the conditional probability of obtaining any value at all of y provided the continuous event u has values between x and x + dx be equal to one.

$$\text{a)} \quad \int_{-\frac{a}{2}}^{\frac{a}{2}} dx\, p(x|y) = 1 \quad \text{and} \quad \text{b)} \quad \int_{-\frac{b}{2}}^{\frac{b}{2}} dy\, p(y|x) = 1 \tag{3.38}$$

That is, the probability that any value at all of the particular variable occurs has to be equal to one provided the condition is kept constant.

The integration of the joint probability density $p(x \cap y)$ that event u has values between x and x + dx and event v has values between y and y + dy over one of the variables only can be more easily performed by expressing the joint probability density in terms of the conditional probabilities, as was done in Equation 3.34 or Equation 3.35.

$$\begin{aligned} &\text{a)} \quad \int_{-\frac{a}{2}}^{\frac{a}{2}} p(x \cap y)\, dx = \int_{-\frac{a}{2}}^{\frac{a}{2}} p(x|y)\, p(y)\, dx \quad \text{and} \\ &\text{b)} \quad \int_{-\frac{b}{2}}^{\frac{b}{2}} p(x \cap y)\, dy = \int_{-\frac{b}{2}}^{\frac{b}{2}} p(y|x)\, p(x)\, dy \end{aligned} \tag{3.39}$$

Note from Equations 3.38a and 3.38b that the integrals of the conditional probability densities $p(x|y)$ and $p(y|x)$ over x or y, respectively, are equal to one. Thus,

$$\text{a)} \quad \int_{-\frac{a}{2}}^{\frac{a}{2}} p(x \cap y)\, dx = p(y) \quad \text{and} \quad \text{b)} \quad \int_{-\frac{b}{2}}^{\frac{b}{2}} p(x \cap y)\, dy = p(x) \tag{3.40}$$

where p(x) and p(y) are the *total probability densities* of the continuous events x and y.

Suppose the values of the conditional probability density $p(x|y)$ of the continuous variable x provided the continuous variable v has values between y and y + dy as well as the values of the probability density p(y) are known. However, suppose one would like to find the value of the conditional probability density $p(y|x)$ of the continuous variable y provided the variable u has values between x and x + dx. This can be calculated by rearranging Equation 3.34.

$$p(y|x) = \frac{p(x \cap y)}{p(x)} \tag{3.41}$$

By substituting Equation 3.35 and Equation 3.40b into Equation 3.41, one obtains, by analogy with the discrete case of Equation 3.14, the expression for *Bayes' theorem of continuous events*:

$$p(y|x) = \frac{p(x|y)\, p(y)}{\int_{-\frac{b}{2}}^{\frac{b}{2}} p(x|y)\, p(y)\, dy} \tag{3.42}$$

A similar expression for $p(x|y)$ in terms of $p(y|x)$ can also be derived.

EXAMPLE 3.4

The conditional probability density per unit length x provided the continuous event v has values between y and y + dy is equal to:

$$p(y|x) = \frac{2xy + a}{a^2(y+1)} \quad \text{for} \quad 0 < x \le a$$

where a is a constant and the probability density per unit length y is

$$p(y) = \frac{2(y+1)}{b^2 + 2b} \quad \text{for} \quad 0 < y \le b$$

One would like to find the conditional probability density p(y|x) per unit length y provided the continuous event u has values between x and x + dx. Substituting the foregoing equations into Bayes' theorem of Equation 3.30:

$$p(y|x) = \frac{\dfrac{2}{a^2(b^2+2b)}(2xy + a)}{\dfrac{2}{a^2(b^2+2b)}\displaystyle\int_0^b (2xy + a)dy}$$

By performing the integration, one obtains for the conditional probability density p(y|x):

$$p(y|x) = \frac{2xy + a}{xb^2 + ba}$$

PROBLEMS

Problem 3.1

Given:

Consider two sets of events $S_A(A_1, A_2, A_3)$ and $S_B(B_1, B_2, B_3)$. The conditional probabilities $P(A_n|B_m)$ that any of the events A_n will occur provided an event B_m has occurred are

$$\begin{bmatrix} P(A_1|B_1) = 0.001 & P(A_1|B_2) = 0.001 & P(A_1|B_3) = 0.988 \\ P(A_2|B_1) = 0.998 & P(A_2|B_2) = 0.005 & P(A_2|B_3) = 0.008 \\ P(A_3|B_1) = 0.001 & P(A_3|B_2) = 0.994 & P(A_3|B_3) = 0.004 \end{bmatrix}$$

The total probabilities Q_m that any of the events B_m will occur are

$$Q_1 = 0.25, Q_2 = 0.6, \text{ and } Q_3 = 0.15$$

Problem:

a) Calculate the three total probabilities P_l that any of the events A_l will occur.
b) Calculate the conditional probability $P(B_3|A_1)$ that the event B_3 will occur provided event A_1 has occurred.

Problem 3.2

Given:

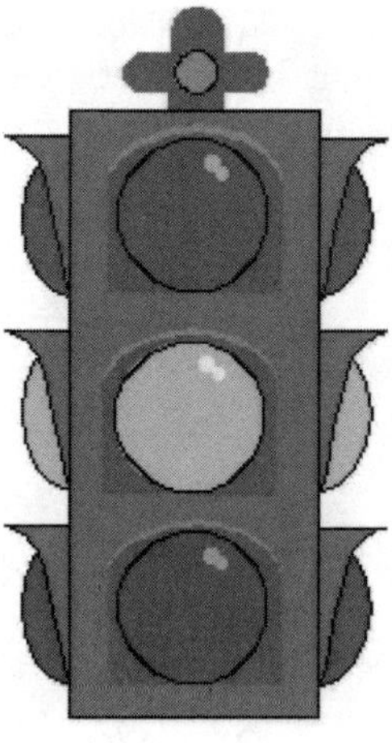

Traffic light.

A traffic light is operated remotely from a control center. The signal to turn on and off the various colors is sent through a long transmission line that is subject to noise. The conditional probabilities P(Color X Received|Color Y Transmitted) that a signal to turn on a particular color Y is transmitted while a signal to turn on color X is received are

P(Green Rec'd.|Green Transm.) = 0.99998
P(Green Rec'd.|Yellow Transm.) = 0.0002
P(Green Rec'd.|Red Transm.) = 0.0003
P(Yellow Rec'd.|Green Transm.) = 0.00001
P(Yellow Rec'd.|Yellow Transm.) = 0.99996
P(Yellow Rec'd.|Red Transm.) = 0.0005
P(Red Rec'd.|Green Transm.) = 0.00001
P(Red Rec'd.|Yellow Transm.) = 0.00002
P(Red Rec'd.|Red Transm.) = 0.9992

The probability P_G that a signal to turn on the green light is transmitted is equal to 0.28. The probability P_Y that a signal to turn on the yellow light is transmitted is equal to 0.44. The probability P_R that a signal to turn on the red light is transmitted is equal to 0.28.

Problem:

a) Calculate the probability P(Yellow Received) of a yellow signal being received.
b) Calculate the conditional probability P(Green Transmitted|Yellow Received) that a signal to turn on the green light was transmitted provided a signal to turn on the yellow light was received.

PROBLEM 3.3

Given:

The conditional P(NM|KL) probabilities that two-digit binary numbers NM were received provided two-digit binary numbers KL were transmitted are

$$\begin{array}{llll} P(00|00)=0.0001 & P(00|01)=0.0003 & P(00|10)=0.002 & P(00|11)=0.9994 \\ P(01|00)=0.9997 & P(01|01)=0.0004 & P(01|10)=0.004 & P(01|11)=0.0003 \\ P(10|00)=0.0001 & P(10|01)=0.9989 & P(10|10)=0.006 & P(10|11)=0.0002 \\ P(11|00)=0.0001 & P(11|01)=0.0004 & P(11|10)=0.988 & P(11|11)=0.0001 \end{array}$$

The probabilities Q_{MN} that the various digital numbers MN were transmitted are

$$Q_{00}=0.2 \quad Q_{01}=0.24 \quad Q_{10}=0.26 \quad Q_{11}=0.3$$

Problem:

a) Calculate the probability P_{10} that the digital number 10 was received.
b) Calculate the conditional probability R(01|10) that the digital number 01 was transmitted provided the digital number 10 was received.

PROBLEM 3.4

Given:

The conditional probability density p(x|y) that the variable u will have values between x and x + Δx provided the variable v will have values between y and y + Δy is

$$p(x|y)=\frac{3}{2a}\frac{x^2+1.8xy+y^2}{a^2+3y^2}$$

for $-a < x \le a$ and $-a < y \le b$.

The probability density p(y) that the variable v will have values between y and y + Δy is

$$p(y) = \frac{1}{4a^3}\left(a^2 + 3y^2\right)$$

Problem:

a) Calculate the probability density p(x) that the variable u will have values between x and x + Δx.
b) Calculate the conditional probability density p(y|x) that the variable v will have values between x and x + Δx provided the variable u will have values between x and x + Δx.

Problem 3.5

Given:

A conditional probability density p(x|y) per unit length is

$$p(x|y) = \frac{x^2 - 2\frac{p}{a^2}xy^3 + \frac{y^6}{a^4}}{\frac{2a^3}{3} + 2\frac{y^6}{a^3}}$$

where −a < x ≤ a and −a < y ≤ a. The probability density q(y) is

$$q(y) = \frac{21}{20a^4}\left[\frac{a^3}{3} + \frac{y^6}{a^3}\right]$$

Problem:

Calculate the conditional probability r(y|x) per unit length y.

Problem 3.6

Given:

An analog voltage is transmitted over a long cable. The transmitted analog voltage V can have values between V = −15 V and V = 15 V. All voltages are equally likely to occur. The conditional probability density p(U Received|V transmitted) that a voltage with a value between U and U + dU was received provided a voltage between V and V + dV was transmitted is

$$p(\text{U Received}|\text{V Transmitted}) = \frac{a}{\pi}\frac{1}{(U-V)^2 + a^2}$$

where −∞ < U ≤ ∞ and −15 V < V ≤ 15 V. Here, a = 5 μV

Problem:

a) Calculate the probability density of receiving a voltage between U and U + dU.
b) Calculate the conditional probability of transmitting a voltage V provided a voltage U has been received.

Problem 3.7

Given:

Mistakes are made in typing someone's handwritten initials AB. Errors are made by typing OB or AP instead of AB. The typing is finished and put through a correction process characterized by the following conditional probabilities:

$$P(AB|AB) = 0.9997 \quad P(AB|OB) = 0.998 \quad P(AB|AP) = 0.998$$
$$P(OB|AB) = 0.0001 \quad P(OB|OB) = 0.001 \quad P(OB|AP) = 0.001$$
$$P(AP|AB) = 0.0002 \quad P(OB|OB) = 0.001 \quad P(AP|AP) = 0.001$$

where P(AB|AB) is the conditional probability that no error was made, AB was typed, provided there was no error, AB was already typed. P(OB|AB) is the conditional probability that an error was made, OB was typed, provided there was no error originally, AB was typed. P(AB|OB) is the conditional probability that an error was corrected, AB was typed provided there was an error, OB was originally typed. P(OB|OB) is the conditional probability that an error was not corrected, OB was typed, provided there was an error, OB was typed originally, etc.

P(AB) = 0.8 is the probability that there was no error initially.
P(OB) = 0.1 is the probability that an error was made initially; OB was typed.
P(AP) = 0.1 is the probability that an error was made initially; AP was typed.

Problem:

a) Calculate the probabilities P(AB), P(OB), and P(AP) after passing the typing an infinite number of times through the correction process.
b) Calculate the probabilities P(AB), P(OB), and P(AP) after passing the typing thrice through the correction process.

Problem 3.8

Given:

Consider three sets of events $S_A\{A_1, A_2, A_3\}$, $S_B\{B_1, B_2, B_3\}$ and $S_C\{C_1, C_2, C_3\}$ where the conditional probabilities are:

$$P(A_1|B_1) = 0.9998 \quad P(A_1|B_2) = 0.0002 \quad P(A_1|B_3) = 0.001$$
$$P(A_2|B_1) = 0.0001 \quad P(A_2|B_2) = 0.9996 \quad P(A_2|B_3) = 0.003$$
$$P(A_3|B_1) = 0.0001 \quad P(A_3|B_2) = 0.0002 \quad P(A_3|B_3) = 0.996$$

$P(B_1\|C_1) = 0.0001$	$P(B_1\|C_2) = 0.0002$	$P(B_1\|C_3) = 0.996$
$P(B_2\|C_1) = 0.9998$	$P(B_2\|C_2) = 0.0002$	$P(B_2\|C_3) = 0.003$
$P(B_3\|C_1) = 0.0001$	$P(B_3\|C_2) = 0.9996$	$P(B_3\|C_3) = 0.001$

and the probabilities that the various events in set S_C will occur are:

$P(C_1) = 0.995$ $\quad P(C_2) = 0.005$ $\quad P(C_3) = 0.003$

Problem:

a) Calculate the probabilities $P(A_1)$, $P(A_2)$ and $P(A_3)$ of the events in set S_A occurring.
b) Calculate the conditional probability $P(B_2|A_2)$ that even B_2 will occur provided event A_2 has occurred.

4 Random Variables and Functions of Random Variables

4.1 CONCEPT OF A RANDOM VARIABLE AND FUNCTIONS OF A RANDOM VARIABLE

Recall that sets of events were discussed in previous chapters. In some cases the labels of the events, and in some other cases the events themselves, are *random variables.* Random variables can be discrete or continuous. A random variable is a continuous or discrete variable when each of its values corresponds to a probability or probability density. For example, the numbers of coins N_H and N_T landing on their heads or tails in the coin-tossing experiment are random variables. In the example where there is a probability of an arrow hitting a target at a radial distance r from its center, the radial distance is a random variable.

The forecasting techniques will also be extended to functions of a random variable. Suppose there is a certain probability P_n of a random variable n occurring. Then, the function f_n of n and the random variable n have the same probability of occurring.

EXAMPLE 4.1

Consider sand being unloaded from a truck as shown in the following figure. Suppose the probability per unit length p(x) of a grain of sand landing between x and x + dx is a Gaussian. Here, x is a continuous random variable. The Gaussian probability density is

$$p(x) = \frac{1}{(2\pi)^{\frac{1}{2}}\sigma}\exp\left[-\frac{x^2}{2\sigma^2}\right]$$

The height y(x) of the sand pile is approximately proportional to the number of grains of sand landing near each position x. Therefore, the height y(x) of the sand pile is approximately proportional to the probability density. The height of the sand at any point x of the sand pile is a random variable x.

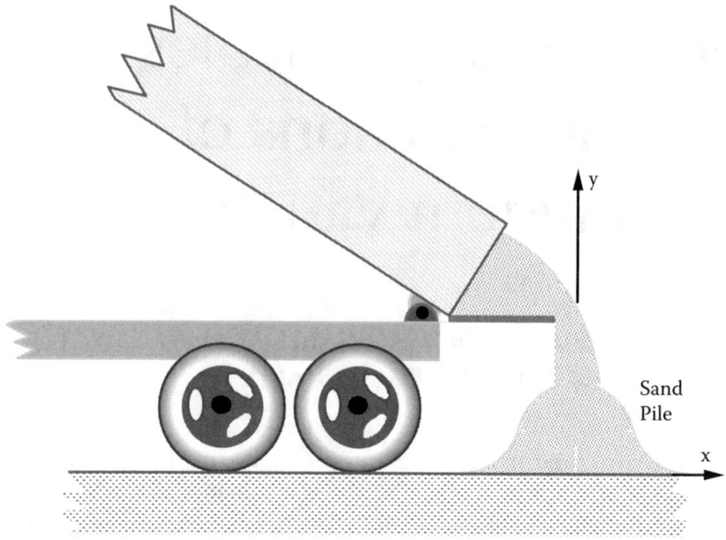

4.2 DISCRETE DISTRIBUTION FUNCTIONS

Consider a random variable n = 0, 1, 2, 3, …, N. There is a probability P_n of the random variable having a particular value n. For example, P_3 is the probability of n being equal to 3. The *distribution function* F_m is the probability $P\{n \leq m\}$ of n having a value less or equal to m:

$$F_m = \sum_{n=1}^{n=m} P_n \tag{4.1}$$

Recall that events discussed in this text are mutually exclusive unless otherwise specified. Because the probabilities P_n are positive and real, note from Equation 4.1 that F_m is a monotonically increasing function with increasing values of the random variable m; that is, $F_m > F_n$ for m > n. For a discrete random variable with a maximum value of N, the distribution function F_N is equal to one; that is it is certain that any one of the N events will occur.

$$F_N = 1 \tag{4.2}$$

The probability of n having values greater than k and less than m is

$$P\{k < n \leq m\} = \sum_{n=k+1}^{n=m} P_n \tag{4.3}$$

or

$$P\{k < n \leq m\} = F_m - F_k \tag{4.4}$$

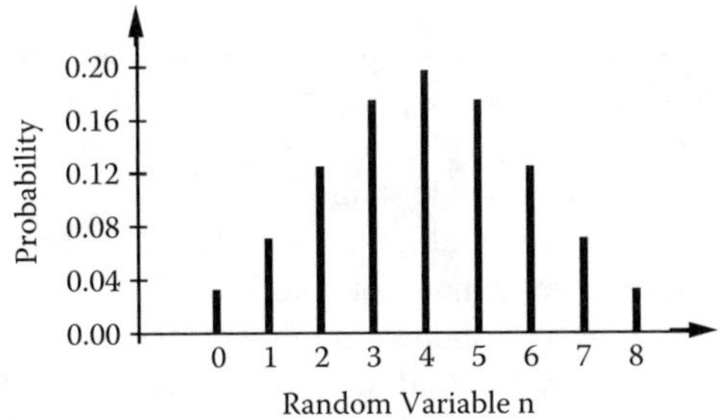

FIGURE 4.1 A plot of the probability that a particular value of the discrete random variable n will occur.

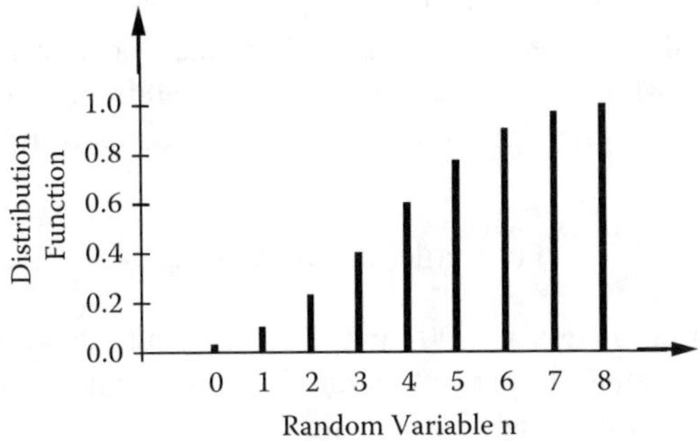

FIGURE 4.2 The distribution function for the discrete random variable of Figure 4.1. Note that the distribution function increases monotonically with n.

An example of the probabilities of certain values of a discrete random variable occurring is shown in Figure 4.1. The corresponding values of the distribution function are shown in Figure 4.2.

An interesting case is the difference between the discrete distribution functions F_n and F_{n-1}.

$$F_n - F_{n-1} = P_n \tag{4.5}$$

where P_n is just the probability that the random variable n will occur. Thus, the probability of the random variable n occurring is equal to the "backward" difference of the distribution functions. A difference of the form $R_{n+1} - R_n$ is called a *forward difference*. The properties of discrete distribution functions can be summarized as follows:

1. The distribution function F_N evaluated at the upper limit of the random variable is equal to one $F_N = 1$ for $0 < n \le N$.
2. For values between the upper and lower limit of the random variable, the distribution function F_n has values between zero and one $0 < F_n \le 1$ for $0 < n \le N$.
3. $F_m < F_n$ for $m < n$.
4. $P\{k < n \le m\} = F_m - F_k$ for $k < m$.
5. $F_m - F_{m-1} = P_m$.

EXAMPLE 4.2

Consider an office building. There is a probability that a certain number of windows will be lit at night.
The probability that 49 or fewer windows are lit is 0.525.
The probability that 50 or fewer windows are lit is 0.550.
The probability that 51 or fewer windows are lit is 0.573.
The probability that 52 or fewer windows are lit is 0.595.
The probability P_{50} that exactly 50 windows are lit is equal to 0.550 – 0.525 = 0.025.
The probability P_{52} that exactly 52 windows are lit is equal to 0.595 – 0.573 = 0.022.

Distribution functions based on probabilities that are expressed in percentages are often used for grouping events or people in categories. Probabilities that are expressed in percentages are defined between 0 and 100 instead of between 0 and 1. The certain event has a probability of 100% instead of a probability of 1 in this case.

Another interesting concept is the *percentile*; it is equal to 100 times one minus the distribution function.

$$\text{Percentile}(n) = 100(1 - F_n) \tag{4.6}$$

That is, the percentile is equal to 100 times the probability that the random variable will have values between the percentile variable, n in Equation 4.6, and the maximum value of the random variable. For example:

Random variable	0	1	2	3	4	5	6	7	8
Probability	0.05	0.10	0.15	0.33	0.12	0.10	0.08	0.05	0.02
Distribution function	0.05	0.15	0.3	0.63	0.75	0.85	0.93	0.98	1.00
Percentile	95	85	70	37	25	15	7	2	0

Another example is as follows:

Grade G	100	96	91	87	80	72	63	53	G < 53
Probability	0.04	0.10	0.12	0.31	0.12	0.11	0.08	0.07	0.05
Distribution function	0.04	0.14	0.26	0.57	0.69	0.80	0.88	0.95	1.00
Percentile	96	86	74	43	25	31	12	5	0

For example, a student who is classified, after taking a test, to be in the 86th percentile has a grade better than 96.

4.3 DISCRETE DISTRIBUTION FUNCTIONS FOR MORE THAN ONE VALUE OF A RANDOM VARIABLE WITH THE SAME PROBABILITY

The cases where some values of a random variable have the same probability must also be analyzed. For example, suppose the following values of a random variable have the same probability:

Random variable	0	1	2	3	4	5	6	7	8
Probability	0.05	0.05	0.1	0.12	0.12	0.2	0.2	0.07	0.07
Subset with same probability	0.05	0.12	0.2	0.07					
Number in subset	2	3	2	2					
Distribution function	0.1	0.46	0.86	1.00					

Consider a set of discrete events $S\{(A_1, A_2, A_3)_1, (A_4, A_5, A_6, A_7)_2, \ldots, (A_{100}, A_{101}, A_{102}, A_{103}, \ldots)_n, \ldots, (A_{562}, A_{563}, A_{564}, A_{565}, \ldots)_N\}$. These events are grouped in subsets that have the same probabilities. Here, n enumerates the number of groups. There are N groups. There are many more individual events than N. The probability that up to a certain number m of groups of events will occur is

$$F_m = \sum_{n=0}^{n=m} G_n P_n \quad 0 < m \leq N \tag{4.7}$$

where the G_m's are the number of events with the same probability. Here, m refers not to an enumeration of individual events but, rather, to an enumeration of groups of events with the same probability. It is true that this is somewhat awkward in the discrete case. With this definition, one cannot obtain the distribution function of an arbitrary number of events. It is restricted to distribution functions of groups of events as defined in Equation 4.7.

As before, consider the difference between the discrete distribution functions F_n and F_{n-1}.

$$F_m - F_{m-1} = \sum_{n=0}^{n=m} G_n P_n - \sum_{n=0}^{n=m-1} G_n P_n \tag{4.8}$$

Summing and dividing by the number of events G_m in group m, one obtains:

$$P_m = \frac{F_m - F_{m-1}}{G_m} \tag{4.9}$$

Obviously, the number of events G_m in group m has to be known.

4.4 CONTINUOUS DISTRIBUTION AND DENSITY FUNCTIONS

Consider a continuous random variable x defined between two limits a and b, $a < x \leq b$. The probability that all values of the random variable x with values less than or equal to u will occur is

$$F(u) = \int_a^u p(x)dx \quad \text{for } a < x \leq b \tag{4.10}$$

where p(x) is the probability density per unit length that the random variable u has values between x and x + dx. Of course, u must be less or equal to the largest value

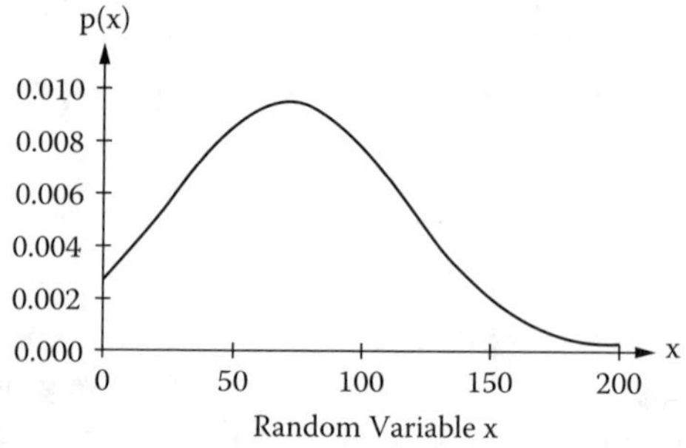

FIGURE 4.3 A plot of the probability density that a particular value of the continuous random variable u will have values between x and x + dx. Here, $0 < x \leq 200$.

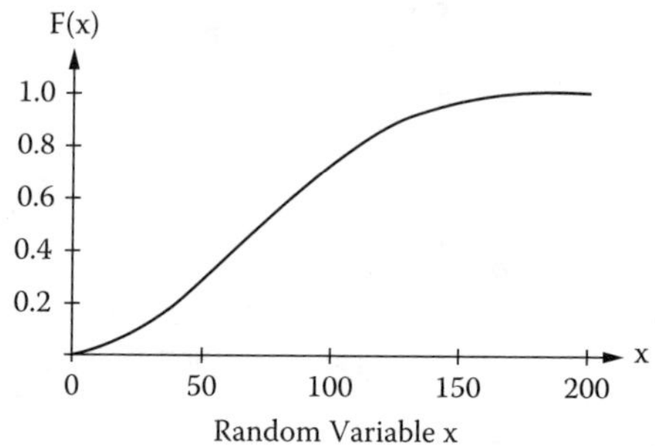

FIGURE 4.4 The continuous distribution function of the random variable x of Figure 4.3. Here, $0 < x \leq 200$. Note that F(x) is a monotonically increasing function of x.

a of the random variable x. The quantity F(u) is the *continuous distribution function*. A typical plot of a probability density as a function of a random variable is shown in Figure 4.3. The corresponding distribution function is plotted in Figure 4.4.

The same concepts in the continuous case were used as in the discrete case. Recall that the probability of n having a value less or equal to M is equal to the distribution function F_m. Similarly, the continuous distribution function F(u) is the probability that values of the random variable less or equal to u will occur. A typical plot of a continuous distribution function as a function of a random variable x is shown in Figure 4.4. As was the case with the discrete distribution function, the continuous distribution function F(u) is also a monotonic increasing function of the random variable; that is, $F(w) > F(u)$ for $w > u$.

For the continuous random variable x defined between two limits $a < x \leq b$

$$F(b) = 1 \tag{4.11}$$

In the continuous case, the distribution function of the lower limit must be equal to zero

$$F(a) = 0 \tag{4.12}$$

As before, the probability that the random variable x has values between w and u for $w > u$ is investigated. In this case, one obtains:

$$P(u < x \le w) = F(w) - F(u) \tag{4.13}$$

Again, of particular interest is the difference between the distribution functions F(u + Δu) and F(u). According to Equation 4.13, this can be written as:

$$F(u + \Delta u) - F(u) = \int_a^{u+\Delta u} p(x)dx - \int_a^u p(x)dx \tag{4.14}$$

The integrals can be combined to form:

$$F(u + \Delta u) - F(u) = \int_u^{u+\Delta u} p(x)dx \tag{4.15}$$

Equation 4.15 can be rewritten for a sufficiently small increment Δu as follows:

$$F(u + \Delta u) - F(u) \approx p(u)\Delta u \tag{4.16}$$

Dividing Equation 4.16 by Δu and taking the limit as Δu goes to zero:

$$\text{a)}\quad p(u) = \lim_{\Delta u \to 0} \frac{F(u + \Delta u) - F(u)}{\Delta u} \quad \text{or} \quad \text{b)}\quad p(u) = \frac{dF(u)}{du} \tag{4.17}$$

Thus, the probability density function p(u) of the random variable x having values between u and u + du is equal to the derivative of the distribution function F(u). The properties of the probability density function can be summarized as follows:

1. p(x) is positive and real.
2. $\int_a^b p(x)dx = 1 \quad \text{for} \quad a < x \le b$
3. $p(u) = \dfrac{dF(u)}{du}$

The properties of the distribution function can be summarized as follows:

1. The distribution function F(a) evaluated at the lower limit of the random variable is equal to zero; F(a) = 0 for $a < x \le b$.
2. The distribution function F(b) evaluated at the upper of the random variable limit is equal to one, F(b) = 1 for $a < x \le b$.
3. For values between the upper and lower limit of the random variable, the distribution function F(x) has values between zero and one; $0 < F(x) \le 1$ for $a < x \le b$.
4. F(w) > F(u) for w > u
5. $P(u < x \le w) = F(w) - F(u)$ for w > u
6. $dF(u)/du = p(u)$

EXAMPLE 4.3

Consider a random variable x defined between $0 < x \le \infty$ having an exponentially varying probability density:

$$p(x) = \frac{1}{\lambda}\exp\left[-\frac{x}{\lambda}\right]$$

The probability that the random variable will have values less than or equal to u is

$$\text{a)}\quad F(u) = \frac{1}{\lambda}\int_0^u \exp\left[-\frac{x}{\lambda}\right]dx \quad \text{b)}\quad F(u) = 1 - \exp\left[-\frac{u}{\lambda}\right]$$

One can check if F(u) indeed has the properties of a distribution function. At the upper limit, at $x = \infty$, the distribution function $F(\infty)$ should be equal to one. This is, indeed, true. The distribution function of the lower limit F(0) should be equal to zero. This is also true. F(u) increases monotonically. This too is true. The derivative of F(u) with respect to u is equal to the previously given probability density function $p(u) = 1/\lambda \exp[-u/\lambda]$. Therefore, F(u) has all the properties of a distribution function.

4.5 CONTINUOUS DISTRIBUTION FUNCTIONS FOR MORE THAN ONE VALUE OF A RANDOM VARIABLE WITH THE SAME PROBABILITY

Consider a set of continuous random variables having values with the same probability. The probability $P\{0 < x \le b\}$ that the random variable x $(0 < x \le a)$ will have values between zero and b $(b < a)$ is:

$$P\{0 < x \le b\} = F(b) \tag{4.18}$$

or

$$F(b) = \int_0^b g(x)\,p(x)\,dx \tag{4.19}$$

As before, consider the difference between the continuous distribution functions $F(u + \Delta u)$ and $F(u)$.

$$F(u+\Delta u) - F(u) = \int_0^{u+\Delta u} g(x)p(x)\,dx - \int_0^u g(x)p(x)dx \tag{4.20}$$

The first integral can be written as two integrals:

$$F(u+\Delta u) - F(u) = \int_0^u g(x)p(x)dx + \int_u^{u+\Delta u} g(x)p(x)dx - \int_0^u g(x)p(x)dx \tag{4.21}$$

Note that the first and third terms on the right-hand side of Equation 4.21 cancel. For sufficiently small Δu, one obtains:

$$F(u+\Delta u)-F(u)\approx \Delta u g(u)p(u) \tag{4.22}$$

Dividing by g(u)Δu and taking the limit as Δu goes to zero:

$$p(u)=\frac{1}{g(u)}\lim_{\Delta u\to 0}\frac{F(u+\Delta u)-F(u)}{\Delta u} \tag{4.23}$$

By taking the limit as Δu goes to zero, one obtains for the probability density p(u):

$$p(u)=\frac{1}{g(u)}\frac{\partial F}{\partial u} \tag{4.24}$$

Obviously, the density of events g(x) with the same probability has to be known.

For example, calculating the distribution function for the arrow hitting the target discussed in Chapter 2:

$$F(u)=2\pi\int_0^u p(r)r\,dr \tag{4.25}$$

where 2πrdr is the region of the target with the same probability density per unit radius, where u is smaller than the radius R of the target.

4.6 DISCRETE DISTRIBUTION FUNCTIONS OF MULTIPLE RANDOM VARIABLES

Consider two sets of events. The events in one set are labeled j and the events in the other set are labeled k. The probability that the discrete random variable j will have values less or equal to m and that the discrete random variable k will have values less or equal to n is a discrete distribution function of two variables F_{nm}.

$$F_{mn}=\sum_{j=1}^{j=m}\sum_{k=1}^{k=n}P(j\cap k) \tag{4.26}$$

where P(j ∩ k) is the joint probability that particular values of the random variables j and k will occur. Here, consistent with the discussion of Chapter 3, the sum of the joint probability over all values of the two random variables must be equal to one.

$$1=\sum_{j=1}^{j=J}\sum_{k=1}^{k=K}P(j\cap k) \tag{4.27}$$

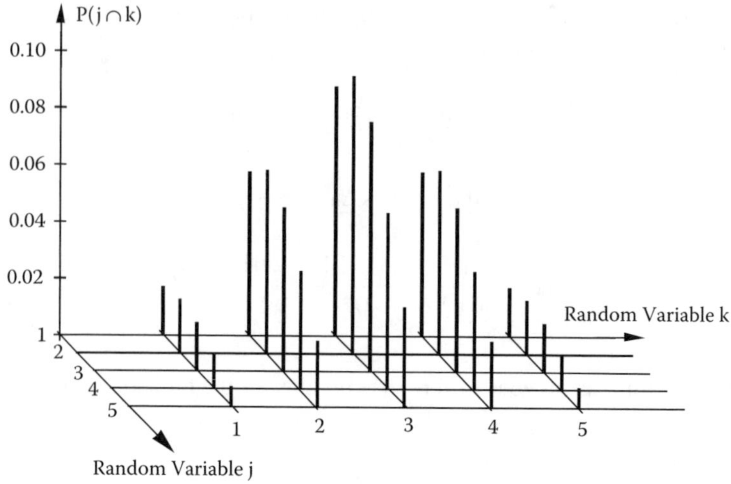

FIGURE 4.5 Probabilities associated with two random variables k and j.

where $0 < j \leq J$ and $0 < k \leq K$. In Equation 4.26, $m \leq J$ and $n \leq K$. A typical plot of the joint probability of two discrete random variables is shown in Figure 4.5, and the corresponding plot of the joint distribution function of the two discrete random variables is shown in Figure 4.6.

One can, of course, construct discrete distribution functions for more than two random variables.

$$F_{mnp\ldots q} = \sum_{j=1}^{j=m}\sum_{k=1}^{k=n}\sum_{\ell=1}^{\ell=p}\cdots\sum_{i=1}^{i=q} P(j \cap k \cap \ell \cap \ldots \cap i) \tag{4.28}$$

Returning to the discrete distribution function of two random variables, in general, the joint probability can be expressed as the product of a conditional probability and a total probability.

$$\text{a)}\quad P(j \cap k) = P(j|k)P_k \quad \text{or} \quad \text{b)}\quad P(j \cap k) = P(k|j)P_j \tag{4.29}$$

As is clear from Equation 4.26 and Figure 4.6, the distribution function F_{mn} is a monotonic increasing function of the random variables m and n; that is, $F_{mn} > F_{qn}$ for $q < m$, $F_{mn} > F_{mr}$ for $r < n$, and $F_{mn} > F_{qr}$ for both $q < m$ and $r < n$.

These rules can be extended to distribution functions of more than two discrete random variables.

It is interesting to investigate the second partial difference $F_{mn} - F_{(m-1)n} - F_{m(n-1)} + F_{(m-1)(n-1)}$.

$$F_{mn} - F_{(m-1)n} - F_{m(n-1)} + F_{(m-1)(n=1)} =$$

$$\sum_{j=1}^{j=m}\sum_{k=1}^{k=n} P(j \cap k) - \sum_{j=1}^{j=m-1}\sum_{k=1}^{k=n} P(j \cap k) - \sum_{j=1}^{j=m}\sum_{k=1}^{k=n-1} P(j \cap k) + \sum_{j=1}^{j=m-1}\sum_{k=1}^{k=n-1} P(j \cap k) \tag{4.30}$$

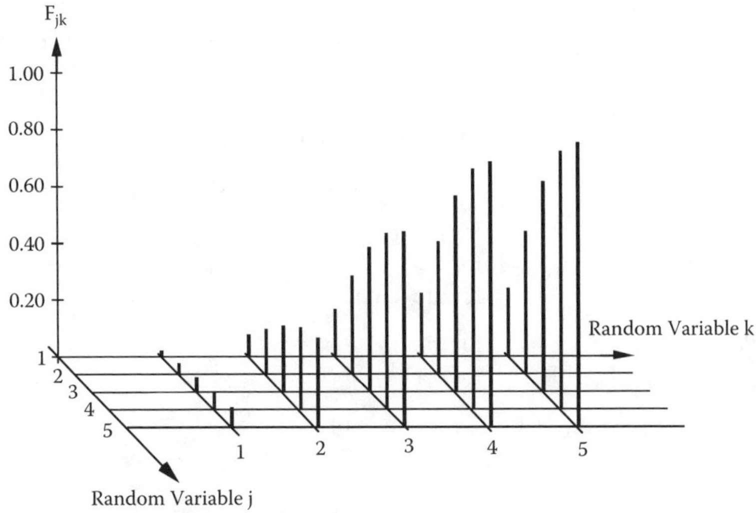

FIGURE 4.6 The distribution function for the two random variables k and j of Figure 4.5.

where $0 < j \le J$, $0 < k \le K$, $m < J$, and $n < K$. The sums can be combined as follows:

$$F_{mn} - F_{(m-1)n} - F_{m(n-1)} + F_{(m-1)(n-1)} = \sum_{k=1}^{k=n} P(m \cap k) - \sum_{k=1}^{k=n-1} P(m \cap k) \tag{4.31}$$

The remaining sums can be combined to form:

$$F_{mn} - F_{(m-1)n} - F_{m(n-1)} + F_{(m-1)(n-1)} = P(m \cap n) \tag{4.32}$$

that is, the joint probability is equal to the second partial backward difference of the distribution function. Another interesting quantity is the partial difference of the distribution function:

$$P_k = F_{Jk} - F_{J(k-1)} \tag{4.33}$$

The quantities F_{Jk} and $F_{J(k-1)}$ appear at the edge or margin of a diagram, as in Figure 4.6. Therefore, one can call P_k a *marginal probability*. Another form for the same marginal probabilities can be obtained by making use of Equation 4.29:

$$P_k = \sum_{j=1}^{j=J} P(j \cap k) \tag{4.34}$$

Of course,

$$P_j = \sum_{k=1}^{k=K} P(j \cap k) \tag{4.35}$$

P_j is also a marginal probability.

Summarizing the properties of distribution functions of two discrete random variables:

1. $F_{JK} = 1$ for $0 \le j \le J$ and $0 \le k \le K$
2. $0 < F_k \le 1$ for $0 \le j \le J$ and $0 \le k \le K$
3. $F_{mn} > F_{qn}$ for $m > q$
   $F_{mn} > F_{mr}$ for $n > r$
   $F_{mn} > F_{qr}$ for $m > q$ and $n > r$
4. $P(k < j \le m) = F_{mn} - F_{kn}$ for $m > k$
   $P(j < n \le r) = F_{mj} - F_{mr}$ for $r > j$
5. $F_{m,n} - F_{m,n-1} - F_{m-1,n} + F_{m-1,n-1} = P(m \cap n)$

EXAMPLE 4.4

Consider the distribution function F_{mn} of two discrete random variables m and n. Here, $0 \le m \le 5$ and $0 \le n \le 4$.

	n = 1	n = 2	n = 3	n = 4
m = 1	0.01	0.03	0.07	0.10
m = 2	0.03	0.09	0.21	0.30
m = 3	0.07	0.21	0.49	0.70
m = 4	0.09	0.27	0.63	0.90
m = 5	0.10	0.30	0.70	1.00

For example:

$$p(m = 3 \cap n = 2) = F_{3,2} - F_{2,2} - F_{3,1} + F_{2,1}$$

$$p(m = 3 \cap n = 2) = 0.21 - 0.09 - 0.07 + 0.03$$

$$p(m = 3 \cap n = 2) = 0.08$$

One can include the effect of more than one value of the discrete random variables having the same probability by following a similar procedure as in Section 4.3.

4.7 CONTINUOUS DISTRIBUTION FUNCTIONS OF MULTIPLE RANDOM VARIABLES

The continuous distribution functions of two random variables x and y can be deduced by extending the concept of a discrete distribution function of two random variables to the continuous case. In this case, the sum becomes an integral. The probability that the random variable x is less than or equal to u, and the random variable y is less than or equal to w is

$$F(u,w)=\int_0^u dx\int_0^w dy\, p(x\cap y) \tag{4.36}$$

where $p(x \cap y)\,dxdy$ is the joint probability that the random variable u will have values between x and x + dx, and the random variable v will have values between y and y + dy. Here, $0 < x \leq a$ and $0 < y \leq b$. Consistent with the discussion of Chapter 3, the integral of the joint probability over all values of the two random variables must be equal to one.

$$1=\int_0^a dx\int_0^b dy\, p(x\cap y) \tag{4.37}$$

As was the case with the discrete distribution function, one can define a distribution function associated with more than two random variables.

$$F(u,v,w,\ldots,q)=\int_0^u dx_1\int_0^v dx_2\int_0^w dx_3\ldots\int_0^q dx_N p(x_1\cap x_2\cap x_3\cap\ldots\cap x_N) \tag{4.38}$$

A method similar to the discussion leading to Equation 4.17 can be used to investigate the continuous version of Equation 4.30.

$$\begin{aligned}&F(u+\Delta u,v+\Delta v)-F(u,v+\Delta v)-F(u+\Delta u,v)+F(u,v)=\\&\int_0^{u+\Delta u} dx\int_0^{v+\Delta v} dy\, p(x\cap y)-\int_0^{u} dx\int_0^{v+\Delta v} dy\, p(x\cap y)\\&-\int_0^{u+\Delta u} dx\int_0^{v} dy\, p(x\cap y)+\int_0^{u} dx\int_0^{v} dy\, p(x\cap y)\end{aligned} \tag{4.39}$$

The first two integrals and last two integrals can be combined as follows:

$$\begin{aligned}&F(u+\Delta u,v+\Delta v)-F(u,v+\Delta v)-F(u+\Delta u,v)+F(u,v)=\\&\int_u^{u+\Delta u} dx\int_0^{v+\Delta v} dy\, p(x\cap y)-\int_u^{u+\Delta u} dx\int_0^{v} dy\, p(x\cap y)\end{aligned} \tag{4.40}$$

For sufficiently small Δu, one can approximate the integrals over x by:

$$\begin{aligned}&F(u+\Delta u,v+\Delta v)-F(u,v+\Delta v)-F(u+\Delta u,v)+F(u,v)\approx\\&\int_0^{v+\Delta v} dy\, p(u\cap y)\Delta u-\int_0^{v} dy\, p(u\cap y)\Delta u\end{aligned} \tag{4.41}$$

These integrals can be further combined to form:

$$F(u+\Delta u, v+\Delta v)-F(u, v+\Delta v)-F(u+\Delta u, v)+F(u, v)\approx\int_{v}^{v+\Delta v} dy\, p(u\cap y)\Delta u \quad (4.42)$$

Similarly, for sufficiently small Δv, the integrals can be approximated over y by:

$$F(u+\Delta u, v+\Delta v)-F(u, v+\Delta v)-F(u+\Delta u, v)+F(u, v)\approx p(u\cap v)\Delta u\Delta v \quad (4.43)$$

Dividing Equation 4.33 by ΔuΔv:

$$p(u\cap v)\approx\frac{\dfrac{F(u+\Delta u, v+\Delta v)-F(u+\Delta u, v)}{\Delta v}-\dfrac{F(u, v+\Delta v)-F(u, v)}{\Delta v}}{\Delta u} \quad (4.44)$$

Taking the limit as Δv goes to zero and keeping Δu constant:

$$p(u\cap v)\approx\frac{\dfrac{\partial F(u+\Delta u, v)}{\partial v}-\dfrac{\partial F(u, v)}{\partial v}}{\Delta u} \quad (4.45)$$

Taking the limit as Δu goes to zero and keeping v constant, it is seen that the joint probability density of the random variables u and v is equal to the second partial derivative of the distribution function F(u, v):

$$p(u\cap v)=\frac{\partial^2 F(u, v)}{\partial u\partial v} \quad (4.46)$$

It is interesting to see what one obtains by integrating Equation 4.46 over all x and y. Recall that $0 < x \le a$ and $0 < y \le b$.

$$\int_0^a dx\int_0^b dy\, p(x\cap y)=\int_0^a dx\int_0^b dy\frac{\partial}{\partial y}\left(\frac{\partial F(x, y)}{\partial x}\right) \quad (4.47)$$

The left-hand side of Equation 4.47 is equal to the probability of obtaining any value of x and y between zero and a and b at all. This probability is equal to one. Integrating over y:

$$1=\int_0^1 dx\left(\frac{\partial F(x, b)}{\partial x}-\frac{\partial F(x, 0)}{\partial x}\right) \quad (4.48)$$

Integrating over x:

$$1=F(a, b)-F(0, b)-F(a, 0)+F(0, 0) \quad (4.49)$$

Note that F(0,b), F(a,0), and F(0,0) are equal to zero and F(a,b) = 1 for $0 < x \le a$ and $0 < y \le b$.

The properties of the joint probability density function $p(x \cap y)$ are as follows:

1. $p(x \cap y)$ is positive and real.
2. $\int_0^a dx \int_0^b dy\, p(x \cap y) = 1$ for $0 < x \le a$, and $0 < y \le b$.
3. $p(x \cap y) = \dfrac{\partial^2 F(x,y)}{\partial x \partial y}$.

Another interesting quantity is

$$p(x) = \frac{\partial F(x,b)}{\partial x} \tag{4.50}$$

Similar to the discrete case, the quantities F(x,b) appears at the edge or margin of the x-y random variable area. Therefore, p(x) can be called a *marginal probability density*. Another way of writing Equation 4.50 is

$$p(x) = \int_0^b dy\, p(x \cap y) \tag{4.51}$$

Of course, p(y), the probability density along the other edge or margin of the x-y area, is also a marginal probability density

$$\text{a)}\quad p(y) = \frac{\partial F(a,y)}{\partial y} \qquad \text{b)}\quad p(y) = \int_0^a dx\, p(x \cap y) \tag{4.52}$$

p(y) is also a marginal probability density. We summarize the properties of the continuous distribution function of two continuous random variables as follows:

1. $F(0,0) = 0$ for $0 < x \le a$ and $0 < y \le b$
2. $F(0,y) = 0$ for $0 < x \le a$ and $0 < y \le b$
3. $F(x,0) = 0$ for $0 < x \le a$ and $0 < y \le b$
4. $F(a,b) = 1$ for $0 < x \le a$ and $0 < y \le b$
5. $0 < F(x,y) \le 1$ for $0 < x \le a$ and $0 < y \le b$
6. $F(x,y) > F(u,y)$ for $u < x$
   $F(x,y) > F(x,v)$ for $v < y$
   $F(x,y) > F(u,v)$ for $u < x$ and $v < y$
7. $P(q < x < u, 0 < y < v) = F(u,v) - F(q,v)$ for $q < u$
   $P(0 < x < u, s < y < v) = F(u,v) - F(u,s)$ for $s < v$
8. $\dfrac{\partial F(u,v)}{\partial u \partial v} = p(u \cap v)$

EXAMPLE 4.5

Consider the function:

$$F(u,v) = 2u^3v - 3u^2v^2 + 2uv^3 \qquad \text{for } 0 < u \le 1 \text{ and } 0 < v \le 1$$

Investigating the properties of F(u,v):

1. At u = 0 and v = 0, F(u,v) = 0
2. At u = 0, F(u,v) = 0
3. At v = 0, F(u,v) = 0
4. At u = 1 and v = 1, F(u,v) = 1
5. F(u,v) < 1 for $0 < u \le 1$ and $0 < v \le 1$
6. F(u,v) is a monotonically increasing function of both u and v.
7. $\partial F(u,v)/\partial u \partial v = 6(u-v)^2$ and $6(u - v)^2$ has all the properties of a joint probability function for $0 < u \le 1$ and $0 < v \le 1$.

Therefore, F(u,v) is indeed a distribution function.

One can include the effect of more than one value of the continuous random variables having the same probability by following a similar procedure as in Section 4.5.

4.8 PHASE SPACE: A SPECIAL CASE OF MULTIPLE RANDOM VARIABLES

There is a special case of multiple random variables. It is the space formed by the position vector components q_μ and momentum vector components p_μ of a physical system. This space is known as *phase space* Γ. The phase space of a one-dimensional physical system is shown in Figure 4.7. Note that the phase space for a one-dimensional system has two dimensions, the position q, and the momentum p. A real three-dimensional particle would have a six-dimensional phase space consisting of the

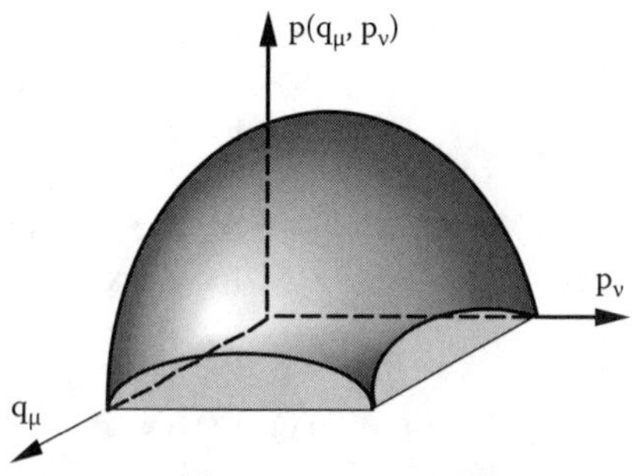

FIGURE 4.7 Phase space Γ of a physical system. Here, $p(q_\mu,p_n)$ is the probability density per unit area in phase space. The phase space here has many dimensions, the position vector components q_μ and the momentum vector components p_n.

three position vector components q_μ and the three momentum vector components p_μ ($\mu = 1, 2, 3$). A system of two particles would have a twelve-dimensional phase space consisting of the three position vector components $q_{\mu a}$ and the three momentum vector components $p_{\mu a}$ ($\mu = 1, 2, 3$ and $a = 1, 2$) of each particle, etc.

The phase space coordinates are random variables, and there is a probability density $p(q_\mu,p_\mu)$ per unit area or unit volume in phase space. The probability of a physical system having position vector components between q_μ and $q_\mu + dq_\mu$ and momentum vector components between p_μ and $p_\mu + dp_\mu$ is $p(q_\mu,p_\mu)d\Gamma$.

$$\begin{aligned} &P\{u_x < q_x \le u_x + du_x, u_y < q_y \le u_y + du_y \ldots v_x < p_x \le v_x + dv_x, \\ &\quad v_y < p_y \le v_y + dv_y \ldots\} = p(u_x < q_x \le u_x + du_x, u_y < q_y \le u_y + du_y \ldots \\ &\quad v_x < p_x \le v_x + dv_x, v_y < p_y \le v_y + dv_y \ldots)d\Gamma \end{aligned} \tag{4.53}$$

The probability must, of course, be normalized.

$$\int_\Gamma p(q_\mu, p_\mu)d\Gamma = 1 \tag{4.54}$$

where a shorthand notation of $p(q_\mu,p_\mu)$ was used instead of listing all position vector and momentum vector components as was done in Equation 4.52.

The average value $\langle U \rangle$ of some quantity such as the energy $U(q_\mu,p_\mu)$ that is a function of the phase space coordinates is

$$\langle U \rangle = \int_\Gamma U(q_\mu, p_\mu)p(q_\mu, p_\mu)d\Gamma \tag{4.55}$$

For quantum mechanical systems, it is sometimes the case that the momentum vector components have discrete values. For example, the momentum vector components $p_\mu = i\hbar 2\pi n_\mu / L_\mu$, $n_\mu = 0, 1, 2, \ldots$, of a free electron in a solid are quantized, whereas its position vector components q_μ have continuous values. Here, $\hbar = 1.054588664 \times 10^{-34}$ J-s is Planck's constant divided by 2π, L_μ is the length of the solid in the μ direction, and $i = \sqrt{-1}$. This is not different from the discrete and continuous random variables discussed previously.

PROBLEMS

Problem 4.1

Given:

Consider three functions G_n, H_n, and K_n of a random variable n having values n = 1, 2, 3, 4, 5, 6, 7, 8, 9, and 10.

n	G_n	H_n	K_n
1	0.08	0.1	0.05
2	0.12	0.15	0.09
3	0.32	0.18	0.13
4	0.48	0.26	0.19
5	0.72	0.48	0.56
6	0.86	0.62	0.72
7	0.72	0.80	0.86
8	0.84	0.91	0.92
9	0.96	0.97	0.96
10	1.00	1.00	0.98

Problem:

a) Which of the functions G_n, H_n, and/or K_n of the random variable n can be a distribution function?
b) Calculate the probabilities associated with function or functions having properties of a distribution function?

Problem 4.2

Given:

In a class of 400 students, 64 students have grades greater than 95, 56 students have a grade higher than 90 but less than 96, 48 students have a grade higher than 83 but less than 91, 52 students have a grade higher than 75 but less than 84, and 72 students have a grade higher than 65 but less than 76. The remaining students have a grade less than 66. The highest grade is 100.

The probability of a number of students with a particular range of grades is proportional to the number of students in that group.

Problem:

a) Calculate the probability of students falling in each grade group.
b) Calculate the distribution function of the students in the grade groups.
c) Calculate the percentile of students having grades greater than the grade in the next lower group.

Problem 4.3

Given:

A particular DNA molecule consists of 320 genes. Thirty-two genes have 16,500 or more nucleotides but less than 20,001 nucleotides, 48 genes have 13,000 or more nucleotides but less than 16,500 nucleotides, 64 genes have 9,500 or more nucleotides but less than 13,000 nucleotides, 28 genes have 6,000 or more nucleotides but less than 9,500 nucleotides, and 28 genes have 2,500 or more nucleotides but less than 6,000 nucleotides.

Problem:

a) Calculate the probability of genes being in a group with a certain range of nucleotides.
b) Calculate the distribution function of the groups of genes.
c) Calculate the percentile of groups of genes having more than a certain number of nucleotides.

Problem 4.4

Given:

Consider three functions of the random variable x. Here, $0 < x \leq 0.8$.

$F_1(x) = 3.2x^2 - 0.67x - x^3$
$F_2(x) = 1.5x^2 + 0.69x - x^3$
$F_3(x) = 0.04 + 2x - x^2$

Problem:

a) Determine which of the given functions are distribution functions.
b) Calculate the probability density for the function that has the properties of a distribution function.

Problem 4.5

Given:

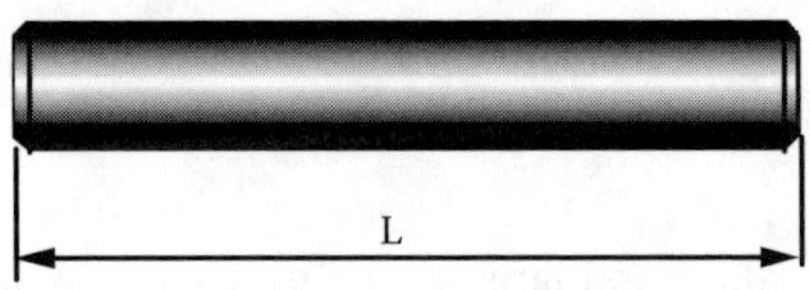

Shaft with a nominal length L.

Shafts for a machine are manufactured within a tolerance ±b. The nominal length of the shaft is L = 300 mm. The probability density p(x) that the shaft will have a length x is

$$p(x) = \frac{\sigma}{2\tan^{-1}\frac{b}{\sigma}} \frac{1}{\sigma^2 + (x-L)^2}$$

where $L - b < x \leq L + b$ and where $b = 0.025$ and $\sigma = 0.0125$.

Note:

$$\int \frac{du}{1+u^2} = \tan^{-1} u$$

Problem:

Calculate and plot the fraction of shafts having various lengths up to values of u. Here, u has values between L – b and L + b.

Problem 4.6

Given:

A tree has roots below the ground of length x. The probability density p(x) per unit length that the roots of the tree will have a length u below the ground having values between x and x + Δx is:

$$p(x) = \frac{\alpha^3}{2} x^2 \exp(-\alpha x) \quad \text{for} \quad 0 < x \leq \infty$$

The tree has a height y above the ground. The probability density q(y) per unit length that the tree will have a height v above ground having values between y and y + Δy is

$$q(y) = \beta^2 y \exp(-\beta y) \quad \text{for} \quad 0 < y \leq \infty$$

where $\alpha > \beta$.

The total height w of the tree, including roots, is

$$w = x + y$$

Problem:

a) Determine the smallest and largest total dimension w of the tree, including roots.
b) Calculate and plot the probability density p(w) per unit length that the tree will have a total dimension s having values between w and w + Δw.

Problem 4.7

Given:

Consider the following function of two discrete random variables:

A_{mn}:

m =	1	2	3	4	5	6	7	8
n = 1	0.01	0.02	0.03	0.03	0.04	0.04	0.04	0.05
n = 2	0.02	0.03	0.04	0.04	0.05	0.06	0.08	0.12
n = 3	0.02	0.03	0.36	0.40	0.42	0.44	0.46	0.48

A_{mn}: **(continued)**

m =	1	2	3	4	5	6	7	8
n = 4	0.03	0.04	0.40	0.60	0.61	0.62	0.63	0.64
n = 5	0.03	0.05	0.42	0.61	0.94	0.95	0.96	0.96
n = 6	0.04	0.06	0.44	0.62	0.95	0.97	0.98	0.98
n = 7	0.04	0.08	0.46	0.63	0.96	0.98	0.99	0.99
n = 8	0.05	0.12	0.48	0.64	0.96	0.98	0.99	1.00

B_{mn}:

m =	1	2	3	4	5	6	7	8
n = 1	0.01	0.02	0.03	0.03	0.04	0.04	0.04	0.05
n = 2	0.02	0.03	0.04	0.04	0.05	0.06	0.08	0.12
n = 3	0.02	0.03	0.36	0.40	0.42	0.44	0.46	0.48
n = 4	0.03	0.04	0.40	0.60	0.61	0.62	0.63	0.64
n = 5	0.03	0.05	0.56	0.88	0.94	0.95	0.96	0.96
n = 6	0.04	0.06	0.48	0.82	0.95	0.97	0.98	0.98
n = 7	0.04	0.08	0.46	0.63	0.96	0.98	0.99	0.99
n = 8	0.05	0.12	0.48	0.64	0.96	0.98	0.99	1.00

Problem:

a) Determine which of the given functions are distribution functions.
b) Calculate the joint probability $P(n = 4 \cap m = 6)$ for the functions that have the properties of a distribution function.

Problems 4.8

Given:

Consider two functions of two continuous random variables x and y. Here, $0 < x \leq 2$ and $0 < y \leq 2$.

$$A(x,y) = 0.0125[2xy^2 + x^2y^2 + xy^3 + x^2y^{32}]$$

$$B(x,y) = 0.0125[2xy^2 + 4x^2 + xy^3 + x^2y^3]$$

Problem:

a) Determine which of the given functions are distribution functions.
b) Calculate the joint probability density $p(x \cap y)$ for the functions that have the properties of a distribution function.

5 Conditional Distribution Functions and a Special Case: *The Sum of Two Random Variables*

5.1 DISCRETE CONDITIONAL DISTRIBUTION FUNCTIONS

Recall from Chapter 4 that the distribution function of two discrete random variables j and k, for $0 < j \le M$ and $0 < k \le N$ has the following form:

$$F_{mn} = \sum_{j=1}^{j=m}\sum_{k=1}^{k=n} P(j \cap k) \tag{5.1}$$

where $P(j \cap k)$ is the joint probability of the two random variables j and k.

$$\sum_{j=1}^{j=M}\sum_{k=1}^{k=N} P(j \cap k) = 1 \tag{5.2}$$

and where $m < M$ and $k < N$. The joint probability $P(j \cap k)$ of the two random variables can be expressed in terms of the conditional probabilities and the total probabilities:

$$\text{a)}\quad P(j \cap k) = P(j|k)P_k \quad \text{or} \quad \text{b)}\quad P(j \cap k) = P(k|j)P_j \tag{5.3}$$

where $P(j|k)$ is the conditional probability that event j will occur, provided event k has occurred. Similarly, $P(k|j)$ is the conditional probability that event k will occur, provided event j has occurred.

Postulating a function $F(m|k)$ defined as:

$$F_{mn} \equiv \sum_{k=1}^{k=n} F(m|k)P_k \tag{5.4}$$

The function F(m|k) is not well defined at this point. Nevertheless, substituting Equation 5.3a into Equation 5.1 and equating the resulting expression to Equation 5.4:

$$\sum_{k=1}^{k=n} F(m|k)P_k = \sum_{j=1}^{j=m}\sum_{k=1}^{k=n} P(j|k)P_k \tag{5.5}$$

Rewriting Equation 5.5:

$$\sum_{k=1}^{k=n}\left[F(m|k) - \sum_{j=1}^{j=m} P(j|k)\right]P_k = 0 \tag{5.6}$$

Because, in general, P_k is not equal to zero, the quantity in the square bracket must be equal to zero.

$$F(m|k) = \sum_{j=1}^{j=m} P(j|k) \tag{5.7}$$

where F(m|k) is the conditional distribution function; it is the probability that any of the values of the random variable j between one and m ($0 < j \le m$) will occur provided event k had occurred. Similarly,

$$F(n|j) = \sum_{k=1}^{k=n} P(k|j) \tag{5.8}$$

Here, F(n|j) is the conditional distribution function; it is the probability that any of the values of the random variable k between one and n ($0 < k \le n$) will occur provided event j had occurred.

The discrete conditional distribution function has properties similar to the distribution function of a single discrete random variable:

1. $F(M|k) = 1$ for $0 < j \le M$
2. $F(m|k) > F(q|k)$ for $m > q$
3. $P(q < j \le m|k) = F(m|k) - F(q|k)$ for $m > q$
4. $p(m|k) = F(m|k) - F([m-1]|k)$

5.2 CONTINUOUS CONDITIONAL DISTRIBUTION FUNCTIONS

Similarly, recall from Chapter 4 that the distribution function of two continuous random variables $0 < x \le a$ and $0 < y \le b$ has the following form:

$$F(u,v) = \int_0^u dx \int_0^v dy\, p(x \cap y) \tag{5.9}$$

where

$$1 = \int_0^a dx \int_0^b dy\, p(x \cap y) \tag{5.10}$$

and where u < a and v < b. The joint probability density $p(x \cap y)$ of the two random variables can be expressed in terms of the conditional probability densities and the total probability densities:

$$\text{a)} \quad p(x \cap y) = p(x|y)p(y) \quad \text{or} \quad \text{b)} \quad p(x \cap y) = p(y|x)p(x) \tag{5.11}$$

where $p(x|y)dx$ is the conditional probability that the random variable x will have values between x and x + dx provided the random variable z has values between y and y + dy. Similarly, $p(y|x)dy$ is the conditional probability that the random variable z will have values between y and y + dy provided the random variable x has values between x and x + dx.

One can postulate a quantity $F(u|y)$:

$$F(u,v) \equiv \int_0^v F(u|y)p(y)dy \tag{5.12}$$

Again, the function $F(u|y)$ is not well defined at this point. Nevertheless, substituting Equation 5.11a into Equation 5.9 and setting the result equal to Equation 5.12:

$$\int_0^v F(u|y)p(y)dy = \int_0^v dy \int_0^u dx\, p(x|y)p(y) \tag{5.13}$$

Equation 5.13 can be rewritten as follows:

$$\int_0^v dy \left[F(u|y) - \int_0^u dx\, p(x|y)\right] p(y) = 0 \tag{5.14}$$

Because, in general, p(y) is not equal to zero, the term in the square bracket must be equal to zero.

$$F(u|y) = \int_0^u dx\, p(x|y) \tag{5.15}$$

where $F(u|y)$ is the conditional distribution function; it is the probability density that any of the values of the random variable x between zero and u ($0 < x \leq u$) will occur provided the random variable z will have values between y and y + dy. Similarly,

$$F(v|x) = \int_0^v dy\, p(y|x) \tag{5.16}$$

Here, $F(v|x)$ is the conditional distribution function; it is the probability density that any of the values of the random variable y between zero and v ($0 < y \leq v$) will occur provided the random variable x will have values between x and x + dx.

The continuous conditional distribution function, which by necessity involves two random variables, has properties similar to the distribution function of a single continuous random variable:

1. $F(0|y) = 0$ for $0 < x \leq a$
2. $F(a|y) = 1$ for $0 < x \leq a$
3. $F(u|y) > F(q|y)$ for $u > q$
4. $P(q < x \leq u|y) = F(u|y) - F(q|y)$ for $u > q$
5. $p(u|y) = \dfrac{\partial F(u|y)}{\partial u}$

This can be extended to the case of more than two random variables.

5.3 A SPECIAL CASE: THE SUM OF TWO STATISTICALLY INDEPENDENT DISCRETE RANDOM VARIABLES

Consider two statistically independent discrete random variables m and n. Here, $0 < m \leq M$ and $0 < n \leq N$, where M is not necessarily equal to N. In this case, the joint probability $P(m \cap n)$ that particular values of the random variables m and n will occur is equal to the product of the individual probabilities P_n and Q_m of the random variables, as shown in Figure 5.1.

$$P(m \cap n) = P_m Q_n \tag{5.17}$$

Note that:

$$\text{a)} \sum_{n=1}^{n=N} P_n = 1 \quad \text{and} \quad \text{b)} \sum_{m=1}^{m=M} Q_m = 1 \tag{5.18}$$

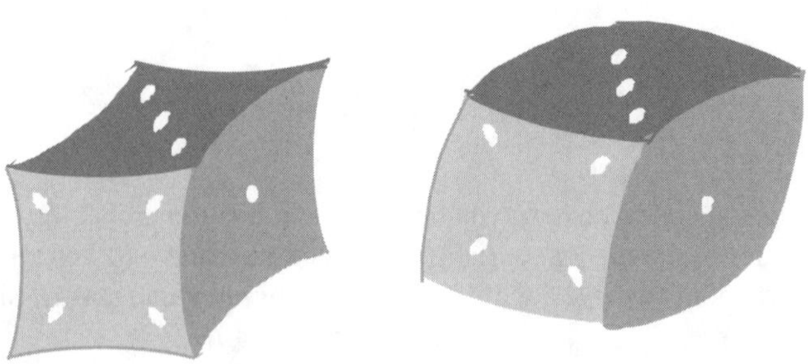

FIGURE 5.1 A distorted pair of dice.

Consider the sum w of the random variables m and n:

$$w \equiv m + n \tag{5.19}$$

where $w \leq M + N$.

This special case is best illustrated by an example. Consider two distorted dice as shown in Figure 5.1. Suppose the probabilities that the light die will be showing a particular face is

$$P_1 = 0.095,\ P_2 = 0.100,\ P_3 = 0.105,\ P_4 = 0.110,\ P_5 = 0.480,\ \text{and}\ P_6 = 0.110$$

where the subscript is equal to the number of points showing. Similarly, suppose the probability that the darker die will be showing a particular face is

$$Q_1 = 0.110,\ Q_2 = 0.510,\ Q_3 = 0.110,\ Q_4 = 0.100,\ Q_5 = 0.090,\ \text{and}\ Q_6 = 0.080$$

Note that the light die will most likely show five dots, while the gray die will most likely show two dots.

To calculate the probabilities $P\{m + n = w\}$ of rolling a particular number $w = m + n$ with both dice, one first organizes the probabilities P_n of rolling a particular number n with the light dice and the probabilities Q_m of rolling a particular number m with the darker dice in two row vectors, as shown in the following text. Note that one can extend the vectors on both ends of the specified probabilities by adding zeros to the respective rows.

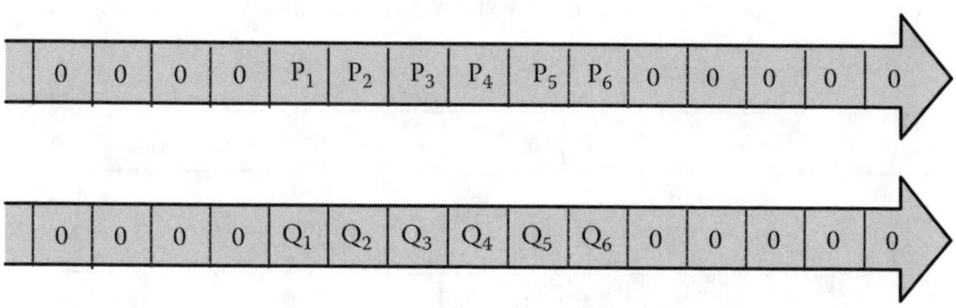

Next, the direction of one of the row probability vectors is reversed as shown below.

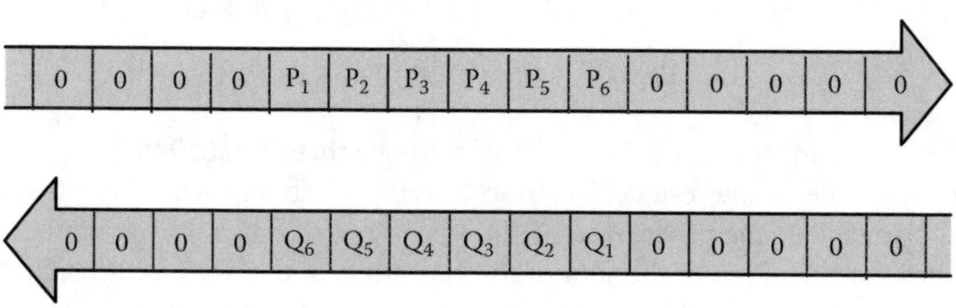

Suppose one is interested in rolling a two with both dice; this can only be achieved in one way, namely, by rolling a one with each die. To calculate the probabilities P{m + n = 2} of rolling a w = 2 with both dice, one slides the vectors until the sum of the subscripts of each pair of probabilities P_n and Q_m that are aligned add to two in this case, as shown in the following figure.

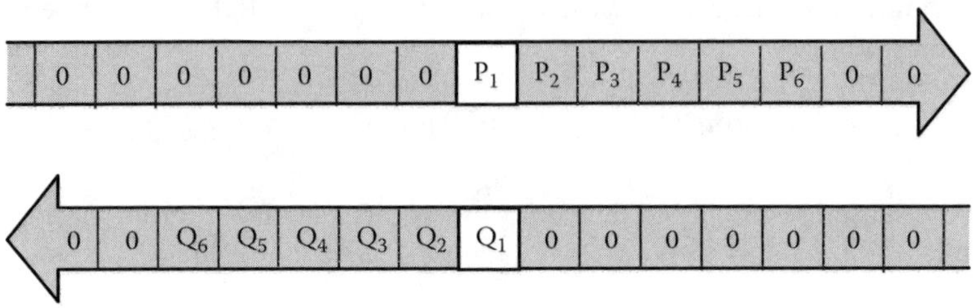

The next step is to multiply the aligned pairs of probabilities and add the results:

$$P\{m + n = 2\} = \ldots + (0 \times Q_6) + (0 \times Q_5) + (0 \times Q_4) + (0 \times Q_3) + (0 \times Q_2) + (P_1 \times Q_1) + (P_2 \times 0) + (P_3 \times 0) + (P_4 \times 0) + (P_5 \times 0) + (P_6 \times 0) + \ldots \quad (5.20)$$

Therefore, the probability P{m + n = 2} of rolling a two is 0.01045.

Suppose that one is interested in rolling a three with both dice. One proceeds as before. To calculate the probabilities P{m + n = 3} of rolling a w = 3 with both dice, one slides the vectors until the sum of the subscripts of each pair of probabilities P_n and Q_m that are aligned add to three, as shown in the following figure.

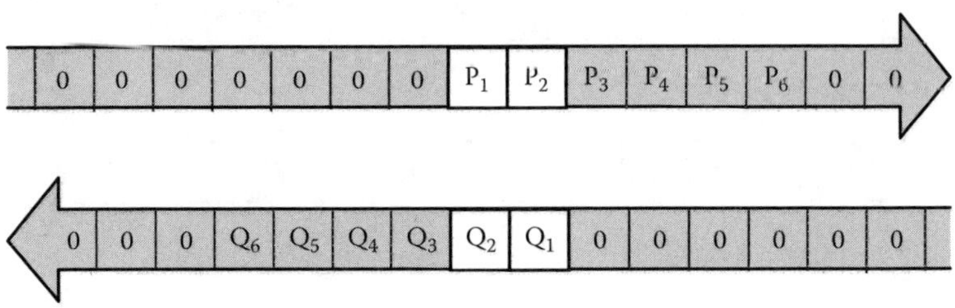

Multiplying the aligned pairs of probabilities and adding the results:

$$P\{m + n = 3\} = \ldots + (0 \times Q_6) + (0 \times Q_5) + (0 \times Q_4) + (0 \times Q_3) + (P_1 \times Q_2) + (P_2 \times Q_1) + (P_3 \times 0) + (P_4 \times 0) + (P_5 \times 0) + (P_6 \times 0) + (0 \times 0) + \ldots \quad (5.21)$$

Therefore, the probability P{m + n = 3} of rolling a three is 0.05945.

Suppose one is interested in rolling a four with both dice. One proceeds as before. To calculate the probabilities P{m + n = 4} of rolling a w = 4 with both dice, one slides the vectors until the sum of the subscripts of each pair of probabilities P_n and Q_m that are aligned add to four as shown in the following figure.

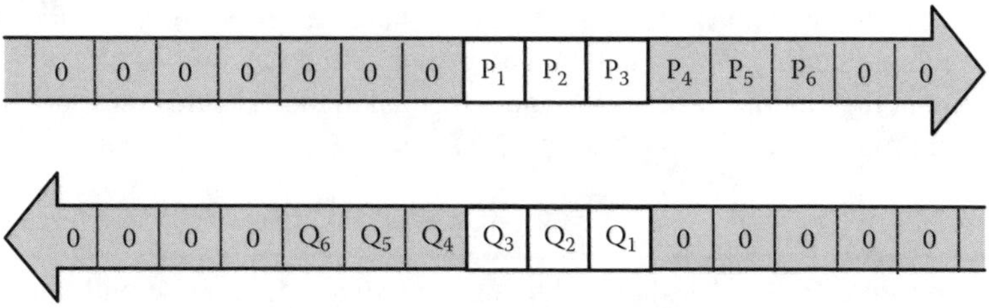

Multiplying the aligned pairs of probabilities and adding the results:

$$P\{m + n = 4\} = \ldots + (0 \times Q_6) + (0 \times Q_5) + (0 \times Q_4) + (P_1 \times Q_3) + (P_2 \times Q_2) + (P_3 \times Q_1) + (P_4 \times 0) + (P_5 \times 0) + (P_6 \times 0) + \ldots \quad (5.22)$$

Therefore, the probability P{m + n = 4} of rolling a four is 0.073, etc. A plot of the probabilities P{w = m + n} of rolling a particular number w with both dice is shown in Figure 5.2.

Note that the preceding process is a *discrete convolution* of the probabilities of the two random variables m and n.

This simple example demonstrates some of the properties of the probability of the sum of the random variables. One obtained the discrete convolution of the probabilities of the two random variables. Equations 5.20, 5.21, and 5.22 can be generalized as follows.

$$P\{w = m + n\} = \sum_{n=-\infty}^{n=\infty} P_n Q_{w-n} \quad (5.23)$$

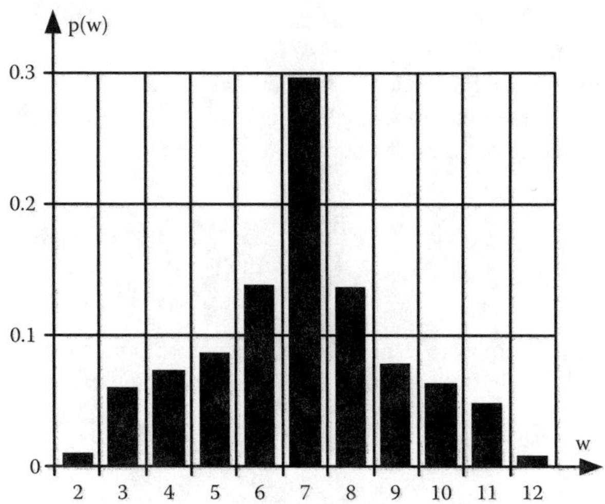

FIGURE 5.2 A typical plot of the probability P{w = m + n} of rolling a particular number with two dice. Observe that for the foregoing values of the probabilities for the dice, one obtained a larger probability of rolling a seven with the two dice.

Thus, for random variables m and n extended to have values between plus and minus infinity, the probability $P\{w = m + n\}$ that a value of the random variable w is equal to the sum of the random variables m and n is equal to the discrete convolution of the individual probabilities P_m and Q_n.

EXAMPLE 5.1

The probabilities P_n and Q_m that n photons are radiated in a mode with an average number of β photons and m photons are radiated in a mode with an average number of β photons are

$$P_n = e^{-\beta} \frac{\beta^n}{n!} \quad \text{and} \quad Q_m = e^{-\beta} \frac{\beta^m}{m!}$$

where m and n are two statistically independent random variables having values $-1 < m \le \infty$ and $-1 < n \le \infty$. The concept of an average value will be discussed in a later chapter.

The total number of photons radiated, k, is equal to the sum of the photons being radiated in the individual modes.

$$k = m + n$$

One can use Equation 5.23 to calculate the probability P_k that a total number k of photons will be radiated.

$$P_k = e^{-2\beta} \sum_{n=0}^{n=k} \frac{\beta^{n+k-n}}{n!(k-n)!}$$

In Equation 5.24 the smallest value of the random variables was equal to one. Here the smallest value of the random variables is equal to zero. By multiplying and dividing by k! and factoring out β^k, one obtains:

$$P_k = e^{-2\beta} \frac{\beta^k}{k!} \sum_{n=0}^{n=k} \frac{k!}{n!(k-n)!}$$

Note that the sum is equal to 2 raised to the k power. Therefore, one obtains for the probability P_k of the total number k of photons being radiated the following expression:

$$P_k = e^{-2\beta} \frac{(2\beta)^k}{k!}$$

Note that, as required, the probabilities P_k are normalized for $-1 < k \le \infty$.

5.4 A SPECIAL CASE: THE SUM OF TWO STATISTICALLY INDEPENDENT CONTINUOUS RANDOM VARIABLES

Consider two statistically independent continuous random variables x and y. Here, $0 < x \leq a$ and $0 < y \leq b$, where a is not necessarily equal to b. In this case, the joint probability density $p(x \cap y)$ of the random variables x and y is equal to the product of the individual probability densities p(x) and q(y) of the random variables.

$$p(x \cap y) = p(x)q(y) \tag{5.24}$$

The integrals of the individual probability densities of the ranges of the respective random variables must be equal to one.

$$\text{a)} \int_0^a p(x)dx = 1 \quad \text{and} \quad \text{b)} \int_0^b q(y)dy = 1 \tag{5.25}$$

Consider the sum u of the random variables x and y:

$$w = x + y \tag{5.26}$$

By solving Equation 5.26 for y and substituting the resulting expression into Equation 5.24, one obtains for the joint probability density $p(x \cap w)$:

$$p(x \cap w) = p(x)q(w - x) \tag{5.27}$$

The probability density s(w) can be obtained from the joint probability density $p(x \cap w)$ by integrating over the random variable x, as was discussed in Chapter 3, Equation 3.28.

$$s(w) = \int_{-\infty}^{\infty} p'(x)q'(w - x)dx \tag{5.28}$$

where $p'(x)$ is equal to p(x) over the range of the random variable x, in this case for values of x between 0 and a. For all other values of x, $p'(x)$ is equal to zero. Here, $q'(y = w - x)$ is equal to $q(y = w - x)$ over the range of the random variable y; in this case $w - x$ lies between 0 and b. This is similar to Equation 5.23 for the discrete case.

Equation 5.28 is the convolution of the probability densities $p'(x)$ and $q'(w - x)$. Note that as w varies, the probability density function $q'(w - x)$ is "sliding through" the probability density function $p'(x)$, similar to the arrows representing the discrete probability densities, as discussed previously.

FIGURE 5.3 A tree has a height x and the roots extend a distance y under the ground. The total height from the bottom of the roots to the top of the tree is the sum w of the tree height x and the root extension y.

As was done in the discrete case, calculating the probability density $s\{w = x + y\}$ of the sum of two statistically independent, continuous random variables is also best illustrated by an example. Consider a forest of trees. Each tree has a height x and the roots extending below the surface to a depth y (see Figure 5.3). Because the heights vary from tree to tree, one can consider that the tree height x is a random variable. Similarly, the depth y to which the roots extend also varies from tree to tree. Thus, y is also a random variable. Of course, the tree height and root depth are not really statistically independent. However, for the purpose of this problem, it is assumed that they are statistically independent random variables. For simplicity, one assumes that the probability densities p(x) and q(y) have the following forms:

$$\text{a)}\quad p(x) = \alpha e^{-\alpha x} \quad \text{and} \quad \text{b)}\quad q(y) = \beta e^{-\beta y} \tag{5.29}$$

where $0 < x \leq \infty$ and $0 < y \leq \infty$. Of course, no tree is infinitely tall. However, the probability densities decay to practically zero long before the random variables reach infinity. In this example, $\alpha = 0.25$ per m and $\beta = 0.125$ per m.

The sum w of the random variables will have a minimum value of zero and a maximum value of infinity, $0 < w \leq \infty$ in this case. Plots of the two probability densities are shown in Figures 5.4 and 5.5.

Next, both p(x) and $q(y = w - x)$ are plotted as a function of x, as shown in Figures 5.6 and 5.7. Increasing values of x correspond to decreasing values of y. Therefore, one first reverses the direction of one of the probability density functions $q'(y)$ or $q'(x - w)$, as shown in Figures 5.6 and 5.7. This is similar to the row vectors of the discrete probabilities.

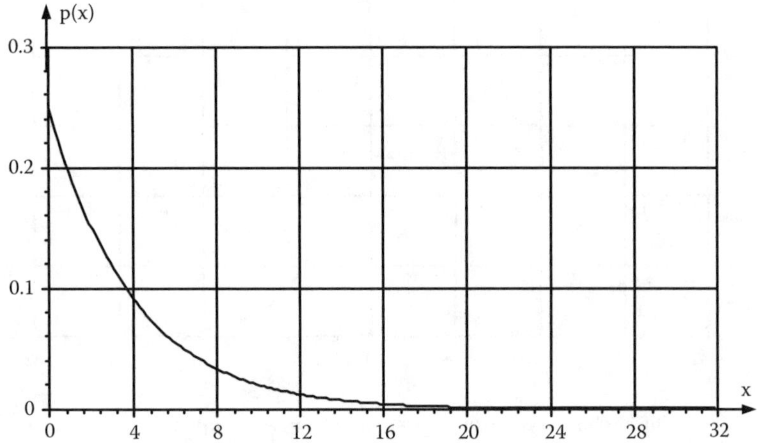

FIGURE 5.4 The probability density p(x) of the random variable x representing the tree height.

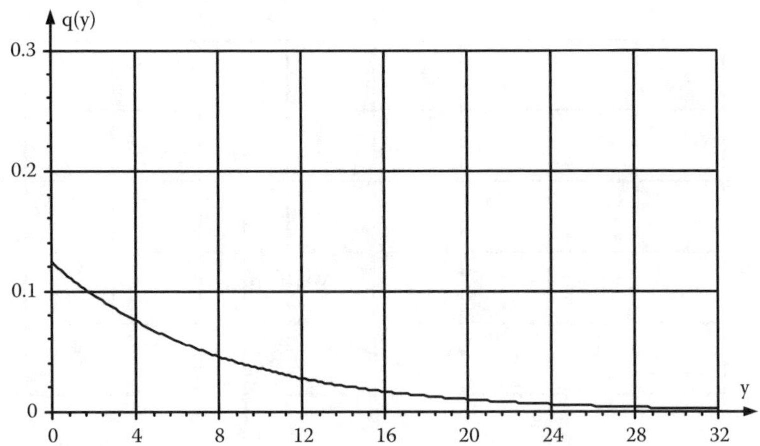

FIGURE 5.5 The probability density q(y) of the random variable y.

In the first case where $w = -8$, that is, $w < 0$, the probability densities p(x) and $q(y = w - x)$ do not overlap and, thus, the probability density p(w) is equal to zero (see Figure 5.6).

In the second case, where $w = 12$, that is, $w > 0$, the probability densities p(x) and $q(y = w - x)$ do overlap for the region of x between zero and w (see Figure 5.7). In this case, from Figure 5.7, the probability density s(w) is

$$s(w) = \alpha\beta e^{-\beta w} \int_0^w e^{-(\alpha-\beta)x} dx \tag{5.30}$$

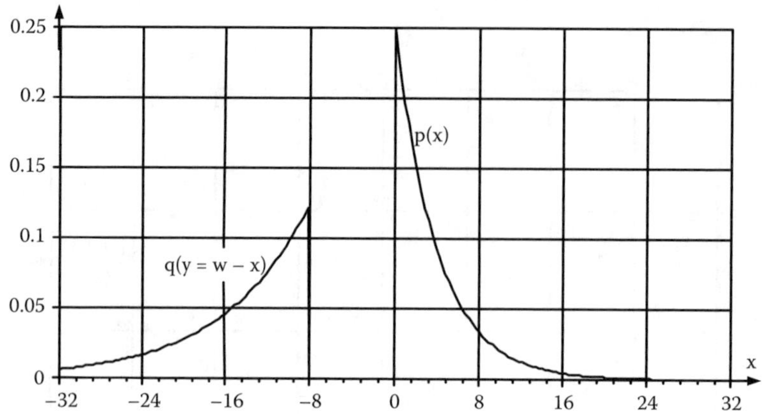

FIGURE 5.6 The probability densities p(x) and q(y = w – x) plotted as a function of the random variable x for w = –8. In this case the probability densities do not overlap and, thus, p(w) is equal to zero.

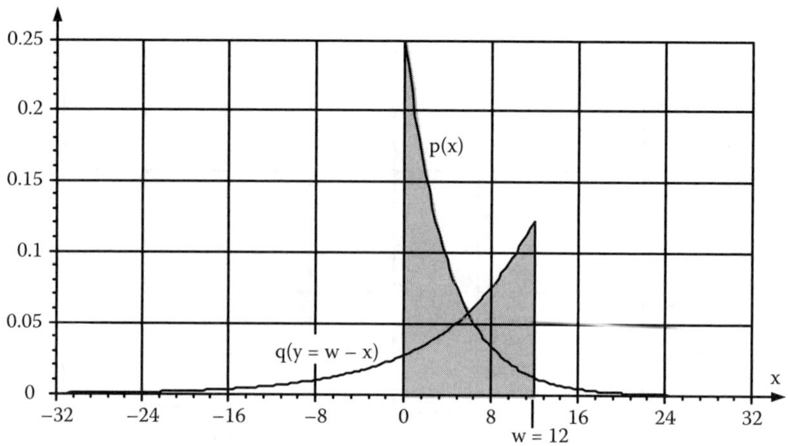

FIGURE 5.7 The probability densities p(x) and q(y = w – x) plotted as a function of the random variable x for w = 12. In this case the probability densities overlap in the region between x = 0 and x = w.

By integrating Equation 5.30, one obtains for the probability density s(w):

$$s(w) = \frac{\alpha\beta}{\alpha - \beta}\left[e^{-\beta w} - e^{-\alpha w}\right] \quad \text{for} \quad 0 < w \le \infty \tag{5.31}$$

A plot of s(w) is shown in Figure 5.8.

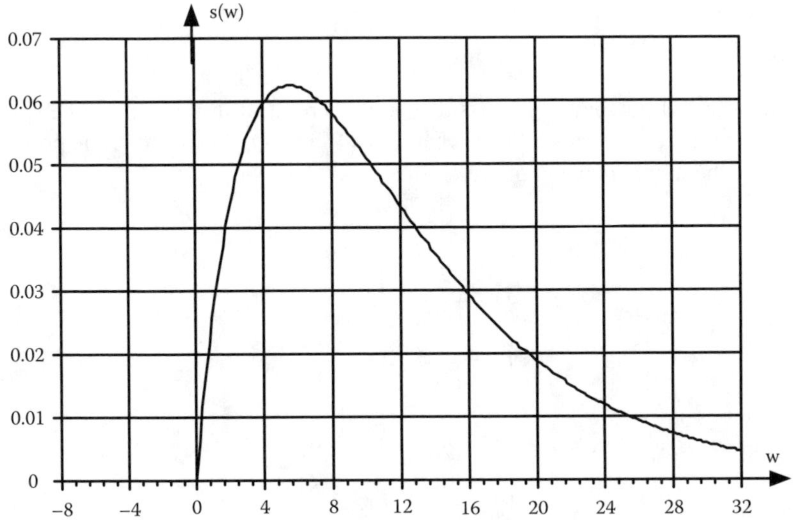

FIGURE 5.8 Plot of the probability density s(w).

PROBLEMS

Problem 5.1

Given:

The conditional probability density $p(x|y)$ that the random variable u will have values between x and $x + \Delta x$ provided the random variable v will have values between y and $y + \Delta y$ is

$$p(x|y) = \frac{3}{2a}\frac{x^2 + 1.8xy + y^2}{a^2 + 3y^2}$$

for $-a < x \le a$ and $-a < y \le b$.

The probability density $p(y)$ that the random variable v will have values between y and $y + \Delta y$ is

$$p(y) = \frac{1}{4a^3}\left(a^2 + 3y^2\right)$$

Problem:

Calculate and plot the conditional distribution function $F(x|y)$.

Problem 5.2

Given:

Two odd-shaped polyhedra, many-sided solids, are used in the game. The sides of the polyhedra are numbered. The probabilities P_n and Q_m that each of the polyhedra will show the numbers n and m are

$$P_1 = \frac{2}{126} \quad Q_1 = \frac{3}{363}$$
$$P_2 = \frac{4}{126} \quad Q_2 = \frac{9}{363}$$
$$P_3 = \frac{8}{126} \quad Q_3 = \frac{27}{363}$$
$$P_4 = \frac{16}{126} \quad Q_4 = \frac{81}{363}$$
$$P_5 = \frac{32}{126} \quad Q_5 = \frac{243}{363}$$
$$P_6 = \frac{64}{126}$$

Here, w = n + m.

Problem:

a) Determine the largest and smallest values of w.
b) Calculate the probabilities P_w of the various values of w being rolled.

Problem 5.3

Given:

Consider two sets of events $S_A\{0 < x \leq \infty\}$ and $S_B\{0 < y \leq \infty\}$. The probability density per unit length is

$$p(x) = \alpha^2 x e^{-\alpha x} \quad q(y) - \beta \in$$

Here, w = x + y.

Problem:

a) Determine the range of w.
b) Calculate the probability density p(w) per unit length.

PROBLEM 5.4

Given:

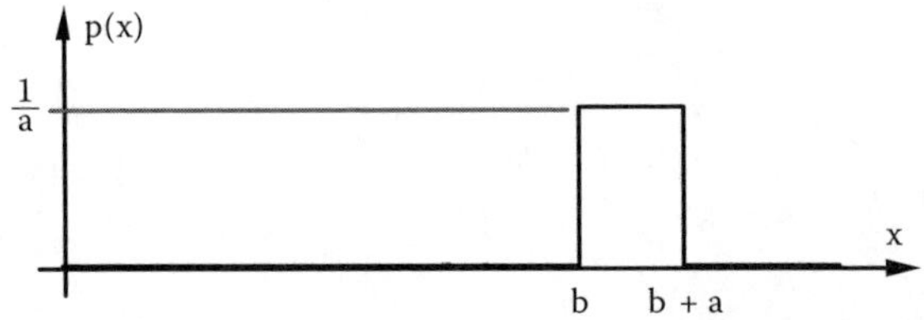

Probability density of javelin being thrown a certain distance x.

The probability density per unit length p(x) that a javelin thrower will toss the javelin a certain distance x is

$$p(x)=\begin{cases}0 \text{ for } x<b \\ \dfrac{1}{a} \text{ for } b<x\le b+a \\ 0 \text{ for } a+b\le x\end{cases}$$

In the competition, the score is counted as the total distance u of three javelin throws. The ratio of b to a is 4.

Problem:

a) Determine the smallest and largest value of the total distance u of three throws (10 points).
b) Calculate and plot the probability density per unit length p(u) of the sum u of the distances of three throws (30 points).

PROBLEM 5.5

Given:

Consider the probabilities p(x) and q(y) per unit length x and y, respectively, that the random variables x and y will have values in an incremental region of area dxdy. Here, $-b < x \le b$ and $-b < y \le b$.

$$p(x)=\frac{3}{4b^3}\left(b^2-x^2\right) \quad \text{and} \quad q(y)=\frac{3}{4b^3}\left(b^2-y^2\right)$$

The random variable w is equal to the sum of the random variables x and y:

$$w = x + y$$

Problem:

a) Determine the largest and smallest values of w.
b) Calculate the probability density p(w) per unit length w.

PROBLEM 5.6

Given:

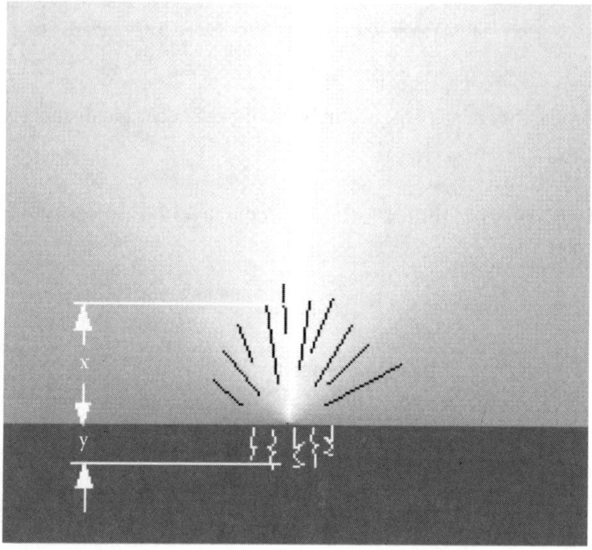

Explosion.

The probability density per unit height x that a particle in an explosion will reach a height x is

$$p(x) = a^2\, x\, e^{-ax} \quad \text{for } 0 < x \le \infty$$

The probability density q(y) per unit depth y that the ground below the explosion will be disturbed up to a distance y is

$$q(v) = 2/b^2 y \quad \text{for } 0 < y \le b$$

Here, w is the total extent of the explosion:

$$w = x + y$$

Problem:

a) Determine the minimum and maximum values of the total extent w of the explosion.
b) Calculate the probability density p(w) per unit length of the total extent of the explosion.
c) Calculate the probability F(u) that the explosion will affect a region up to w = u; that is, calculate the distribution function F(u) as a function of u.

6 Average Values, Moments, and Correlations of Random Variables and of Functions of Random Variables

6.1 THE MOST LIKELY VALUE OF A RANDOM VARIABLE

The *most likely value* of a random variable is the value at which the probability has the largest value. For example, consider a discrete random variable m with the following associated probabilities:

	m = 1	m = 2	m = 3	m = 4	m = 5
$P_m =$	0.01	0.19	0.80	0.19	0.01

The value of the random variable m that is most likely to occur in the preceding example is 3 because it has the largest value of the probability P_m. One can predict that there is an 80% chance that m will be equal to 3.

Similarly, the most likely value of the continuous random variable x with a probability density $p(x) = 1/\sqrt{2\pi\sigma}\exp\left[(x-a)^2/2\sigma^2\right]$ is at x = a. The probability density has a maximum of $p(x = a) = 1/\sqrt{2\pi\sigma}$ at x = a. One can predict that it is most likely that the value of x equal to a will occur. As will be shown, the most likely value is not necessarily equal to the average value of the random variable.

6.2 THE AVERAGE VALUE OF A DISCRETE RANDOM VARIABLE AND OF A FUNCTION OF A DISCRETE RANDOM VARIABLE

The average value of $\langle f_n \rangle$, a function f_n of a discrete random variable n, is equal to the sum of the products of the functions f_n and the probabilities P_n:

$$\langle f \rangle = \sum_{n=0}^{n=N} f_n P_n \tag{6.1}$$

where $0 \le n \le N$ and P_m is the probability of the random variable n having a particular value m. The values of the random variable in this case are mutually exclusive. Recall that events considered in this text are mutually exclusive unless otherwise specified.

Here, the bracket $\langle \ldots \rangle$ denotes the average value operation. In the literature other formulations such as $E(f_n)$, the expected value of f_n and $\overline{f_n}$, are also used to express the average value. Values of the function f_n for which the random variable has large probabilities will contribute to the average value more than other values of the function f_n.

EXAMPLE 6.1

Consider a discrete random variable n having the following probabilities P_n of occurring:

n	1	2	3	4	5	6
P_n	0.01	0.02	0.04	0.08	0.16	0.69

There exists a function $f_n = (7 - n)^3$ of the random variable n.

n	1	2	3	4	5	6
f_n	216	125	64	27	8	1

The average value $\langle f_n \rangle$ is

$$\langle f_n \rangle = (216 \times 0.01) + (125 \times 0.02) + (64 \times 0.04) \\ + (27 \times 0.08) + (8 \times 0.16) + (1 \times 0.69)$$

That is, one can predict that, on the average, the function f_n will have a value:

$$\langle f_n \rangle = 11.35$$

The average value is a *macroscopic parameter.* A stochastic system can, of course, be described by the probabilities of the individual values of the random variable. At times this might be a very large number of values. It is sometimes more convenient to describe a stochastic system with macroscopic parameters such as the average value. Macroscopic parameters usually convey less information than the large number of microscopic parameters. However, in many cases the description of a stochastic system by macroscopic parameters is sufficient.

6.3 AN OFTEN-USED SPECIAL CASE

In the special case when nothing is known about the stochastic nature of a discrete random variable n, the probabilities P_n are uniform, as will be shown in Chapter 8. For such a random variable n having values 1, 2, …, N, the probabilities have uniform values:

$$P_n = \frac{1}{N} \tag{6.2}$$

In this case, according to Equations 6.1 and 6.2, the average value $\langle f_n \rangle$ of a function f_n of the random variable is simply:

$$\langle f_n \rangle = \frac{1}{N}\sum_{n=1}^{n=N} f_n \tag{6.3}$$

That is, in the case when nothing is known about the stochastic nature of a discrete random variable n, the average value of a function of this random variable is simply equal to the sum of the functions for each value of the random variable divided by the number of such values.

EXAMPLE 6.2

Consider a class with 10 students. The grades on an exam were as follows:

Student number (n):	1	2	3	4	5	6	7	8	9	10
Grade g_n:	94	86	84	98	96	100	68	100	98	96

If all students are considered to be equal, then, according to Equation 6.3 the average grade $\langle g_n \rangle$ is simply:

$$\langle g_n \rangle = \frac{94+86+84+98+96+100+68+100+98+96}{10} \quad \text{or} \quad \langle g_n \rangle = 92$$

6.4 THE PROBABILISTIC MATHEMATICAL MODEL OF DISCRETE QUANTUM MECHANICS

The particular stochastic model discussed here is useful whenever the probing of a stochastic system alters the system. It is mainly used to formulate the quantum mechanical model of nature. However, it can be used to describe other stochastic systems.

In quantum mechanics, the motion of particles follows probabilistic rules. However, the probability itself propagates in conformity with the law of causality; that is, only the probability that a particle will have a certain position and will have a certain momentum can be calculated according to quantum mechanics. However, the probability changes according to an entirely predictive model.

Here, only the rules of quantum mechanics are derived. The actual application of the rules of quantum mechanics to construct a mathematical model of real physical systems require a great deal of practice with these rules. Quantum mechanics courses, which are usually two semesters in duration, attempt to give students some practice in using the rules of quantum mechanics.

Historically, quantum mechanics was developed first by Werner Heisenberg (1901–1976) as a vector and matrix theory in 1925.[1,2] Heisenberg developed his mathematical model from an earlier empirical model of quantized natural phenomena by Niels Bohr (1885–1962) and Arnold Johannes Wilhelm Sommerfeld (1868–1951). Bohr and Sommerfeld were trying to develop a mathematical model that described the peaks observed in the emission and absorption of light by various substances. The then-known mathematical models of nature, Newton's classical mechanics and Maxwell's electromagnetic theory, proved inadequate to model light absorption satisfactorily. Because the mathematical models of nature such as classical mechanics and electromagnetic theory used continuous mathematics, the scientists of the day did not welcome Heisenberg's vector and matrix theory with much enthusiasm.

In 1926 Erwin Schrödinger (1887–1961) developed a wave theory of quantum mechanics. Schrödinger was the first to recognize that both Heisenberg's matrix quantum mechanics theory and his own wave theory were different mathematical descriptions of the same physical phenomena; that is, both particles and waves can sometimes have particle and at other times have wavelike properties. Max Born (1886–1970) soon realized that the square of the components of the vectors used in Heisenberg's matrix theory and the square of the magnitude of the wave in Schrödinger's model at each position could be interpreted as probabilities and probability densities, respectively. The square of the magnitude of the wave is the probability of the particle described by the wave being at that location. It was Max Born who coined the term *quantum mechanics*.

Niels Bohr

Particles are often confined to a particular region, such as a room. If it is assumed that the particles cannot penetrate into the wall, the probability of the particle being in the wall has to be zero. The particle can be represented by a wave. Thus, the amplitude of the wave has to be equal to zero at the wall. A wave goes to zero after each half-wavelength. Therefore, it is necessary to accommodate an integer number of half-waves between the walls of the room. This implies that only waves with certain wavelengths can exist in the room (see Figure 6.1); that is, the particle in the room can only be described by certain discrete waves. This applies to protons, neutrons, electrons, and other particles whose motion is described by quantum mechanics. The quantum mechanical description also worked well for vibrational modes in solids, liquids, and gases. The average value of the vibration of the atoms in various materials determine

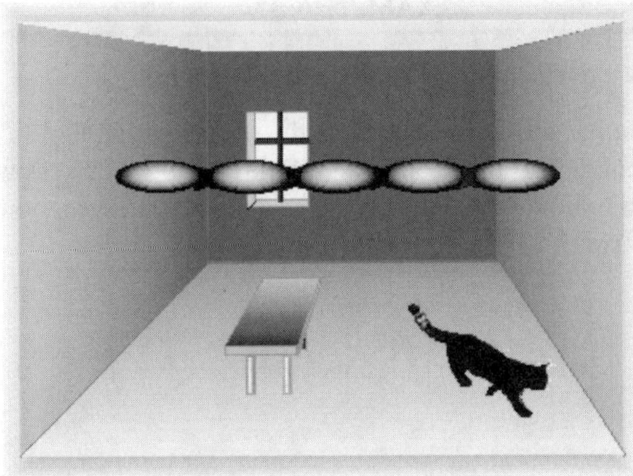

FIGURE 6.1 Room with standing wave. It is necessary to have an integer number of half-waves between the walls of the room. Schrödinger's cat played by Leon Travis; the author's cat is walking around, but that is another story.

their thermodynamic properties. To describe the interaction of electrons, protons, etc., with electromagnetic waves required that one make a quantum mechanical model of the electromagnetic waves. This was not done until about 32 years after the development of the original quantum mechanics model. Richard Feynman (1918–1988), at that time at the California Institute of Technology, developed quantum electrodynamics (QED), a quantum mechanical mathematical model of electromagnetism in 1960.

Werner Heisenberg

EXAMPLE 6.3

Heisenberg's Uncertainty Principle

Actually, a wave that has a finite extended as the one shown in Figure 6.1 does not have a single wavelength. It is made up of an infinite number of waves with different wavelengths λ and amplitudes b_λ. The proper mathematical expression in quantum mechanics for the wave shown in Figure 6.1. is:

$$\psi(x) = \sqrt{\frac{2}{L}} \cos\left[\frac{\pi n}{L} x\right] \quad \text{for} \quad -\frac{L}{2} < x \le \frac{L}{2} \qquad \text{EX6.3.1}$$

where L is the length of the room and n = 1, 2, 3, …. is an integer. The wave function $\psi(x)$ of Equation EX6.3.1, and as shown in Figure 6.1, has an integer number of half wavelength. As required $\psi(x)$ is equal to zero at x equal to $\pm L/2$, at the position of the walls of the room; that is, the particle has zero probability of being in the wall. As will be shown below, it is required by the rules of quantum mechanics that the function $\psi(x)$ has to be normalized; that is, the integral of the probability density $\psi^*(x)\psi(x)$ has to be equal to one.

$$1 = \int_{-\frac{L}{2}}^{\frac{L}{2}} \psi^*(x)\psi(x)dx \qquad \text{EX6.3.2}$$

The amplitude b_k of one of the waves with wavevector $k = 2\pi/\lambda$ that is a solution of the Schrödinger's wave of equation, Equation 6.51 is:

$$\text{a)} \quad b_k = \sqrt{\frac{2}{L}} \int_{-\frac{L}{2}}^{\frac{L}{2}} \cos\left[\frac{\pi n}{L} x\right] \exp[-jkx]dx$$

or EX6.3.3

$$\text{b)} \quad b_k = \sqrt{\frac{1}{2L}} \left\{ \int_{-\frac{L}{2}}^{\frac{L}{2}} \exp\left[-jkx + j\frac{\pi n}{L} x\right]dx + \int_{-\frac{L}{2}}^{\frac{L}{2}} \exp\left[-jkx - j\frac{\pi n}{L} x\right]dx \right.$$

By integrating, one obtains:

$$b_k = \sqrt{\frac{2}{L}} \left\{ \frac{\sin\left[\frac{kL}{2} - \frac{\pi n}{2}\right]}{k - \frac{\pi n}{L}} + \frac{\sin\left[\frac{kL}{2} + \frac{\pi n}{2}\right]}{k + \frac{\pi n}{L}} \right\} \qquad \text{EX6.3.4}$$

Thus, for even values of n equation EX6.3.4 can be rewritten as:

$$b_k = \sqrt{\frac{L}{2}}(-1)^n \sin\frac{kL}{2}\left[\frac{1}{\frac{kL}{2}-\frac{n\pi}{2}} + \frac{1}{\frac{kL}{2}+\frac{n\pi}{2}}\right] \qquad \text{EX6.3.5}$$

or

$$b_k = \sqrt{\frac{L}{2}}(-1)^n \frac{kL\sin\frac{kL}{2}}{\left(\frac{kL}{2}\right)^2 - \left(\frac{n\pi}{2}\right)^2} \qquad \text{EX6.3.6}$$

The smallest value of n is n = 0.

$$b_k = (2L)^{\frac{1}{2}} \frac{\sin\frac{kL}{2}}{\frac{kL}{2}} \qquad \text{EX6.3.7}$$

This sin k function has a width

$$\frac{kL}{2} = \pi \qquad \text{EX6.3.8}$$

That is, its zero occurs at $kL/2 = \pi$, where n = 1, 2, 3, …. The momentum p in quantum mechanics is equal to $p = \hbar k$ where $\hbar = 1.054588664 \times 10^{-34}$ Joule seconds is Planck's constant divided by 2π. By multiplying both sides of equation EX6.3.8 by $2\hbar$ one obtains:

$$\text{a)}\ \ \hbar kL \geq 2\pi\hbar \quad \text{or} \quad \text{b)}\ \ pL \geq h \qquad \text{EX6.3.9}$$

where $h = 6.626175999 \times 10^{-34}$ Joule seconds is Planck's constant. Equation EX6.3.9 is *Heisenberg's Uncertainty Principle*. It states that if the position L is known, the momentum can only be known to within a value of $p \geq h/L$, and if the momentum p is known the position can only be known to within a value of $L \geq h/p$; that is, the position and momentum cannot be known to any arbitrary accuracy simultaneously. The same is true of the time and energy.

Erwin Schrödinger

In 1935 Schrödinger published a three-part essay on quantum mechanics in which his famous Schrödinger's cat paradox appears. This was a thought experiment in which a cat in a closed box either lived or died according to whether a quantum event occurred. The paradox is that both universes, one with a dead cat and one with a live one, seem to exist in parallel until an observer opens the box.

Heisenberg was the only major physicist to remain in Germany during World War II. He was in charge of the German nuclear weapons effort. Heisenberg's last contact with the physics community was his secret visit to Niels Bohr in Copenhagen at the beginning of the war. The Germans never did develop a working nuclear weapon. To this day it is not known if Heisenberg was unable or unwilling to produce a German nuclear bomb.

As modern circuit elements shrink to smaller and smaller dimensions, the motion of the charge carriers in these devices has to be described by quantum mechanics. At the same time, new areas such as quantum encoding, quantum communication, and quantum computing are in the process of being developed. For example, there are encryption schemes that are based on either very large integers or computer-generated random numbers. However, these computer-generated random numbers are not truly random. It is always possible to build a larger computer to decipher such codes. It is possible to generate quantum mechanical functions that are truly random and can be used for quantum encryption.

Part of the mathematical model of quantum mechanics is derived here from basic probabilistic concepts. The discussion starts with a discrete model because it is simpler to derive. Recall that Heisenberg's first model of quantum mechanics, too, used a discrete matrix theory.

Recall that P_n, the probability that any particular value of the random variable n will occur, must be positive and real. One can express this mathematically as:

$$P_n = \psi_n^* \psi_n \tag{6.4}$$

where * denotes the complex conjugate and where the function ψ_n does not have to be positive or even real. The sum of the probabilities P_n must be equal to one.

$$1 = \sum_{n=1}^{n=N} \psi_n^* \psi_n \tag{6.5}$$

This is similar to the dot product of two vectors, **V** and **W**.

$$\mathbf{V} \cdot \mathbf{W} = V_x W_x + V_y W_y + V_z W_z \tag{6.6}$$

The vectors represented in Equation 6.5 can, of course, have more than three components. The generalized dot product is called an *inner product* of two vectors. The inner product of the two vectors of Equation 6.5 can also be written as follows:

$$1 = \begin{pmatrix} \psi_1^* & \psi_2^* & \psi_3^* \ldots \psi_n^* \ldots \psi_N^* \end{pmatrix} \begin{pmatrix} \psi_1 \\ \psi_2 \\ \psi_3 \\ \ldots \\ \psi_n \\ \ldots \\ \psi_N \end{pmatrix} \tag{6.7}$$

The individual vectors are traditionally known as wave functions. This terminology is derived from the description of quantum mechanical particles as waves by Schrödinger. Perhaps a better name for these vectors would be information functions. Because the magnitude squared $\psi_n^* \psi_n$ of each information function or wave function vector component ψ_n is equal to a probability, the inner product of the row and column vectors of Equation 6.5 has to be equal to one; that is, the sum of all probabilities is equal to one. Thus, the information vectors have to be normalized. The column information vector can be written in shorthand notation as:

$$|\psi\rangle = \begin{pmatrix} \psi_1 \\ \psi_2 \\ \psi_3 \\ \ldots \\ \psi_n \\ \ldots \\ \psi_N \end{pmatrix} \tag{6.8}$$

Note from Equation 6.7 that a row vector that is the complex conjugate transpose of the column vector, or the Hermitian transpose $\langle\psi|$, of the vector $|\psi\rangle$ is also used. The row wave function or information function vector $\langle\psi|$ can also be written in shorthand notation as:

$$\langle\psi| = \left(\psi_1^* \; \psi_2^* \; \psi_3^* \ldots \psi_n^* \ldots \psi_N^*\right) \tag{6.9}$$

Here, $|\psi\rangle$ is called a cket vector, and $\langle\psi|$ is called a bra vector. This notation was developed by the physicist Paul Adrian Maurice Dirac[3,4] (1902–1984), and it is therefore known as Dirac notation. The N-dimensional vectors form an N-dimensional space. This space can be thought of as a conceptual extension of physical three-dimensional space. Equation 6.5 can be written in shorthand or Dirac notation as follows:

$$1 = \langle\psi|\psi\rangle \tag{6.10}$$

that is, the length of the $|\psi\rangle$ vectors is unity. In general, the inner product of two information function vectors $|\phi\rangle$ and $|\psi\rangle$ is a conceptual extension of the dot product of two vectors in real space:

$$\langle\psi|\phi\rangle = \psi_1^*\phi_1 + \psi_2^*\phi_2 + \ldots + \psi_n^*\phi_n + \ldots + \psi_N^*\phi_N \tag{6.11}$$

All vectors in the physical three-dimensional space can be formed out of linear combinations of three unit vectors. Ordinary three-dimensional vectors can be represented using unit vectors $\hat{\mathbf{a}}_x$, $\hat{\mathbf{a}}_y$, and $\hat{\mathbf{a}}_z$ as follows:

$$\mathbf{V} = \hat{\mathbf{a}}_x V_x + \hat{\mathbf{a}}_y V_z + \hat{\mathbf{a}}_z V_z \tag{6.12}$$

Similarly, the information function vectors can be represented by the use of information function basis vectors $|n\rangle$ and $\langle n|$. The basis function vectors act much like unit vectors in this N-dimensional space formed by the information function vectors.

$$\begin{aligned} &\text{a)} \quad |\psi\rangle = b_1|1\rangle + b_2|2\rangle + b_3|3\rangle + \ldots + b_n|n\rangle + \ldots + b_N|N\rangle \\ &\text{b)} \quad \langle\psi| = \langle 1|b_1^* + \langle 2|b_2^* + \langle 3|b_3^* + \ldots \langle n|b_n^* + \ldots + \langle N|b_N^* \end{aligned} \tag{6.13}$$

The unit vectors in the real three-dimensional space are taken to be at 90° to each other; that is, they are *orthogonal.* The dot product of two orthogonal vectors is equal to zero. Just as in the case of the real three-dimensional space, the unit vectors in the information function space are also orthogonal.

$$\langle n|m\rangle = \begin{cases} 1 \text{ if } n = m \\ 0 \text{ if } n \neq m \end{cases} \tag{6.14}$$

David Hilbert

Just as the three unit vectors of a real three-dimensional space represent a complete set, the basis function vectors of an information function also represent a complete set; that is, no additional unit vectors are necessary to describe a vector in this space. This N-dimensional space is known as a *Hilbert space*, named after the German mathematician David Hilbert (1868–1943). It is often convenient to take the eigen vectors of an operator as the unit vectors in the information function Hilbert space. Eigenvectors will be discussed later.

In a system described by a classical mathematical model of mechanics, an individual particle can only have one energy, one velocity, or one momentum at a time. This is not true for systems described by a quantum mechanical mathematical model of nature. It has probabilities of having any number of values of the physical quantities. A simple example of such an information function is a so-called qubit, a unit of information in a quantum computing scheme.

$$|\text{Qubit}\rangle = a_1|1\rangle + a_0|0\rangle \tag{6.15}$$

where, before any measurement on the qubit is performed, there is a probability $|a_1|^2$ that the qubit represents a binary ONE and a probability $|a_0|^2$ that it represents a binary ZERO. Recall that this is similar to Schrödinger's cat in the box. The $|1\rangle$ and $|0\rangle$ could, for example, represent the polarization states of a light energy quanta, a photon. $|1\rangle$ could represent a horizontal polarization and $|0\rangle$ a vertical polarization. Thus, $|a_1|^2$ is the probability that the photon is horizontally polarized.

By substituting Equation 6.4 into Equation 6.1, one obtains for the average value $\langle f_n \rangle$ of a function f_n of a random variable n that has values 1, 2, …, N:

$$\langle f_n \rangle = \sum_{n=1}^{n=N} \psi_n^* f_n \psi_n \tag{6.16}$$

Observe that the components of the two vectors of Equations 6.8 and 6.9 appear also in the equation for the average value, Equation 6.16. The foregoing expression can be written in matrix form. To conveniently illustrate this, one can use a small matrix with N equal to only 3.

$$\langle f_m \rangle = \begin{pmatrix} \psi_1^* & \psi_2^* & \psi_3^* \end{pmatrix} \begin{pmatrix} f_1 & 0 & 0 \\ 0 & f_2 & 0 \\ 0 & 0 & f_3 \end{pmatrix} \begin{pmatrix} \psi_1 \\ \psi_2 \\ \psi_3 \end{pmatrix} \tag{6.17}$$

Note that the functions f_n in this case became the elements of a diagonal matrix and ψ_m as well as ψ_n^* are components of vectors as discussed in Equations 6.8 and 6.9. In the example of Equation 6.17 the vectors are specified in a three-dimensional space similar to the physical three-dimensional space. If the space is rotated, that is, the unit vectors of the three-dimensional space are chosen in different directions, each component of the vectors would consist of linear combinations of the vector components of Equation 6.17.

A rotation of a coordinate system in two dimensions is shown in Figure 6.2. In the rotated space, the matrix appearing in Equation 6.17 would have off-diagonal components. A similar situation will occur in multidimensional space. Equation 6.17 can be written as follows in the rotated coordinate system:

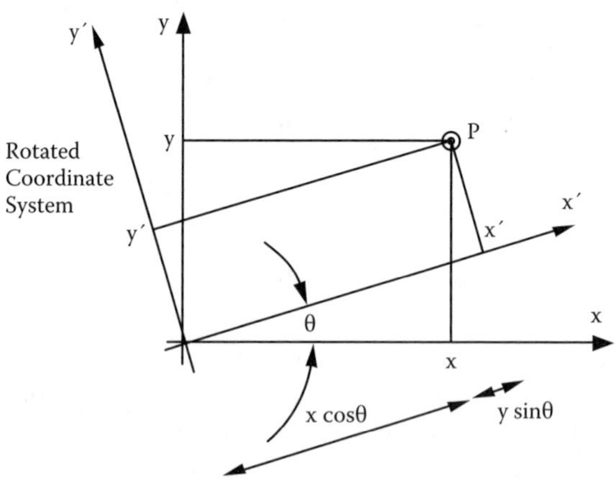

FIGURE 6.2 The location of point P specified in an original and rotated coordinate system. Observe that $x' = x\cos\theta + y\sin\theta$.

$$\langle f_{mn}\rangle = \sum_{n=1}^{n=N} \psi_n^* \left[\sum_{m=1}^{m=N} f_{nm}\psi_m \right]_n \tag{6.18}$$

where m also has values between 1 and N, m = 1, 2, 3, …, N. Observe that f_{nm} is the nm-th component of a matrix. Here, f_{nm} has to be a component of a square matrix, as will be shown later. The square bracket in Equation 6.18 is only a function of the random variable n because the summation over the random variable m occurs inside the square bracket. Equation 6.18 can also be rewritten in matrix form as follows for, say, N = 3:

$$\langle f_{mn}\rangle = \begin{pmatrix} \psi_1^* & \psi_2^* & \psi_3^* \end{pmatrix} \begin{pmatrix} f_{11} & f_{12} & f_{13} \\ f_{21} & f_{22} & f_{12} \\ f_{31} & f_{32} & f_{33} \end{pmatrix} \begin{pmatrix} \psi_1 \\ \psi_2 \\ \psi_3 \end{pmatrix} \tag{6.19}$$

Note that the quantities f_{nm} are components of a matrix operator and that ψ_m as well as ψ_n^* are components of vectors. The foregoing multiplication of vectors and the matrix is carried out in two steps. First, the multiplication of the matrix with components f_{nm} by the vector with components ψ_m, or the operation of the matrix with components f_{nm} on the vector with components ψ_m as this multiplication is called, is performed:

$$\begin{pmatrix} f_{11} & f_{12} & f_{13} \\ f_{21} & f_{22} & f_{12} \\ f_{31} & f_{32} & f_{33} \end{pmatrix} \begin{pmatrix} \psi_1 \\ \psi_2 \\ \psi_3 \end{pmatrix} = \begin{pmatrix} f_{11}\psi_1 + f_{12}\psi_2 + f_{13}\psi_3 \\ f_{21}\psi_1 + f_{22}\psi_2 + f_{12}\psi_3 \\ f_{31}\psi_1 + f_{32}\psi_2 + f_{33}\psi_3 \end{pmatrix} \tag{6.20}$$

The mathematical expression on the right-hand side of Equation 6.20 is a column vector. However, recall that for the square of the magnitude of the components $\eta_n^*\eta_n$ of the resulting vector to be probabilities, the sum of all the magnitude squares of the vector components must be equal to one; that is, the vector has to be normalized. This can be accomplished by multiplying and dividing the right-hand side of Equation 6.20 by a constant A.

$$\begin{pmatrix} f_{11} & f_{12} & f_{13} \\ f_{21} & f_{22} & f_{12} \\ f_{31} & f_{32} & f_{33} \end{pmatrix} \begin{pmatrix} \psi_1 \\ \psi_2 \\ \psi_3 \end{pmatrix} = A\frac{1}{A} \begin{pmatrix} \eta_1 \\ \eta_2 \\ \eta_3 \end{pmatrix} \tag{6.21}$$

The normalization constant A can be calculated by requiring that the sum of the magnitude squared of the components of the vector on the right-hand side of Equation 6.21, which now includes the reciprocal of A, be equal to one:

$$1 = \frac{1}{A^2}\left[\eta_1^*\eta_1 + \eta_2^*\eta_2 + \eta_3^*\eta_3\right] \tag{6.22}$$

Thus, one can now define a normalized vector as follows:

$$\begin{pmatrix} \phi_1 \\ \phi_2 \\ \phi_3 \end{pmatrix} = \frac{1}{A} \begin{pmatrix} \eta_1 \\ \eta_2 \\ \eta_3 \end{pmatrix} \tag{6.23}$$

Substituting Equation 6.23 into Equation 6.21:

$$\begin{pmatrix} f_{11} & f_{12} & f_{13} \\ f_{21} & f_{22} & f_{12} \\ f_{31} & f_{32} & f_{33} \end{pmatrix} \begin{pmatrix} \psi_1 \\ \psi_2 \\ \psi_3 \end{pmatrix} \begin{matrix} \text{State 1} \\ \text{State 2} \\ \text{State 3} \end{matrix} = A \begin{pmatrix} \phi_1 \\ \phi_2 \\ \phi_3 \end{pmatrix} \begin{matrix} \text{State 1} \\ \text{State 2} \\ \text{State 3} \end{matrix} \tag{6.24}$$

or, using the shorthand notation:

$$f|\psi\rangle = A|\phi\rangle \tag{6.25}$$

The ψ_m's are components of an information function vector. Equations 6.24 and 6.25 are interpreted as follows. The information function vector contains all the information about a particular system, and the matrix operator is used to read some of the information A. In general, in the process of reading the information, the operator will change the information contained in the information function vector. The matrix operator usually corresponds to some physical property of the quantum mechanical system such as its energy, linear momentum, angular momentum, position, etc. Thus, operating, say, with the energy operator on the system information vector yields the energy of the system. One can think of the information function as a computer disk that stores all the information about some system. The computer and the printer take the place of the operator. They read and display the information (see Figure 6.3). In the process of reading this information, the computer can change the information on the disk.

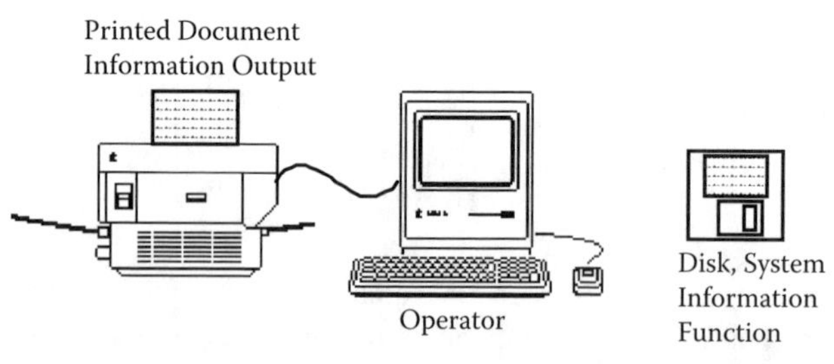

FIGURE 6.3 One can think of the information function as a computer disk that stores all the information about some system. The computer and the printer take the place of the operator. They read and display the information. In quantum mechanics, in general, the information on the disk is altered in the process of reading it.

Each component ψ_n of the information function vector corresponds to a state of the quantum mechanical system (see Equation 6.24). The magnitude squared $\psi_n^*\psi_n$ of each component of the information function vector is the probability P_n of the system being in that particular state. Objects whose motion is described by quantum mechanics, for example, can have discrete energy values. Each value of the energy would then be a state of the system. The angular momentum of quantum mechanical objects can have discrete values. Each value of the angular momentum is then a state of the system, etc. A state can also be described by several parameters such as the energy, plus the value of the vector components of the linear momentum, plus a spin angular momentum value. The laws of quantum mechanics suggest that objects can literally exist in two or more states or places simultaneously until perturbed in some way, after which they collapse out of this "superposition" to just one outcome. Indeed, it is observed that after a measurement is performed on a set of stochastic events, one finds that a particular event occurred. The probability of that event occurring is then equal to one and the probabilities of any other event occurring become zero.

For example, Schrödinger's cat, shown in Figure 6.1, might have nine lives but under the quirky laws of quantum mechanics, it can be simultaneously alive and dead; that is, until the cat is observed, which then determines if it is alive or dead.

In general, reading the information in quantum mechanics changes the information vector. Thus, the next time the information is read, a different result will be obtained. For example, by reading the position of the system, one obtains the value of the position and an altered information vector. By next reading the momentum, one obtains the value of the momentum of the system, which has been disturbed by first reading its position. If this had been done in the reverse order, different results for the energy and momentum would have been obtained. The various quantum mechanical operators, in general, do not commute. Indeed, the matrix description of quantum mechanical operators is appropriate because matrices, in general, do not commute.

EXAMPLE 6.4

The spin angular momentum of a particle such as an electron can only have one magnitude and two states. It could either spin in the clockwise or counterclockwise direction. The operators corresponding to the three components of the spin angular momentum are

$$\mathbb{S}_x = \frac{1}{2}\left(\begin{array}{c|c} 0 & 1 \\ \hline 1 & 0 \end{array}\right) \quad \mathbb{S}_y = \frac{1}{2}\left(\begin{array}{c|c} 0 & -j \\ \hline j & 0 \end{array}\right) \quad \mathbb{S}_z = \frac{1}{2}\left(\begin{array}{c|c} 1 & 0 \\ \hline 0 & -1 \end{array}\right)$$

These are also known as the spin matrices. First, the commutation relation of, say, the first two spin angular momentum operators will be calculated:

$$\mathbb{S}_x\mathbb{S}_y - \mathbb{S}_y\mathbb{S}_x = \frac{1}{4}\left[\left(\begin{array}{c|c} 0 & 1 \\ \hline 1 & 0 \end{array}\right)\left(\begin{array}{c|c} 0 & -j \\ \hline j & 0 \end{array}\right) - \left(\begin{array}{c|c} 0 & -j \\ \hline j & 0 \end{array}\right)\left(\begin{array}{c|c} 0 & 1 \\ \hline 1 & 0 \end{array}\right)\right]$$

Multiplying out:

$$\mathbb{S}_x\mathbb{S}_y - \mathbb{S}_y\mathbb{S}_x = j\frac{1}{2}\left(\begin{array}{c|c} 1 & 0 \\ \hline 0 & -1 \end{array}\right)$$

or

$$\mathbb{S}_x\mathbb{S}_y - \mathbb{S}_y\mathbb{S}_x = j\mathbb{S}_Z$$

Indeed, the spin angular momentum component operators do not commute. Consider the information vector:

$$|\psi\rangle = \frac{1}{\sqrt{3}}\left(\frac{1}{1+j}\right) \begin{array}{ll} \text{State of clockwise spin.} & P_{CW} = \frac{1}{3} \\ \text{State of counterclockwise spin.} & P_{CCW} = \frac{2}{3} \end{array}$$

Operating with the operator $\mathbb{S}_y$ on the information vector:

$$\frac{1}{2}\left(\begin{array}{c|c} 0 & -j \\ \hline j & 0 \end{array}\right)\frac{1}{\sqrt{3}}\left(\frac{1}{1+j}\right) = \frac{1}{2}\frac{1}{\sqrt{3}}\left(\frac{1-j}{j}\right)$$

The resulting information function vector from this operation is different from the original information function vector. The resulting vector is normalized and is thus an information vector. The information one obtained by operating with the operator corresponding to the y component of the spin angular momentum is equal to ½; that is, the y component of the spin angular momentum is equal to ½. The new information function vector $|\phi\rangle$ corresponds to the following probabilities:

$$|\phi\rangle = \frac{1}{\sqrt{3}}\left(\frac{1-j}{1}\right) \begin{array}{ll} \text{State of clockwise spin.} & P_{CW} = \frac{2}{3} \\ \text{State of counterclockwise spin.} & P_{CCW} = \frac{1}{3} \end{array}$$

There is yet another interesting aspect to the information vector. Consider the following example:

$$\begin{pmatrix} 4 & -2 \\ -2 & 3 \end{pmatrix}\frac{1}{\sqrt{2}}\begin{pmatrix} 1 \\ 1 \end{pmatrix} = \sqrt{\frac{5}{2}}\frac{1}{\sqrt{5}}\begin{pmatrix} 2 \\ 1 \end{pmatrix} \tag{6.26}$$

Both the initial and resulting information vectors are normalized as required; that is, the information vectors have a length of one. The length of the information vector is not altered by operating on it. Thus, operating on an information vector corresponds

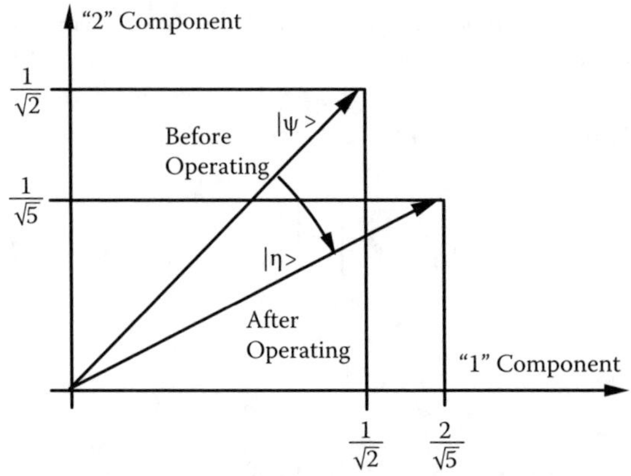

FIGURE 6.4 The effect of operating on the initial information function vector $|\psi\rangle$ of Equation 6.24 is to rotate the vector. The magnitude of the information function vectors remains equal to one.

to a rotation of the information vector, as shown in Figure 6.4. The cosine of the angle between the vectors is equal to the inner product of the information function vectors.

$$\cos\theta = \langle \psi | \phi \rangle \tag{6.27}$$

There are very important special cases of information functions that are not altered by a particular operator. Such information functions are said to "belong" to the particular operator. "Belong to" in German is *eigen*. Thus, information functions that are not altered by a particular operator are called eigen functions of this operator. The information obtained in this case is called an eigen value.

$$\sum_{m=1}^{m=N} f_{nm}\psi_n = f\psi_n \tag{6.28}$$

where ψ_n is a component of the eigen vector and f is the eigen value of the matrix operator with components f_{nm}. For example:

$$\begin{pmatrix} 4 & -1 \\ -1 & 4 \end{pmatrix} \frac{1}{\sqrt{2}} \begin{pmatrix} 1 \\ -1 \end{pmatrix} = 5 \frac{1}{\sqrt{2}} \begin{pmatrix} 1 \\ -1 \end{pmatrix} \tag{6.29}$$

The information vector in Equation 6.29 is not changed by operating with the matrix. Therefore, this information vector is an eigen vector of this matrix, and the information 5 is an eigen value of the matrix.

The operators that represent real physical quantities are *hermitian*. A hermitian matrix is equal to its hermitian conjugate. The hermitian conjugate of a matrix is obtained by taking the complex conjugate of all its matrix elements and then transposing the matrix about its diagonal, as shown in Equations 6.30 and 6.31. Such a matrix can be put in a form where it has only diagonal elements. These diagonal elements are real and are called *eigen values.*

a) Complex conjugate of $\begin{bmatrix} f_{11} & f_{12} \\ f_{21} & f_{22} \end{bmatrix} = \begin{bmatrix} f_{11}^* & f_{12}^* \\ f_{21}^* & f_{22}^* \end{bmatrix}$

and

b) Complex conjugate transpose of $\begin{bmatrix} f_{11} & f_{12} \\ f_{21} & f_{22} \end{bmatrix} = \begin{bmatrix} f_{11}^* & f_{21}^* \\ f_{12}^* & f_{22}^* \end{bmatrix}$ (6.30)

Thus, hermitian matrices must have the following property:

$$\begin{bmatrix} f_{11} & f_{12} \\ f_{21} & f_{22} \end{bmatrix} = \begin{bmatrix} f_{11}^* & f_{21}^* \\ f_{12}^* & f_{22}^* \end{bmatrix} \tag{6.31}$$

For Equation 6.31 to hold, the diagonal elements of a hermitian matrix must be real, $f_{11} = f_{11}^*$ and $f_{22} = f_{22}^*$.

EXAMPLE 6.5

Consider an energy matrix operator $\mathbb{E}$ of the form

$$\mathbb{E} = \begin{bmatrix} E_A & j\Delta \\ -j\Delta & E_B \end{bmatrix}$$

Indeed, the energy operator matrix is hermitian. It is equal to its complex conjugate transpose. Suppose a system is described by the following information function vector $|\psi\rangle$:

$$|\psi\rangle = \begin{bmatrix} \frac{1+j}{2} \\ \frac{1-j}{2} \end{bmatrix}$$

One can check if the information function vector $|\psi\rangle$ is normalized; that is, is $\langle\psi|\psi\rangle = 1$?

$$\langle\psi|\psi\rangle = \begin{bmatrix} \frac{1-j}{2} & \frac{1+j}{2} \end{bmatrix} \begin{bmatrix} \frac{1+j}{2} \\ \frac{1-j}{2} \end{bmatrix}$$

$$\langle\psi|\psi\rangle=\left[\frac{1+j}{2}\frac{1-j}{2}+\frac{1-j}{2}\frac{1+j}{2}\right]$$

$$\langle\psi|\psi\rangle=\frac{1}{2}+\frac{1}{2}$$

$$\langle\psi|\psi\rangle=1$$

That is, the information function vector is normalized. This result in itself is interesting. The system has a probability of 50% of being in one state, and it has a 50% probability of simultaneously being in the other state. No classical system can do this.

Next, the average value of the energy is calculated.

$$\langle\psi|\mathbb{E}|\psi\rangle=\begin{pmatrix}\frac{1-j}{2} & \frac{1+j}{2}\end{pmatrix}\begin{pmatrix}E_A & j\Delta\\ -j\Delta & E_B\end{pmatrix}\begin{pmatrix}\frac{1+j}{2}\\ \frac{1-j}{2}\end{pmatrix}$$

Multiplying the energy operator matrix by the cket vector $|\psi\rangle$:

$$\langle\psi|\mathbb{E}|\psi\rangle=\begin{pmatrix}\frac{1-j}{2} & \frac{1+j}{2}\end{pmatrix}\begin{pmatrix}E_A\frac{1+j}{2}+j\Delta\frac{1-j}{2}\\ -j\Delta\frac{1+j}{2}+E_B\frac{1-j}{2}\end{pmatrix}$$

The new information vector is now:

$$|\phi\rangle=A\begin{bmatrix}E_A\frac{1+j}{2}+j\Delta\frac{1+j}{2}\\ -j\Delta\frac{1+j}{2}+E_B\frac{1-j}{2}\end{bmatrix}$$

where A is a normalization constant. For $|\phi\rangle$ to be an information vector, it has to be normalized.

$$|\phi\rangle=\frac{\sqrt{2}}{\sqrt{(E_B+\Delta)^2+(E_A+\Delta)^2}}\begin{pmatrix}E_A\frac{1+j}{2}+j\Delta\frac{1-j}{2}\\ -j\Delta\frac{1+j}{2}+E_B\frac{1-j}{2}\end{pmatrix}$$

Finally, the calculation for the average energy $\langle \psi|\mathbb{E}|\psi\rangle$ is completed.

$$\langle \psi|\mathbb{E}|\psi\rangle = \frac{1}{2}(E_A + E_B) + \Delta$$

Observe that, as required, the average energy is real.

A method of finding the eigen values and eigen vectors of a matrix will be demonstrated here. By operating with an operator f on one of its eigen functions $|\psi_n\rangle$, one obtains:

$$f|\psi_n\rangle = f_n|\psi_n\rangle$$

where f_n is the corresponding eigen function. This can be rewritten as follows:

$$f|\psi_n\rangle = f_n\mathbb{E}|\psi_n\rangle$$

where $\mathbb{E}$ is the unity matrix. By subtracting the left-hand side of the preceding equation, one obtains:

$$[f - f_n\mathbb{E}]|\psi_n\rangle = 0$$

This requires that the matrix resulting from the subtraction be equal to zero. By a matrix equal to zero, it is meant that its determinant is equal to zero. This can best be illustrated by an example. Consider the following hermitian matrix operator:

$$f = \begin{bmatrix} 5 & 2j \\ -2j & 2 \end{bmatrix}$$

Subtracting the unit matrix multiplied by one of the eigen values from the matrix operator as was done earlier:

$$\left[\begin{bmatrix} 5 & 2j \\ -2j & 2 \end{bmatrix} - f_n \begin{bmatrix} 1 & 0 \\ 0 & 1 \end{bmatrix}\right]\begin{bmatrix} a_n \\ b_n \end{bmatrix} = 0$$

The matrices in the square brackets can be combined into a single matrix.

$$\begin{bmatrix} 5 - f_n & 2j \\ -2j & 2 - f_n \end{bmatrix} = 0$$

The determinant of this matrix is:

$$(5 - f_n)(2 - f_n) - 4 = 0 \quad \text{or} \quad f_n^2 - 7f_n + 10 - 4 = 0$$

By solving for the eigen values f_n, one obtains:

$$f_1 = 6 \quad \text{and} \quad f_2 = 1$$

Next, the eigen vectors are calculated. To this end, one multiplies the matrix that is used for calculating the eigen value by a two-dimensional vector. This will be the eigen vector. Next, the values of these vector components are calculated.

$$\begin{bmatrix} 5-f_n & 2j \\ -2j & 2-f_n \end{bmatrix}\begin{bmatrix} a_n \\ b_n \end{bmatrix} = 0$$

By multiplying out, one obtains:

a) $(5 - f_n)a_n + 2jb_n = 0$
b) $-2ja_n + (2 - f_n)b_n = 0$

These are two homogeneous equations that do not have unique solutions for the components of the eigen vectors. Therefore, consider one less equation than there are above. In this two-dimensional case, this leaves us with only one equation, say, equation **a**. First, the eigen value $f_1 = 6$ is chosen. One obtains from equation **a**:

$$a_1 = 2ib_1$$

Of course, the eigen vector must be normalized:

$$1 = b_1^2(4+1) \quad \text{or} \quad a_1 = 2i\frac{1}{\sqrt{5}} \quad \text{and} \quad b_1 = \frac{1}{\sqrt{5}}$$

If equation **b** is chosen instead of equation **a** one obtains for the same eigen value $f_1 = 6$:

$$2ia_1 = -4b_1 \quad \text{or} \quad a_1 = 2ib_1$$

Of course, the eigen vector must be normalized:

$$1 = b_1^2(4+1) \quad \text{or} \quad a_1 = 2i\frac{1}{\sqrt{5}} \quad \text{and} \quad b_1 = \frac{1}{\sqrt{5}}$$

This result is identical to that obtained from using equation **a**.

Similarly, for the eigen value $f_2 = 1$, one obtains:

$$a_2 = \frac{i}{\sqrt{5}} \quad \text{and} \quad b_2 = -\frac{2}{\sqrt{5}}$$

One can test if the calculation has been correct by operating with the original matrix on the two eigen vectors:

$$\begin{bmatrix} 5 & 2j \\ -2j & 2 \end{bmatrix} \frac{1}{\sqrt{5}} \begin{bmatrix} 2j \\ 1 \end{bmatrix} = 6 \frac{1}{\sqrt{5}} \begin{bmatrix} 2j \\ 1 \end{bmatrix} \quad \text{and} \quad \begin{bmatrix} 5 & 2j \\ -2j & 2 \end{bmatrix} \frac{1}{\sqrt{5}} \begin{bmatrix} j \\ -2 \end{bmatrix} = 1 \frac{1}{\sqrt{5}} \begin{bmatrix} j \\ -2 \end{bmatrix}$$

Indeed, it is found that by operating with the original matrix on the first eigenvector, an eigenvalue of 5 is obtained and by operating with the original matrix on the second eigenvector, an eigen value of 1 is obtained as expected. The two eigenvectors are orthogonal as required.

$$\frac{1}{\sqrt{5}} \begin{bmatrix} -2j & 1 \end{bmatrix} \frac{1}{\sqrt{5}} \begin{bmatrix} j \\ -2 \end{bmatrix} = 0$$

There is a special case where two or more eigenvalues have the same numerical value. Eigen values that have the same numerical value are said to be degenerate eigen values. In this case it is not possible to obtain N – 1 linear independent equations. Here, N is the dimensionality of the matrix. The simplest method is to, first, find all the eigen vectors that can readily be calculated by the preceding illustrated method and then find a vector of unit length that is orthogonal to all the eigen vectors that were readily calculated.

The eigen vectors of the operator can form the unit vectors of the information function Hilbert space.

6.5 THE AVERAGE VALUE OF A CONTINUOUS RANDOM VARIABLE AND OF A FUNCTION OF A CONTINUOUS RANDOM VARIABLE

The *average value* of a function f(x) of a continuous random variable x is

$$\langle f(x) \rangle = \int_0^a f(x) p(x) dx \tag{6.32}$$

where $0 < x \leq a$ and p(x) is the probability density of the random variable x. It is here assumed that x has values between 0 and a. Similar to the discrete case in the literature, other formulations such as E[f(x)], the expected value of f(x) and $\overline{f}(x)$, are also used to express the average value.

EXAMPLE 6.6

Consider a continuous random variable θ having a probability per unit angle of:

$$p(\theta) = \frac{1}{2} \sin\theta \quad \text{for} \quad 0 < \theta \leq \pi$$

The average value of the random variable is:

$$\langle\theta\rangle = \frac{1}{2}\int_0^{\pi} \theta \sin\theta \, d\theta$$

This can be integrated by parts.

$$\langle\theta\rangle = -\frac{1}{2}\left[\theta\cos\theta\right]_0^{\pi} + \frac{1}{2}\int_0^{2\pi} \cos\theta \, d\theta$$

The integral when evaluated at the limits is equal to zero, and the quantity in the square brackets when evaluated at the limits yields:

$$\langle\theta\rangle = \frac{\pi}{2}$$

6.6 THE PROBABILISTIC MODEL OF CONTINUOUS QUANTUM MECHANICS

In Section 6.2 a discrete stochastic model that is used in the quantum mechanical model of nature is discussed. Next, a continuous stochastic model of the quantum mechanical model of nature is described. It too can be used to describe other stochastic systems in which a measurement of the stochastic system changes it.

The probability density p(x) of the random variable x must be positive and real. This is expressed mathematically as follows:

$$p(x) = \psi^*(x)\psi(x) \tag{6.33}$$

The normalization condition of the probability density p(x) of the random variable x requires:

$$1 = \int_0^{a} p(x)\,dx \tag{6.34}$$

where the random variable x has values between x = 0 and x = a. By substituting Equation 6.33 into Equation 6.34, one obtains an expression analogous to Equation 6.5 of the discrete case.

$$1 = \int_0^{a} \psi^*(x)\psi(x)\,dx \tag{6.35}$$

Similar to the discrete case, ψ(x) as well as ψ*(x) are components of *information function vectors*. Recall that there was an information function vector component

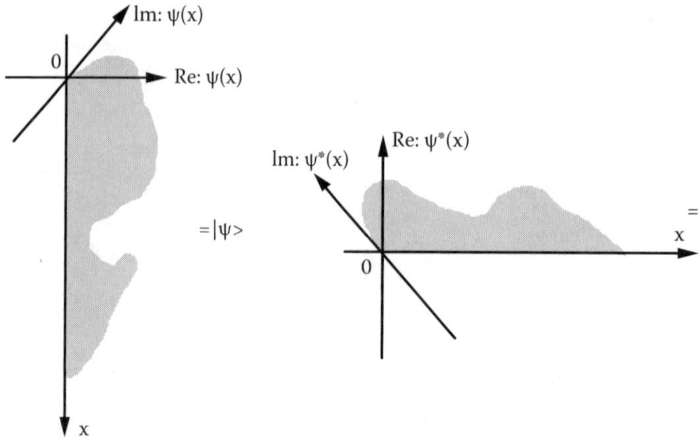

FIGURE 6.5 Similar to the discrete case, there is a vector component of the continuous information function vectors $|\psi\rangle$ and $\langle\psi|$ for every value x of the continuous random variables. The values of the various vector components are plotted as a continuous curve in this case. It was assumed here that $0 < x \leq a$.

for every value n of the discrete random variable. This concept is extended to the continuous case. Similarly, one can say that there is a vector component of the information function vector for every value x of the continuous random variable. Of course, because there are an infinite number of such vector components, they cannot be listed. The continuous information function vectors are schematically illustrated in Figure 6.5. It is conventional to use the same shorthand notation $|\psi\rangle$ and $\langle\psi|$ for the continuous information function vector and its complex conjugate transpose, as for the discrete versions. As has been shown in the discrete case, the information function vectors exist in a space with dimension equal to the number of components in the vector, a so-called Hilbert space. Thus, the continuous information function vectors exist in Hilbert space with an infinite number of dimensions. It, too, can be thought of as a conceptual extension of the three-dimensional vectors of our real three-dimensional space. Because there are as many unit vectors as there are dimensions of the space, this information function Hilbert space must have an infinite number of unit vectors. These unit vectors are also orthogonal to one another. Using the Dirac notation, Equation 6.35 can be written in shorthand notation as follows:

$$1 = \langle\psi|\psi\rangle \tag{6.36}$$

Just as the information function vector of the discrete variables had to have a length of unity, the length of the information function vector of the continuous random variables has to be equal to one.

Substituting Equation 6.33 into Equation 6.32:

$$\langle f(x)\rangle = \int_0^a dx\psi^*(x)f(x)\psi(x) \tag{6.37}$$

As was done in the discrete case, one can extend this result by a method analogous to the one used to go from Equations 6.16 to 6.18.

$$\langle f(x,y)\rangle = \int_0^a dx \int_0^a dy\psi^*(x)f(x,y)\psi(y) \tag{6.38}$$

The kernel f(x,y) is the xy′-th component of an *operator*. This operator is similar to the matrix operator with components f_{nm}, where there is a matrix component f(x,y) for every value of x and y. The operation of the kernel operator on the information function vector with components ψ(x) can be written as follows:

$$\int_0^a dyf(x,y)\psi(y) = A\theta(x) \tag{6.39}$$

where A is an information and θ(x) is the x-th component of a new information vector. By using the shorthand or Dirac notation for the continuous information function vectors, Equation 6.39 can be written as follows:

$$\mathbb{f}|\psi\rangle = A|\theta\rangle \tag{6.40}$$

where $\mathbb{f}$ is the integral kernel operator.

For the special case where the operator $\mathbb{f}$ does not change the information function vector, the information function vector is an eigen function of the kernel operator and the information F is an eigen value of this operator.

$$\mathbb{f}|\psi\rangle = F|\psi\rangle \tag{6.41}$$

where $|\psi\rangle$ is an eigen vector and F is an eigen value of the operator $\mathbb{f}$. Writing Equation 6.41 in integral form:

$$\int_0^a dyf(x,y)\psi(y) = F\psi(x) \tag{6.42}$$

Equation 6.42 is a homogeneous Fredholm equation.

The integral operator $\mathbb{L}$ can be defined as:

$$\mathbb{L} = \int_0^a dyf(x,y) \tag{6.43}$$

where $\mathbb{L}$ has the continuous equivalent of the matrix operator of Equation 6.21. Equation 6.42 can then be expressed in operator form:

$$\mathbb{L}\psi(y) = \mathbb{f}\psi(x) \tag{6.44}$$

Multiplying Equation 6.44 by the inverse $\mathbb{L}^{-1}$ of the operator $\mathbb{L}$:

$$\frac{1}{F}\psi(x) = \mathbb{L}^{-1}\psi(x) \tag{6.45}$$

The inverse $\mathbb{L}^{-1}$ of the integral operator $\mathbb{L}$ is a differential operator $\mathbb{D}$. Unlike in the discrete mathematical model of quantum mechanics, in the continuous model the kernel operator discussed earlier is seldom used. Corresponding to each kernel operator f(x,y), there is a differential operator $\mathbb{D}$. Equation 6.45 is a differential equation:

$$\text{a)}\quad \mathbb{D}\psi(x) = A\psi(x) \quad \text{where} \quad \text{b)}\quad A = \frac{1}{F} \tag{6.46}$$

Indeed, in the continuous model of quantum mechanics, the differential operator model is used instead of the integral equation model. One can, again, interpret Equation 6.46 as follows: The information function vector contains all the information about a particular system, and the operator $\mathbb{D}$ is used to read some of the information.

The one-dimensional momentum operator $(\hbar/j)(d/dx)$, and the one-dimensional kinetic energy operator $-(\hbar^2/2m)(d^2/dx^2)$ are examples of differential operators $\mathbb{D}$. Here, $\hbar = 1.054588664 \times 10^{-34}$ J-s is Planck's constant divided by 2π, and m is the mass of a particle.

It is of special interest to return to the case where the information function vector is an eigen function of the operator, Equation 6.34. Of special interest in quantum mechanics is the eigen value equation of the energy operator $\mathbb{H}$.

$$\mathbb{H}|\psi\rangle = E|\psi\rangle \tag{6.47}$$

where the energy operator in this case is equal to the sum of the kinetic energy operator $\mathbb{K}$(x,y,z) and the potential energy operator $\mathbb{V}$(x,y,z):

$$\mathbb{H} = \mathbb{K}(x,y,z) + \mathbb{V}(x,y,z) \tag{6.48}$$

and where the kinetic energy operator as discussed earlier is equal to the square of the momentum operator divided by twice the mass m of a particle. For the one-dimensional case:

$$\mathbb{K}(x) = -\frac{\hbar^2}{2m}\frac{d^2}{dx^2} \tag{6.49}$$

The potential energy operator in one dimension is simply equal to the classical potential energy:

$$\mathbb{V}(x) = V(x) \tag{6.50}$$

Substituting Equations 6.49 and 6.50 into Equation 6.48 and the result into Equation 6.47:

$$\left[-\frac{\hbar^2}{2m}\frac{d^2}{dx^2}+V(x)\right]\psi_n(x)=E_n\psi_n(x) \tag{6.51}$$

This is the well-known Schrödinger equation in one dimension. It is a differential equation that usually only has solutions of $\psi_n(x)$ for specific values of the eigen value E_n. The operator in the square brackets of the Schrödinger equation is an operator of the probabilistic model of continuous quantum mechanics.

EXAMPLE 6.7

The kinetic energy operator $\mathbb{E}$ for a particle moving only in the x direction is given by:

$$\mathbb{E}=-\frac{\hbar^2}{2m}\frac{\partial^2}{\partial x^2}$$

where $0 < x \leq \infty$. The following form can be assumed for the information function:

$$|\psi\rangle=(2\alpha)^{\frac{1}{2}}\exp\left[\left(j\frac{2\pi}{\lambda}-\alpha\right)x\right]$$

One should have used an information function that consisted of the one used here plus its complex conjugate. However, for simplicity the specified information function is used. The information function has the form of a decaying wave with wavelength λ and decay constant α. First, it has to be checked whether the information function is normalized. This can be accomplished by substituting into Equation 6.35.

$$2\alpha\int_0^\infty \exp\left(-j\frac{2\pi}{\lambda}-\alpha\right)\exp\left(j\frac{2\pi}{\lambda}-\alpha\right)dx=?$$

or

$$2\alpha\int_0^\infty \exp(-2\alpha x)dx=1$$

Thus, the information function is, indeed, normalized. Next, the information function is substituted into Equation 6.38 to calculate the average value of the energy:

$$\langle\psi|\mathbb{E}|\psi\rangle=-\frac{\alpha\hbar^2}{m}\int_0^\infty \exp\left(-j\frac{2\pi}{\lambda}-\alpha\right)\frac{\partial^2}{\partial x^2}\exp\left(j\frac{2\pi}{\lambda}-\alpha\right)dx$$

or

$$\langle\psi|\mathbb{E}|\psi\rangle = \frac{\left(\frac{4\pi^2}{\lambda^2} - j\frac{4\pi}{\lambda}\alpha - \alpha^2\right)\hbar^2}{2m}$$

The average energy has to be real. Therefore, one should have used an information function that consisted of the one used here plus its complex conjugate. In this case, one would obtain a real average energy. In the limit, as the decay constant goes to zero, one obtains the average energy of a free quantum mechanical particle. Note that λ is the wavelength of the particle. In quantum mechanics, particles have wave properties, and waves have particle properties.

As was shown earlier, the quantum mechanical matrix operators of the discrete case did not, in general, commute. Neither do the quantum mechanical operators of continuous systems commute. Consider, for example, the one-dimensional position operator $\mathbb{x} = x$ and the one-dimensional momentum operator $\mathbb{p} = (\hbar/j)(d/dx)$. One can calculate the difference between first operating with the position operator x on an information function $|\psi\rangle$ and next operating with the momentum operator $\mathbb{p}$, and first operating with the momentum operator $\mathbb{P}$ on an information function $|\psi\rangle$ and next operating with the position operator $\mathbb{x}$.

$$\text{a)}\quad \mathbb{p}\mathbb{x}|\psi\rangle - \mathbb{x}\mathbb{p}|\psi\rangle = \frac{\hbar}{j}\frac{d}{dx}(x\psi) - x\frac{\hbar}{j}\frac{d\psi}{dx} \quad \text{or}$$

$$\text{b)}\quad \mathbb{p}\mathbb{x}|\psi\rangle - \mathbb{x}\mathbb{p}|\psi\rangle = \frac{\hbar}{j}\left[\psi + x\frac{d\psi}{dx} - x\frac{d\psi}{dx}\right] \quad \text{or} \quad \text{c)}\quad \mathbb{p}\mathbb{x}|\psi\rangle - \mathbb{x}\mathbb{p}|\psi\rangle = \frac{\hbar}{j}\psi \tag{6.52}$$

This can be written as:

$$\text{a)}\quad \mathbb{p}\mathbb{x}|\psi\rangle - \mathbb{x}\mathbb{p}|\psi\rangle = \frac{\hbar}{j}\psi \quad \text{or} \quad \text{b)}\quad [\mathbb{p},\mathbb{x}]|\psi\rangle = \frac{\hbar}{j}|\psi\rangle \tag{6.53}$$

A continuous information function can be a stepwise continuous function. This occurs when the potential energy ϕ varies stepwise in space. Consider a particle subject to a potential energy, as shown in Figure 6.6.

For a stepwise continuous potential energy, there will be a stepwise continuous information function. This potential energy can be obtained by solving Schrödinger's equation in each region where the potential energy is continuous. For the example illustrated in Figure 6.6, Schrödinger's equation has to be solved in two regions, region 1 for $-\infty < x \le 0$ and region 2 for $0 < x \le \infty$, where the potential energy is discontinuous. Schrödinger's equation for region 1, for $-\infty < x \le 0$, where the potential energy is equal to zero is

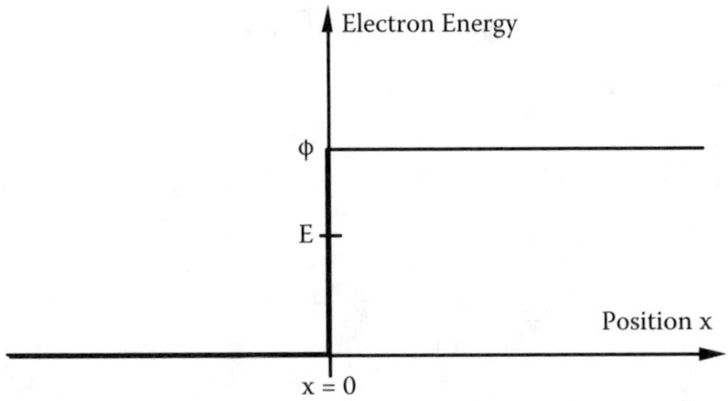

FIGURE 6.6 Stepwise continuous electron potential energy.

$$-\frac{\hbar}{2m}\frac{\partial^2\psi_1(x)}{\partial x^2} = E\psi_1(x) \tag{6.54}$$

where $\hbar$ is Planck's constant divided by 2π and m is the mass of the particle. It is assumed in this case that the information function $\psi_1(x)$ consists of an incoming wave and a reflected wave:

$$\psi_1(x) = \exp[-jkx] + R\exp[jkx] \tag{6.55}$$

As will be shown later, the derivative of the information function $\psi_1(x)$ will also be required.

$$\frac{\partial\psi_1(x)}{\partial x} = -jk\exp[-jkx] + jkR\exp[jkx] \tag{6.56}$$

Substituting each term of Equation 6.55 separately into Equation 6.54 and canceling the exponential and any constants:

$$E = \frac{\hbar^2 k^2}{2m} \tag{6.57}$$

Schrödinger's equation for region 2, for $\infty < x \le 0$, where the potential energy is equal to ϕ is

$$-\frac{\hbar}{2m}\frac{\partial^2\psi_2(x)}{\partial x^2} + \phi\psi_2(x) = E\psi_2(x) \tag{6.58}$$

The following form is assumed for the information function $\psi_2(x)$:

$$\psi_2(x) = T\exp[-\alpha x] \tag{6.59}$$

Similarly, as will be shown later, the derivative of the information function $\psi_2(x)$ will also be required.

$$\frac{\partial \psi_2(x)}{\partial x} = -\alpha T\exp[-\alpha x] \tag{6.60}$$

At the discontinuity, at x = 0 in this case, both the information functions and the derivative of the information function must be continuous. These are known as *boundary conditions.*

$$\text{a)}\quad \psi_1(x=0) = \psi_2(x=0) \quad \text{or} \quad \text{b)}\quad 1+R = T \tag{6.61}$$

and

$$\text{a)}\quad \left.\frac{\partial \psi_1(x)}{\partial x}\right]_{x=0} = \left.\frac{\partial \psi_2(x)}{\partial x}\right]_{x=0} \quad \text{or} \quad \text{b)}\quad -jk + jkR = -\alpha T \tag{6.62}$$

Equations 6.61b and 6.62b are two equations with two unknowns, the reflection coefficient R and the transmission coefficient T. These equations can be solved for R and T:

$$\text{a)}\quad R = \frac{1+j\dfrac{\alpha}{k}}{1-j\dfrac{\alpha}{k}} \quad \text{and} \quad \text{b)}\quad T = \frac{2}{1-j\dfrac{\alpha}{k}} \tag{6.63}$$

6.7 MOMENTS OF RANDOM VARIABLES

The random variable itself can be considered to be a function of itself; $f_n = n$. Therefore, the average value of the random variable is

$$\langle n \rangle = \sum_{n=1}^{n=N} nP_n \tag{6.64}$$

where $0 < n \le N$. Similarly, the average value of a continuous random variable x is

$$\langle x \rangle = \int_0^a xp(x)dx \tag{6.65}$$

where $0 < x \le a$.

EXAMPLE 6.8

Another important example is as follows:

$$\langle x \rangle = 0.25$$

$$\langle x^2 \rangle = 0.125$$

$$\langle x^3 \rangle = 0.09375$$

Suppose the values of, say, the average value $\langle x \rangle$, the average value of the square $\langle x^2 \rangle$, and the average value of the cube $\langle x^3 \rangle$ of the random variable x are known.

Consider the following function f(x) of the random variable x:

$$f(x) = x^3 + 6x^2 - 3x + 2$$

By taking the average value of f(x) one obtains:

$$\langle f(x) \rangle = \langle x^3 \rangle + 6\langle x^2 \rangle - 3\langle x \rangle + 2$$

The average value of the function f(x) of the random variable is

$$\langle f(x) \rangle = 0.09375 + (6 \times 0.125) - (3 \times 0.25) + 2$$

or

$$\langle f(x) \rangle = 2.09375$$

The average value of the various powers of a random variable are known as *moment of a random variable.* For example, the m-th moment of the continuous random variable x is equal to the average value of x^m.

$$\langle x^m \rangle = \int_0^a x^m p(x)\,dx \quad \text{for} \quad 0 < x \le a \tag{6.66}$$

Here, x could also have been a discrete random variable.

$$\langle n^m \rangle = \sum_{n=1}^{n=N} n^m p_n \quad \text{for } 0 < n \le N \tag{6.67}$$

The reason these quantities are called moments is that they bear some similarity to mechanical moments of real physical systems. The first mechanical moment is equal to the product of the force times the length of a lever arm, as shown in Figure 6.7. Consider, for example, a distributed mechanical system shown in Figure 6.7. The

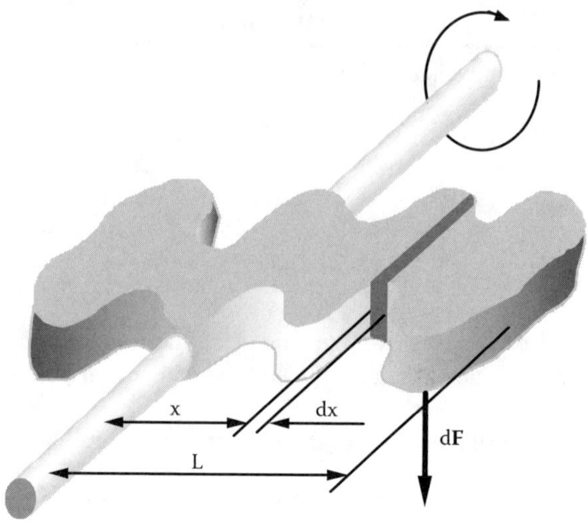

FIGURE 6.7 A lever where the mass density, the mass per unit length, varies with the distance x from the axis.

force, here, is due to the gravitational attraction of the mass of the lever. The incremental force dF due to an incremental mass $\rho(x)dx$ is equal to:

$$dF = g\rho(x)dx \tag{6.68}$$

where g is the acceleration due to gravity and $\rho(x)$ is the mass per unit length of the lever arm. The incremental torque dT or incremental first moment due to this incremental force is equal to:

$$\text{a)}\quad dT = xdF \quad \text{or} \quad \text{b)}\quad dT = xg\rho(x)dx \tag{6.69}$$

The total torque exerted by the lever arm on the shaft is

$$T = \int_0^L xg\rho(x)dx \tag{6.70}$$

where L is the length of the lever (see Figure 6.7). Indeed, the equation of the first mechanical moment bears a resemblance to the first stochastic moment of a continuous random variable. There exists a second mechanical moment that is equal to the square of the length of the lever arm times the mass. By integrating over the incremental second moments $dI = x^2 g\rho(x)dx$, one obtains:

$$I = \int_0^L x^2 g\rho(x)dx \tag{6.71}$$

which is equal to the moment of inertia of the system. Its equation resembles the second stochastic moment of a continuous random variable. This concept can be extended to higher-order moments. In the stochastic case, p(x) takes the place of gρ(x).

EXAMPLE 6.9

Consider a continuous random variable x with a probability density p(x) given by:

$$p(x) = a\exp(-ax) \quad \text{for} \quad 0 < x \le \infty$$

The fifth moment of the random variable x is

$$\langle x^5 \rangle = \frac{1}{\alpha^5}\int_0^\infty \alpha^5 x^5 \exp(-\alpha x)\alpha\, dx$$

Note that:

$$\int_0^\infty u^n \exp(-u)\, du = n!$$

By integrating, one obtains:

$$\langle x^5 \rangle = \frac{5!}{\alpha^5}$$

For the case where there is a region of the random variable x that has the same probability, the m-th moment is

$$\langle x^n \rangle = \int_0^a x^n g(x) p(x)\, dx \tag{6.72}$$

where g(x) is the density of events with the same probability. A similar expression exists for discrete random variables.

6.8 CONDITIONAL AVERAGE VALUE OF A RANDOM VARIABLE AND OF A FUNCTION OF A RANDOM VARIABLE

The average value of a random variable m subject, say, to event B occurring is known as the conditional average value $\langle m|B \rangle$:

$$\langle m|B \rangle = \sum_{n=1}^{n=N} mp(m|B) \quad \text{for} \quad 0 < n \le N \tag{6.73}$$

where $p(m|B)$ is the conditional probability that the random variable m will have a certain value m provided event B has occurred.

This can be extended to a continuous random variable:

$$\langle x|B\rangle = \int_0^a x p(x|B)\,dx \quad \text{for} \quad 0 < x \le a \tag{6.74}$$

where $p(x|B)$ is the conditional probability density. This, of course, can be extended to functions of a random variable.

$$\langle f|B\rangle = \int_0^a f(x)p(x|B)\,dx \quad \text{for} \quad 0 < x \le a \tag{6.75}$$

Because moments are average values of powers of a random variable and the power of a random variable is a function of a random variable, conditional moments of a random variable are similar to the conditional average value of a function of a random variable.

6.9 CENTRAL MOMENTS

The central moment μ_n of a random variable is equal to the average value of a power of the difference between a random variable, x, and its average value, $\langle x \rangle$. For example, for a continuous random variable x, the n-th central moment is

$$\mu_n = \int_0^a \left[x - \langle x \rangle\right]^n p(x)\,dx \quad \text{for} \quad 0 < x \le a \tag{6.76}$$

A similar expression can be derived for a discrete random variable. For the case when there is more than one value of the random variable n that has the same probability, the central moment, for example, for a discrete random variable is

$$\mu_n = \sum_{k=1}^{k=K} \left[k - \langle k \rangle\right]^n G_k P_k \tag{6.77}$$

where the G_n's are the number of values of the random variable with the same probability.

6.10 VARIANCE AND STANDARD DEVIATION

The *variance* σ^2 is the average value of the square of the change Δx from the average $\langle x \rangle$ of a random variable x (see Figure 6.8); that is the variance is the second central moment of the random variable. The first central moment of any random variable is equal to zero.

$$\sigma^2 = \left\langle \left[x - \langle x \rangle\right]^2 \right\rangle \tag{6.78}$$

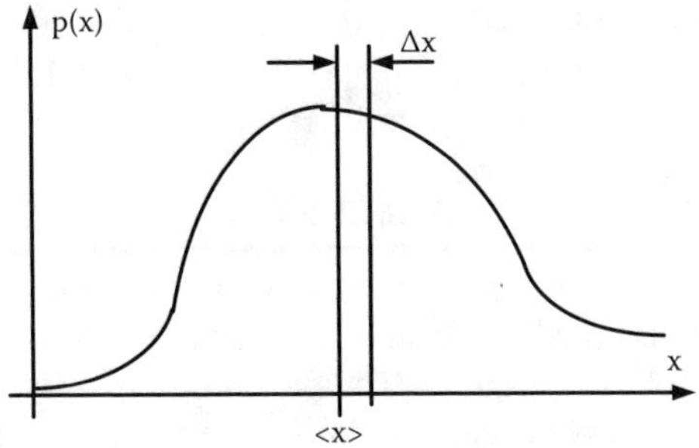

FIGURE 6.8 The change Δx of a random variable from its average value $\langle x \rangle$.

where, recall that the brackets $\langle \ldots \rangle$ denote the average value operation. The quantity σ is the standard deviation. Multiplying out the quantity in the square brackets:

$$\sigma^2 = \left\langle \left[x^2 - 2x\langle x \rangle + \langle x \rangle^2 \right] \right\rangle \tag{6.79}$$

Finally, the average value operation on the random variable x is performed to obtain for the variance σ^2:

$$\sigma^2 = \langle x^2 \rangle - \langle x \rangle^2 \tag{6.80}$$

Thus, the variance σ^2 is equal to the difference between the average value of the square of the random variable $\langle x^2 \rangle$ and the square of the average value of the random variable $\langle x \rangle^2$. For example, the variance σ^2 of the discrete random variable n, $0 < n \le N$, is

$$\sigma^2 = \sum_{n=1}^{n=N} n^2 P_n - \left[\sum_{n=1}^{n=N} n P_n \right]^2 \tag{6.81}$$

The variance of a complex random variable $z = x + jy$ is

$$\sigma^2 = \left\langle \left[z - \langle z \rangle \right]^* \left[z - \langle z \rangle \right] \right\rangle \tag{6.82}$$

Multiplying out the quantity in the square brackets:

$$\sigma^2 = \left\langle |z|^2 \right\rangle - \left| \langle z \rangle \right|^2 \tag{6.83}$$

Similarly, the variance σ^2 is equal to the difference between the average value of the square of the magnitude of the random variable $\langle |z|^2 \rangle$ and the square of the magnitude of the average value of the random variable $|\langle z \rangle|^2$.

EXAMPLE 6.10

One can calculate the variance σ^2 of the complex random variable $z = r\exp[j\phi]$. The random variable z is specified within and on a circle of radius R ($|z| \le R$) in the complex plane (see the following figure). The probability density p(z) that the random variable z will have values between z and z + dz in this example is

$$p(z) = \frac{1}{\pi R^2}$$

Indeed, as required, the probability density p(z) is real and positive because the radius R is a real and positive number. Observe that the probability density is independent of the angle ϕ.

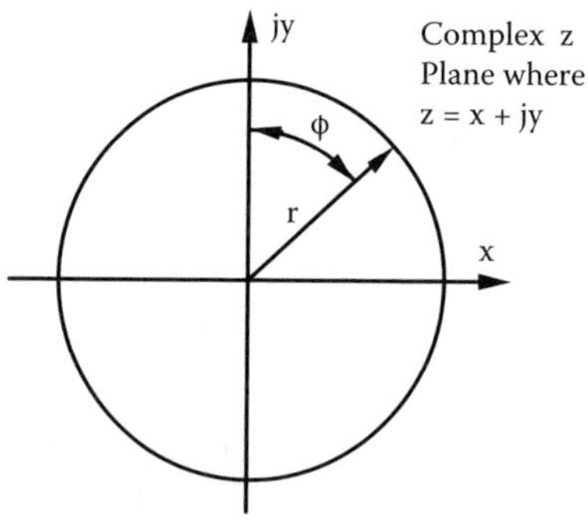

First, one can calculate the average value of z:

$$\langle z \rangle = \frac{1}{\pi R^2}\int_0^{2\pi} \exp[j\phi]\,d\phi \int_0^R r^2 dr$$

Integrating:

$$\langle z \rangle = 0$$

Observe that the average value of the complex random variable in this case is zero. One next calculates the average value of the square of the magnitude of the random variable z:

$$\langle |z|^2 \rangle = \frac{1}{\pi R^2}\int_0^{2\pi} d\phi \int_0^R r^3 dr$$

Integrating:

$$\sigma^2 = \frac{R^2}{2}$$

Using Equation 6.80 to calculate the variance:

$$\sigma^2 = \frac{R^2}{2}$$

6.11 CORRELATIONS OF TWO RANDOM VARIABLES AND OF FUNCTIONS OF RANDOM VARIABLES

Consider the correlation of two functions f(x) and h(y) of two random variables x and y.

$$R_{fh} = \langle f^*(x)h(y) \rangle \tag{6.84}$$

where R_{fh} is the *correlation* of the functions f(x) and h(y) of the two random variables x and y, and * denotes the complex conjugate if the functions are complex. Here, x and y can either be discrete {n = 1, 2, 3, …, m = 0, 1, 2, 3, …} or continuous random variables. Here, R_{fh} can also be the correlation between two functions f(x) and h(x) of the random variable x. Indeed, the latter case is the more common one. The two functions f(x) and h(y) can also be equal to the random variables x and y themselves.

$$R_{xy} = \langle xy \rangle \tag{6.85}$$

There are several special cases of the correlation R_{fh}: If the correlation R_{fh} of two functions of two random variables or the same random variable is equal to the product of the average value of the individual functions, the functions are *uncorrelated.*

$$\text{a)}\quad R_{fh} = \langle f^*(x) \rangle \langle h(y) \rangle \quad \text{or} \quad \text{b)}\quad R_{fh} = \langle f^*(x) \rangle \langle h(x) \rangle \tag{6.86}$$

If, as before, the functions are equal to the individual random variables x and y and the correlation R_{xy} is equal to the product of the average value of the random variables, the random variables x and y are *uncorrelated.*

$$R_{xy} = \langle x \rangle \langle y \rangle \tag{6.87}$$

If two random variables x and y are statistically independent, then the random variables will be uncorrelated. However, two statistically uncorrelated random variables are not necessarily statistically independent.

If the correlation of two functions f(x) and h(y) of two random variables x and y, or two functions of f(x) and h(x) of the same random variable x is equal to zero, the two functions are *orthogonal.*

$$R_{fh} = 0 \tag{6.88}$$

Similarly, if the functions are equal to the individual random variable x and y and the correlation R_{xy} is equal to zero, the two random variables are *orthogonal.*

$$R_{xy} = 0 \tag{6.89}$$

This is best illustrated by a discrete example. Consider two functions $f_1, f_2, f_3, f_3, \ldots$ and $h_1, h_2, h_3, h_3, \ldots$ of two random variables n = 1, 2, 3, … and m = 1, 2, 3, …. The correlation of these two functions:

$$R_{fh} = \sum_{n=1}^{n=N}\sum_{m=1}^{m=N} f_n^* h_m P_{n\cap m} \tag{6.90}$$

where $P_{m\cap n}$ is the joint probability that the values n and m of the random variables will occur. Because only an example is considered, a joint probability that is equal to zero if m is not equal to n and is equal to $P_{m\cap n} = 1/N$ if m is equal to n is chosen. With this assumption, Equation 6.90 can be rewritten as follows:

$$R_{fh} = \frac{1}{N}\sum_{n=1}^{n=N} f_n h_n \tag{6.91}$$

One can interpret f_n and h_n as components of two vectors. The correlation R_{fh} of Equation 6.91 has the form of a scalar or dot product of the two vectors. The scalar product of two vectors in real space that are at 90° to each other, or "orthogonal," is equal to zero. One can extend this concept and say that if the correlation of two random variables or functions of random variables is equal to zero, the two random variables or functions of the random variables are orthogonal.

For the case where the joint probability $P_{m\cap n}$ is not restricted to the preceding example, one can interpret the space in which the vectors with components f_n and h_n exist as a *curved space.* Here, the joint probability $P_{m\cap n}$ scales the various vector components. The joint probabilities in this case become the components of a *metric tensor* of sorts.

$$R_{fh} = \sum_{n=1}^{n=N} f_n^* P_{n\cap m} h_m \tag{6.92}$$

For Equation 6.92 to represent the inner product of two vectors with components f_n and h_m in a curved space, one should divide the joint probabilities by a constant so that the determinant of the joint probability tensor is equal to one. The determinant of a metric tensor has to be equal to one. Of course, Equation 6.92 has to also be multiplied by the same constant in order not to change it.

The *autocorrelation* R_{ff} of a function f(x) of a random variable x is equal to the average value of the square of the magnitude of the function f(x).

$$R_{ff} = \langle |f(x)|^2 \rangle \tag{6.93}$$

For the case when the function f(x) of the random variable is equal to the random variable x, the autocorrelation of the random variable R_{xx} is equal to its second moment.

$$R_{xx} = \langle x^2 \rangle \tag{6.94}$$

Consider next the joint central average value of the two functions f(x) and h(y) of two random variables x and y.

$$C_{fh} = \langle [f(x) - \langle f(x) \rangle][h(y) - \langle h(y) \rangle] \rangle \tag{6.95}$$

where C_{fh} is the *covariance* of the two functions of the random variables x and y. Multiplying out and taking the average value:

$$C_{fh} = \langle f(x)h(y) \rangle - \langle f(x) \rangle \langle h(y) \rangle \tag{6.96}$$

or

$$C_{fh} = R_{fh} - \langle f(x) \rangle \langle h(y) \rangle \tag{6.97}$$

Similarly, the joint central average value of the two functions f(x) with itself of the same random variable x is equal to:

$$C_{ff} = R_{ff} - \langle f(x) \rangle^2 \tag{6.98}$$

In the case where the functions f(x) and h(y) of the two random variables are equal to the random variables themselves, the joint central average value of the two functions reduces to:

$$C_{xy} = \langle [x - \langle x \rangle][y - \langle y \rangle] \rangle \tag{6.99}$$

where C_{xy} is the covariance of the random variables x and y. Multiplying out and taking the average value:

$$C_{xy} = R_{xy} - \langle x \rangle \langle y \rangle \tag{6.100}$$

The covariance of two uncorrelated random variables is equal to zero because in this case

$$R_{xy} = \langle x \rangle \langle y \rangle \tag{6.101}$$

The covariance of two orthogonal random variables or of orthogonal functions of the two random variables is

$$\text{a)}\quad C_{xy} = -\langle x \rangle \langle y \rangle \quad \text{and} \quad \text{b)}\quad C_{fh} = -\langle f(x) \rangle \langle h(y) \rangle \tag{6.102}$$

The *autocovariance* C_{ff} of a function f(x) of a random variable x is equal to:

$$C_{ff} = R_{ff} - \langle f^2(x) \rangle \tag{6.103}$$

The *autocovariance* C_{xx} of the random variable itself is equal to the variance σ^2. This result can readily be obtained by letting y equal x in Equation 6.103.

$$C_{xx} = \sigma^2 \tag{6.104}$$

The correlation and autocorrelation are macroscopic parameters. In the previous discussion, one did not require knowledge of the probabilities or probability densities. Indeed, one did not even have to know whether the random variables were discrete or continuous.

EXAMPLE 6.11

Suppose $\langle x \rangle = 4$, $\langle x^2 \rangle = 18$, and $\langle x^3 \rangle = 0.1$

and

$$f(x) = [2 + x]^2 \text{ and } h(x) = 3 - x.$$

The correlation R_{fh} of the two functions can be calculated:

$$R_{fh} = \langle [2 + x]^2[3 - x] \rangle$$

By squaring the first square bracket:

$$R_{fh} = \langle [4 + 4x + x^2][3 - x] \rangle$$

By multiplying out and taking average values:

$$R_{fh} = 12 + 12\langle x \rangle + 3\langle x^2 \rangle - 4\langle x \rangle - 4\langle x^2 \rangle - \langle x^3 \rangle$$

By substituting numerical values for the various moments of the random variable x, one finally obtains for the correlation:

$$R_{fh} = 12 + (8 \times 4) - 18 - 0.1$$

or

$$R_{fh} = 25.9$$

EXAMPLE 6.12

Consider two functions f(x) and h(x) of the same random variable x defined between plus and minus infinity, $-\infty < x \leq \infty$:

$$\text{a)}\quad f(x) = \exp\left[-\frac{(x+b)^2}{2\sigma^2}\right], \quad \text{b)}\quad h(x) = \exp\left[-\frac{(x-a)^2}{2\sigma^2}\right],$$

$$\text{c)}\quad p(x) = \frac{1}{(2\pi)^{\frac{1}{2}}\sigma}\exp\left[-\frac{x^2}{2\sigma^2}\right]$$

Here, p(x) is the probability density of the random variable x. The correlation of the two functions can be calculated:

$$R_{fh} = \frac{1}{(2\pi)^{\frac{1}{2}}\sigma}\int_{-\infty}^{\infty} dx \exp\left[-\frac{(x+b)^2 + (x-a)^2 + x^2}{2\sigma^2}\right]$$

Multiplying out the exponent and arranging it to contain a complete square involving the random variable x:

$$\text{exponent} = -\frac{x^2 - \frac{2}{3}x(a-b) + \frac{(a-b)^2}{9} + \frac{2a^2 + 2ab + 2b^2}{9}}{\frac{2}{3}\sigma^2}$$

By substituting this exponent back into the integral:

$$R_{fh} = \frac{1}{(2\sigma)^{\frac{1}{2}}\sigma}\exp\left[-\frac{a^2 + b^2 + ab}{3\sigma^2}\right]\int_{-\infty}^{\infty} dx \exp\left[-\frac{\left(x - \frac{a-b}{3}\right)^2}{\frac{2}{3}\sigma^2}\right]$$

Integrating:

$$R_{fh} = (3)^{\frac{1}{2}}\exp\left[-\frac{a^2 + b^2 + ab}{3\sigma^2}\right]$$

Note that the correlation R_{fg} is a maximum when both a and b are equal to zero. This will occur when both functions f(x) and g(x) overlap. The further apart the maxima of the functions f(x) and g(x) become the smaller the correlation R_{fg}.

6.12 A SPECIAL CASE: THE AVERAGE VALUE OF e^{-jkx}

It is interesting to calculate the average value of the complex function e^{-jkx} of the real continuous random variable x. Here, $-\infty < x \le \infty$, and k is a constant. As will be shown, this is quite useful.

$$\phi(k) = \int_{-\infty}^{\infty} p(x) e^{-jkx} dx \tag{6.105}$$

Observe that the average value f(ϕ) of the complex function e^{-jkx} is equal to the Fourier transform of the probability density. Of course, one can recover the probability density by calculating the inverse Fourier transform of the function f(ϕ).

$$p(x) = \frac{1}{2\pi} \int_{-\infty}^{\infty} \phi(k) e^{jkx} dk \tag{6.106}$$

Besides being the Fourier transform of the probability density, the function f(ϕ) has another useful property, namely, the m-th moment of the random variable is equal to the limit as k approaches zero of the product of j^n and the m-th derivative of f(ϕ) with respect to k.

$$\frac{\lim}{k \to 0} j^n \frac{\partial^n \phi(k)}{\partial k^n} = \int_{-\infty}^{\infty} x^n e^{-jkx} p(x) dx \tag{6.107}$$

or

$$\langle x^n \rangle = \frac{\lim}{k \to 0} j^n \frac{\partial^n \phi(k)}{\partial k^n} \tag{6.108}$$

An important example of this is the n-th moment of a Gaussian. For simplicity, consider the probability density of a zero-mean Gaussian probability density:

$$p(x) = \frac{1}{\sqrt{2\pi\sigma^2}} \exp\left(-\frac{x^2}{2\sigma^2}\right) \tag{6.109}$$

The Fourier transform ϕ(k) of the zero-mean Gaussian can be obtained by substituting Equation 6.109 into Equation 6.105:

$$\phi(k) = \frac{1}{\sqrt{2\pi\sigma^2}} \int_{-\infty}^{\infty} \exp\left(-\frac{x^2}{2\sigma^2}\right) e^{-jkx} dx \tag{6.110}$$

One can combine the exponents of the exponentials to form a complete square:

$$\text{exponent} = -\frac{\left(x + jk\sigma^2\right)^2}{2\sigma^2} - \frac{k^2\sigma^2}{2} \tag{6.111}$$

Substituting the exponent of Equation 6.111 into Equation 6.110:

$$\phi(k)=\frac{1}{\sqrt{2\pi\sigma^2}}\exp\left(-\frac{k^2\sigma^2}{2}\right)\int_{-\infty}^{\infty}\exp\left[-\frac{\left(x+jk\sigma^2\right)^2}{2\sigma^2}\right]dx \tag{6.112}$$

Because the Gaussian probability density is normalized, the following value for the Fourier transform of the Gaussian probability density is obtained:

$$\phi(k)=\exp\left(-\frac{k^2\sigma^2}{2}\right) \tag{6.113}$$

Indeed, the Fourier transform of a Gaussian is also a Gaussian.

The first moment of the random variable x is

$$\text{a)}\ \langle x\rangle=\frac{\lim}{k\to 0}\left[-jk\sigma^2\exp\left(-\frac{k^2\sigma^2}{2}\right)\right]\quad\text{or}\quad\text{b)}\ \langle x\rangle=0 \tag{6.114}$$

As expected, because the zero-mean Gaussian is a symmetric function and the random variable x is an antisymmetric function, the average value $\langle x\rangle$ is equal to zero. Because the form of the Gaussian was chosen in such a way that the average value $\langle x\rangle$ of the random variable x is equal to zero, the second moment of the random variable x is equal to the variance σ^2:

$$\text{a)}\ \langle x^2\rangle=\frac{\lim}{k\to 0}(-1)\left[-\sigma^2+k^2\sigma^4\right]\exp\left(-\frac{k^2\sigma^2}{2}\right)\quad\text{or}\quad\text{b)}\ \langle x^2\rangle=\sigma^2 \tag{6.115}$$

The third moment of the random variable x is

$$\text{a)}\ \langle x^3\rangle=\frac{\lim}{k\to 0}(-j)\left[3k\sigma^4-k^3\sigma^6\right]\exp\left(-\frac{k^2\sigma^2}{2}\right)\quad\text{or}\quad\text{b)}\ \langle x^3\rangle=0 \tag{6.116}$$

Similarly, because the zero-mean Gaussian is a symmetric function and the cube x^3 of the random variable is an antisymmetric function, the average value $\langle x^3\rangle$ is equal to zero. Finally, the fourth moment of the random variable x is

$$\begin{aligned}&\text{a)}\ \langle x^4\rangle=\frac{\lim}{k\to 0}(-j)\left[3\sigma^4-6k^2\sigma^6+k^4\sigma^8\right]\exp\left(-\frac{k^2\sigma^2}{2}\right)\quad\text{or}\\&\text{b)}\ \langle x^4\rangle=3\sigma^4\end{aligned} \tag{6.117}$$

This result is difficult to obtain by other methods.

REFERENCES

1. *Uber quantentheoretische Umdeutung kinematischer und mechanischer Beziehungen* von W. Heisenberg Zeitschrift für Physik, Vol. 33, pp 879–893, 1925. Received 29 July 1925. An English translation of the paper may be found in: Van Der Waerden, B.L. Ed. and translator, *Sources of Quantum Mechanics* (New York: Dover, 1967), 261–276.
2. Zeitschrift für Physik, 43 (1927), 172–198, received 23 March 1927. Heisenberg's uncertainty paper has been translated into English by John Archibald Wheeler and Hubert Zurek, In *Quantum Theory and Measurement*, Wheeler and Zurek, Eds. (Princeton: Princeton University Press, 1983), pp. 62–84.
3. *Lectures on Physics* Vol. III (Quantum Mechanics) by Richard P. Feynman, Robert B. Leighton, and Matthew Sands, Addison-Wesley Publ., Reading, MA, Though written in 1963 it is still the simplest and best explanation of quantum mechanics.
4. *Principles of Quantum Mechanics*, 2nd ed., Ed. Shankar, R., Springer, London, 1994.

PROBLEMS

Problem 6.1

Given:

Two quantum mechanical operators have the following form:

$$
\mathbb{a} = \begin{pmatrix} 0 & 1 & 0 & 0 & 0 & \\ 0 & 0 & \sqrt{2} & 0 & 0 & \\ 0 & 0 & 0 & \sqrt{3} & 0 & \\ 0 & 0 & 0 & 0 & \sqrt{4} & \\ & & & & & \end{pmatrix}
$$

and

$$
\mathbb{a}^{\dagger} = \begin{pmatrix} 0 & 0 & 0 & 0 & 0 & \\ 1 & 0 & 0 & 0 & 0 & \\ 0 & \sqrt{2} & 0 & 0 & 0 & \\ 0 & 0 & \sqrt{3} & 0 & 0 & \\ & & & & & \end{pmatrix}
$$

The energy operator $\mathbb{H}$ of a type of a simple harmonic oscillator (SHO) is

$$
\mathbb{H} = \hbar\omega\left[\mathbb{a}^{\dagger}\mathbb{a} + \frac{1}{2}\mathbb{I}\right]
$$

where Planck's constant divided by 2π is $\hbar = 1.05458872 \times 10^{-34}$ J-s, ω is the oscillating frequency of the oscillator, and where $\mathbb{a}^\dagger$ is a raising operator, $\mathbb{a}$ is a lowering operator, and $\mathbb{I}$ is the unit matrix:

$$\mathbb{I} = \begin{pmatrix} 1 & 0 & 0 & 0 & 0 & \\ 0 & 1 & 0 & 0 & 0 & \\ 0 & 0 & 1 & 0 & 0 & \\ 0 & 0 & 0 & 1 & 0 & \\ & & & & & \end{pmatrix}$$

For example, if there were an information function vector $|2\rangle$ with a 100% probability of being in state 2. Operating with $\mathbb{a}$ on this information function vector would result in an information function vector $|1\rangle$ with a 100% probability of being in state 1 multiplied by $\sqrt{2}$. The information function vector describing this particular system is

$$|\psi\rangle = \frac{1}{\sqrt{8}} \begin{pmatrix} 1+j \\ 1-j \\ 2 \\ 0 \\ \ \end{pmatrix} \begin{matrix} \text{State 0} \\ \text{State 1} \\ \text{State 2} \\ \text{State 3} \\ \ \end{matrix}$$

The interpretation of this information function is that the probability P_n of the system being in the n's state is equal to $y_n^* y_n$; that is

$$P_o = \frac{1}{4}, \; P_1 = \frac{1}{4}, \quad \text{and} \quad P_2 = \frac{1}{2}$$

Problem:

a) Calculate the difference matrix obtained by first operating with $\mathbb{a}$ and then with $\mathbb{a}^\dagger$ on an information function $|n\rangle$ and first operating with $\mathbb{a}^\dagger$ and then with $\mathbb{a}$ on an information function $|n\rangle$.
b) Calculate the new information function vector $|\eta\rangle$ and the information obtained by operating with the energy operator $\mathbb{K}$ on the information function vector $|\psi\rangle$.
c) Calculate the average value $\langle \mathbb{H} \rangle$ of the operator $\mathbb{H}$.
d) Calculate the angle θ between the original information function vector $|\psi\rangle$ and the new information function vector $|h\rangle$ obtained by operating with the operator $\mathbb{H}$ on the information function vector $|\psi\rangle$.

PROBLEM 6.2

Given:

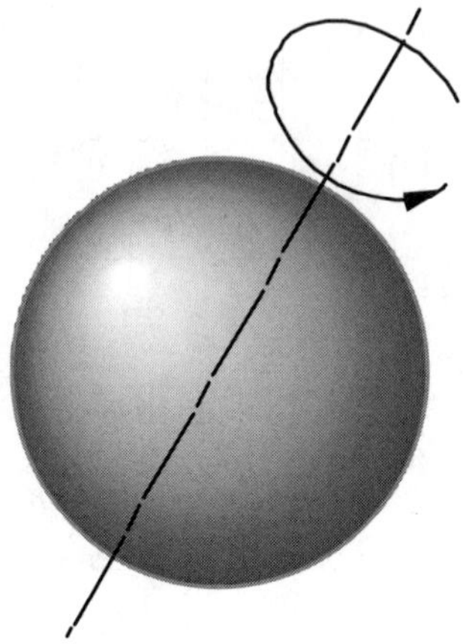

In the quantum mechanical model of nature, a particle's angular momentum associated with the particle spinning about its own axis can have only two values. The two values depend to some extent on the angle between the spin axis of the particle and the chosen coordinate system. The operator representing a "spin" angular momentum s of an electron spinning about its own axis is

$$\mathbb{S} = \frac{\hbar}{2}\left(\begin{array}{c|c} 1 & ja \\ \hline -ja & 1 \end{array}\right)$$

where Planck's constant divided by 2π is h = $1.05458872 \times 10^{-34}$ J-s. Consider two different information function vectors $|\psi_1\rangle$ and $|\psi_2\rangle$ describing the electron at different times:

$$|\psi_1\rangle = \frac{1}{\sqrt{2}}\left(\begin{array}{c} 1 \\ \hline j \end{array}\right)$$

and

$$|\psi_2\rangle = \frac{1}{\sqrt{2}}\left(\begin{array}{c} 1 \\ \hline -j \end{array}\right)$$

Problem:

a) Calculate the new information function vectors and information obtained operating with the spin angular momentum operator $\mathbb{S}$ on each of the information function vectors $|\psi_1\rangle$ and $|\psi_2\rangle$.
b) Calculate the angle θ between the information function vectors $|\eta_1\rangle$ and $|\eta_2\rangle$ obtained by operating with the spin angular momentum operator $\mathbb{S}$ on

the information function vectors $|\psi_1\rangle$ and $|\psi_2\rangle$. Does any special relationship exist between the spin angular momentum operator $\mathbb{S}$, the information function vectors $|\psi_1\rangle$ and $|\psi_2\rangle$, the resulting information function vectors $|\eta_1\rangle$ and $|\eta_2\rangle$, and the information?

Problem 6.3

Given:

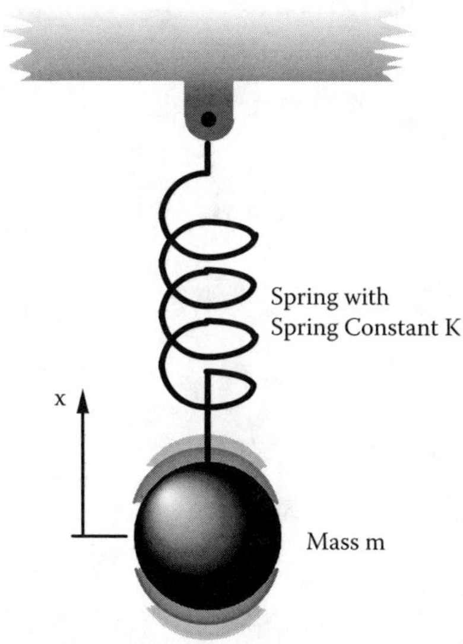

The total energy E of a classical simple harmonic oscillator (SHO) is

$$E = \frac{p^2}{2m} + \frac{1}{2}Kx^2$$

The position $\mathbb{x}$ and momentum $\mathbb{p}$ operators in one dimension have the following forms:

$$\mathbb{x} = x, \quad \text{and} \quad \mathbb{p} = \frac{\hbar}{j}\frac{d}{dx}$$

The quantum mechanical information function y_o corresponding to the lowest energy quantum is

$$y_o = \frac{1}{\left[2\pi\sigma^2\right]^{\frac{1}{4}}}\exp\left[-\frac{x^2}{4\sigma^2}\right]$$

Problem:

a) Calculate the difference between operating first with $\mathbb{x}$ on an information function ψ and then with $\mathbb{p}$ on the resulting information function, and operating first with $\mathbb{p}$ on ψ and then with $\mathbb{x}$ on the resulting information function (20 points).
b) Formulate the quantum mechanical energy operator $\mathbb{H}$ using the form of the position operator $\mathbb{x}$ and the momentum operator $\mathbb{p}$ (10 points).
c) Calculate the lowest value of the energy eigen function E_0 and the value of the constant s^2 obtained by operating with the energy operator $\mathbb{H}$ on the information function ψ_0 (20 points).

Problem 6.4

Given:

The potential energy of an electron in a very thin semiconductor layer has the following form:

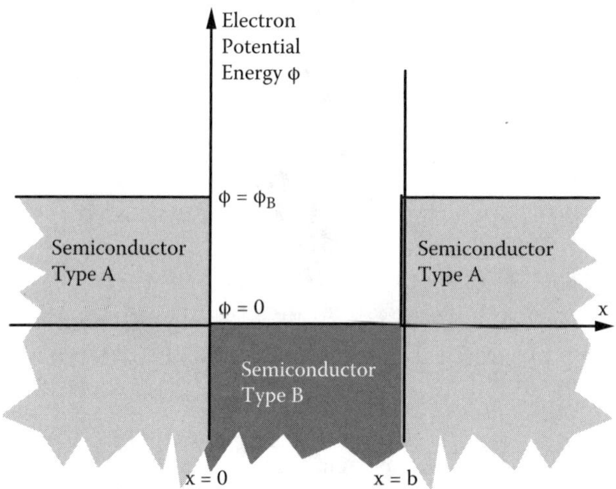

The thin semiconductor layer has a thickness b = 10 nm. As the absolute value of the electron potential energy is arbitrary, a potential energy of the thin semiconductor layer is chosen as zero potential energy. Relative to this potential energy, the potential energy in the adjacent semiconductor layers has a value f_B = 0.5 eV (0.5 × e), where the charge of an electron e =1.60217733 × 10^{-19} C. The electrons are mainly confined in the thin semiconductor region. However, there is a small probability that an electron can wander into the adjacent semiconductors in the region adjacent to the thin layer. Of course, the probability that an electron penetrates far from the thin semiconductor layer is very small. The probability distribution of the electron has to reflect this. Of course, the information function y(x) has to be a solution of Schrödinger's equation:

$$-\frac{\hbar^2}{2m}\frac{\partial^2\psi(x)}{\partial x^2}+\phi(x)\psi(x)=E\psi(x)$$

Problem:

a) Calculate the information function $\psi(x)$ in each region of the semiconductor layer.
b) Calculate the energy levels of an electron in this "quantum well."

Problem 6.5

Given:

The various moments of the random variable x are

$$\langle x\rangle = 0.6,\ s^2 = 0.12,\ \langle x^3\rangle = 0.48,\quad \text{and}\quad \langle x^4\rangle = 0.576$$

Consider two functions:

$$f(x) = u^2 - 4xu + x^2 \quad \text{and} \quad g(x) = 4x - 2x^2$$

Given:

a) Calculate the correlation R_{fg} of the two functions as a function of the parameter u.
b) Calculate the values of the parameter u for which the functions f(x) and g(x) are orthogonal.
c) Calculate the covariance C_{fg} of the two functions as a function of the parameter u.

7 Randomness and Average Randomness

7.1 THE CONCEPT OF RANDOMNESS OF DISCRETE EVENTS

As was shown before, the stochastic nature of a set of events can be described by specifying the probabilities of each event occurring. For example, there corresponds a probability P_n to each event A_n of a set $S\{A_1, A_2, A_3, ..., A_N\}$ of events.

Another way of describing a stochastic system is to compare it to a random system with well-known properties. This is best illustrated with a specific example. For example, consider a set of events with the following probabilities P_n:

Event:	A_1	A_2	A_3	A_4
Probability:	$P_1 = \frac{1}{8}$	$P_2 = \frac{1}{8}$	$P_3 = \frac{1}{4}$	$P_4 = \frac{1}{2}$
Randomness in Coins:	$R_1 = 3$	$R_2 = 3$	$R_3 = 2$	$R_4 = 1$

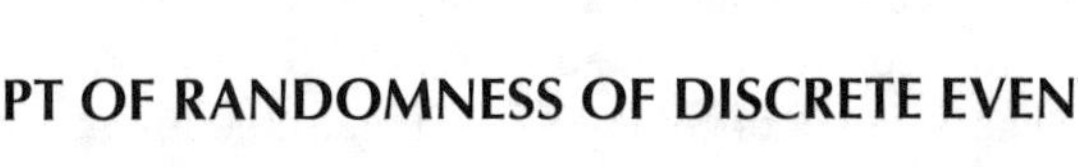

Each probability can be compared to the probability of a set of distinct coins. By distinct coins, it is meant that each coin is separately identifiable. For example, a 1 cent coin, a 5 cent coin, and a 25 cent coin form a set of distinct coins. In this example, the probability $P_1 = 1/8$ is equal to the probability $1/2 \times 1/2 \times 1/2 = 1/8$ of three coins landing in any particular arrangement. Recall that the probability of a coin landing on, say, its head side, is equal to ½. In this case, one would say that the event A_1 is as random as a set of three coins; that is, event A_1 has a randomness R_1 of three coins. Similarly, event A_3 having a probability $P_3 = 1/4$ has a randomness of two coins, etc.

There is a mathematical relation between a probability P_n and the corresponding randomness R_n in coins:

$$\text{a)}\quad R_n = -\ln_2 P_n \quad \text{or} \quad \text{b)}\quad R_n \equiv \frac{\ln P_n}{\ln\left(\frac{1}{2}\right)} \tag{7.1}$$

where $\ln_2 P_n$ is the logarithm to the base 2 of P_n. Another way of writing Equation 7.1a is by the use of the natural logarithms, as in Equation 7.1b. Here, ½ is the probability of a coin landing on, say, the head side, and $\ln P_n$ is the natural logarithm of P_n. Indeed, the negative logarithm to the base 2 of the probability $P_1 = 1/8$ is equal to 3. In this example, a stochastic system is compared to systems of *coins* or *binary bits*. Another way to look at this is that one binary bit is the information required to make a YES or NO decision. Binary-bit-based randomness is widely used in information theory.

The probabilities must not necessarily be equal to the reciprocal of a power of 2, as was the case in the previous simple example. For example, an event having a probability $P_n = 0.3$ of occurring has a randomness equal to $-\ln_2 0.3 = 1.736965594$ coins or binary bits. Indeed, in the general case, one will seldom obtain a randomness equal to an integer number of coins. One can interpret this as a particular probability that is, on the average, equal to a noninteger number of coins. One can compare stochastic systems to probabilistic systems with well-known properties other than coins. For example, one can compare the foregoing event to a set of dice. The probability of any face of a die appearing is equal to 1/6. In this case, the randomness R_n is

$$\text{a) } R_n = -\ln_6 P_n \quad \text{or} \quad \text{b) } R_n = \frac{\ln P_n}{\ln\left(\frac{1}{6}\right)} \tag{7.2}$$

where $\ln_6 P_n$ is the logarithm to the base 6 of the probability P_n of event A_n occurring. For $P_n = 1/8$, the randomness R_n is equal to 1.60558422; that is, event A_n has a randomness, on the average, of 1.60558422 dice. All the randomnesses discussed so far involved a logarithm. However, there are, of course, non-logarithmic-based randomnesses.

A completely certain event, which has a probability equal to one, has a randomness equal to ($\ln 1 = 0$) zero; that is, this event is not random at all. An event with a very small probability of occurring has a large randomness; that is, such an event is very random. The paradox is that an event with zero probability of occurring has an infinite randomness. It might be difficult to find a physical interpretation of the fact that a probability equal to zero corresponds to an infinite randomness, but, nevertheless, it is a useful mathematical quantity.

Other logarithmic randomnesses can be derived by comparing the probability under consideration to the probabilities of a particular side of a polyhedron appearing, where the polyhedron has numbered or otherwise marked sides and is rolled like a die.

Leonhard Euler (1707–1783) a Swiss mathematician and physicist who lived in the middle of the 18th century, discovered the relationship between the number of sides, edges, and corners of a polyhedron:

$$\text{Number of faces} - \text{Number of edges} + \text{Number of vertices} = 2$$

For example, a tetrahedron, a solid having 4 faces each of which is an equilateral triangle, has 6 edges and 4 vertices.

$$4 \text{ equal faces} - 6 \text{ edges} + 4 \text{ vertices} = 2$$

A cube has 6 equal faces, 12 edges, and 8 vertices.

6 faces − 12 edges + 8 vertices = 2

A hexagonal prism has 8 unequal faces, 18 edges, and 12 vertices.

8 faces − 18 edges + 12 vertices = 2, etc.

The properties of the logarithm-based randomness will be investigated here. As was shown earlier, the joint probability $P_{m\cap n}$ of two statistically independent events is equal to the individual probabilities P_n and P_m.

$$P_{m\cap n} = P_m P_n \tag{7.3}$$

The logarithmic joint randomness $R_{m\cap n}$ of two statistical events is

$$R_{m\cap n} = \frac{\ln P_{m\cap n}}{\ln Q} \tag{7.4}$$

where Q is the probability of an event of a system to which the stochastic system is being compared. Substituting the joint probability of two statistically independent events shown in Equation 7.3 into Equation 7.4:

$$\text{a)}\quad R_{m\cap n} = \frac{\ln P_m}{\ln Q} + \frac{\ln P_n}{\ln Q} \quad \text{or} \quad \text{b)}\quad R_{m\cap n} = R_m + R_n \tag{7.5}$$

Thus, the joint randomness of two statistically independent events is equal to the sum of the individual randomnesses.

In general, the joint probability $P_{m\cap n}$ of events A_m and A_n is equal to the product of a conditional probability and a total probability.

$$\text{a)}\quad P_{m\cap n} = P\left(A_m \middle| A_n\right) P_n \quad \text{or} \quad \text{b)}\quad P_{m\cap n} = P\left(A_n \middle| A_m\right) P_m \tag{7.6}$$

Substituting Equation 7.6 into Equation 7.4:

$$\begin{aligned} &\text{a)}\quad R_{m\cap n} = \frac{\ln P\left(A_m \middle| A_n\right)}{\ln Q} + \frac{\ln P_n}{\ln Q} \quad \text{or} \quad \text{b)}\quad R_{m\cap n} = R\left(A_m \middle| A_n\right) + R_n \\ &\text{c)}\quad R_{m\cap n} = \frac{\ln P\left(A_n \middle| A_m\right)}{\ln Q} + \frac{\ln P_m}{\ln Q} \quad \text{or} \quad \text{d)}\quad R_{m\cap n} = R\left(A_n \middle| A_m\right) + R_m \end{aligned} \tag{7.7}$$

where $R(A_m|A_n)$ and $R(A_n|A_m)$ are conditional randomnesses.

A randomness has the following properties:

1. The randomness is a real positive number; $R_n = |\psi|^2$.
2. The randomness of a certain discrete event is equal to zero; $R_n(\text{certain}) = 0$.
3. The randomness increases monotonically as the probability of the event occurring decreases; $R_m(P_m) > R_n(P_n)$ if $P_n > P_m$.
4. The joint randomness $R_{m\cap n}$ of two statistically independent discrete events is equal to the sum of the randomnesses of the individual events $R_m(P_m)$ and $R_n(P_n)$; $R_{m\cap n} = R_m(P_m) + R_n(P_n)$.
5. The joint randomness of two discrete statistically dependent events $R_{m\cap n}$ is equal to the sum of a conditional randomness $R(m|n)$ and the randomness associated with the individual events R_n; $R_{m\cap n} = R(m|n) + R_n$.

7.2 THE CONCEPT OF RANDOMNESS OF CONTINUOUS EVENTS

The randomnesses of discrete events or random variables are based on easy, understandable concepts. The concept of a randomnesses of continuous random variables is an extension of the concept of the discrete randomnesses. One can define the randomness r(x) of a continuous random variable x in a similar manner to the randomness of the discrete random variable of Equation 7.1.

$$r(x) \equiv \frac{\ln\left[p(x)\lambda\right]}{\ln Q} \tag{7.8}$$

where λ is a characteristic length. Here, Q is the probability of the system to which the random variable is being compared. The concept of a "probability of the system to which the random variable is being compared" is the same as in the discrete case. For example, for a randomness r(x) measured in coins or binary bits, Q = ½.

Because Q is taken to be less than one (Q < 1), the logarithm of Q is negative. However, the probability density p(x) can have values greater than one, which could result in a negative randomness. One would then have to say this system is as random as, say, −8 coins? This does not make sense. The randomness has to be positive and real. Therefore, λ is chosen in such a way as to make the randomness positive. Also, p(x)λ is a dimensionless number.

EXAMPLE 7.1

Consider a zero-mean Gaussian probability density p(x), where $-\infty < x \le \infty$.

$$p(x) = \frac{1}{\sqrt{2\pi\sigma^2}}\exp\left[-\frac{x^2}{2\sigma^2}\right]$$

The corresponding randomness is in coins or binary bits:

$$r(x) = \frac{\frac{1}{2}\ln\left[2\pi\frac{\sigma^2}{\lambda^2}\right]}{\ln 2} + \frac{x^2}{2\sigma^2\ln 2}$$

Since the randomness here is measured in coins or binary bits, Q = ½. The minimum value of r(x) will occur at x equal to zero. In order for the randomness r(x) to be positive for all values of x, λ has to satisfy the following inequality.

$$\text{a)}\ \ 2\pi\frac{\sigma^2}{\lambda^2} < 1 \quad \text{or} \quad \text{b)}\ \ \lambda < \sqrt{2\pi^2}$$

Similar to the discrete case, a joint randomness $r(x \cap y)$ of two statistically independent random variables x and y can be defined as:

$$\text{a)}\ \ r(x \cap y) \equiv \frac{\ln p_1(x)\lambda_1 p_2(y)\lambda_2}{\ln Q} \quad \text{or} \quad \text{b)}\ \ r(x \cap y) = \frac{\ln p_1(x)\lambda_1}{\ln Q} + \frac{\ln p_2(y)\lambda_2}{\ln Q} \tag{7.9}$$

$$\text{or} \quad \text{c)}\ \ r(x \cap y) = r(x) + r(y)$$

where λ_1 and λ_2 are characteristic lengths. Again, similar to the discrete case, the joint randomness $r(x \cap y)$ of the product of a conditional probability density $p(x|y)$ and a total probability density p(y) of continuous random variables x and y is

$$\text{a)}\ \ r(x \cap y) \equiv \frac{\ln p(x|y)\lambda_1 p(y)\lambda_2}{\ln Q} \quad \text{or} \quad \text{b)}\ \ r(x \cap y) = \frac{\ln p(x|y)\lambda_1}{\ln Q} + \frac{\ln p(y)\lambda_2}{\ln Q} \tag{7.10}$$

$$\text{or} \quad \text{c)}\ \ r(x \cap y) = r(x|y) + r(y)$$

where $r(x|y)$ is a conditional randomness. The same joint randomness $r(x \cap y)$ is also given by:

$$\text{a)}\ \ r(x \cap y) \equiv \frac{\ln p(y|x)\lambda_1 p(x)\lambda_2}{\ln Q} \quad \text{or} \quad \text{b)}\ \ r(x \cap y) = \frac{\ln p(y|x)\lambda_1}{\ln Q} + \frac{\ln p(x)\lambda_2}{\ln Q} \tag{7.11}$$

$$\text{or} \quad \text{c)}\ \ r(x \cap y) = r(y|x) + r(x)$$

where $r(y|x)$ is a conditional randomness.

The properties of a continuous randomness are listed as follows:

1. The randomness of a continuous random variable is a real positive number; $r(x) = |\psi|^2$.
2. The randomness increases monotonically as the probability of the event occurring decreases; $r(x_1) > r(x_2)$ if $p(x_2) > p(x_1)$.
3. The joint randomness $r(x \cap y)$ of two statistically independent events is equal to the sum of the randomnesses of the individual events r(x) and r(y); $r(x \cap y) = r[p(x)] + r[p(y)]$.

4. The joint randomness $r(x \cap y)$ of two statistically dependent events is equal to the sum of a conditional randomness $r(x|y)$ and a randomness associated with the individual events $r(y)$, $r(x \cap y) = r(x|y) + r(y)$.

7.3 THE AVERAGE RANDOMNESS OF DISCRETE EVENTS

The average value of the randomness of Equation 7.1 of a discrete random variable is

$$\mathbb{H} = \frac{1}{\ln Q} \sum_{n=1}^{n=N} P_n \ln P_n \tag{7.12}$$

where $\mathbb{H}$ is the average randomness or entropy. This quantity is also known as the Shannon entropy. In 1948, Claude E. Shannon (1916–2001),[1,2] in his famous paper "A Mathematical Theory of Communication," suggested that this quantity can be used to describe both randomness and information. Here, $0 < n \le N$, and $Q < 1$ is the probability of a stochastic system to which the random variable n is being compared. For the case where the probabilities P_n are compared to coins or binary bits:

$$\mathbb{H} = -\sum_{n=1}^{n=N} P_n \ln_2 P_n \tag{7.13}$$

where the average randomness in this case is measured in binary bits.

Claude E. Shannon

For the completely certain case when one of the probabilities P_m is equal to one and the others are equal to zero, the average randomness has the following form:

$$\mathbb{H} = \frac{\ln 1}{\ln Q} + \frac{N-1}{\ln Q} \lim_{u \to 0} u \ln u \tag{7.14}$$

where u stands for the probabilities that are equal to zero. In the limit, as u approaches zero, the quantity ulnu too approaches zero because u approaches zero faster than lnu. Therefore, for the completely certain case, the average randomness is equal to zero.

$$\lim_{P_m \to 1} \lim_{P_{n \neq m} \to 0} \mathbb{H} \to 0 \tag{7.15}$$

Indeed, the system is not random at all in this case.

EXAMPLE 7.2

Consider a set of events A_n with the following corresponding probabilities:

Events:	A_1	A_2	A_3	A_4	A_5
Probabilities:	0.05	0.1	0.2	0.4	0.25

Let us calculate the average randomness in binary bits.

$$\mathbb{H} = -1/\ln 2\ [(0.05\ln 0.05) + (0.1\ln 0.1) + (0.2\ln 0.2) + (0.4\ln 0.4) + (0.25\ln 0.25)]$$

$$\mathbb{H} = 2.04144607 \text{ binary bits.}$$

The average randomness is used in information theory. As unlikely as this may seem, the average randomness is equal to the information measured in binary bits. Suppose the values of a function A_n of the random variable n correspond to various pieces of information that are to be conveyed. The pieces of information are measured in binary bits. Here, P_n's are probabilities that a piece of information corresponding to the value A_n of the function of the random variable will be conveyed. Conveying only one piece of information corresponds to one of the P_n's being equal to one, and the other P_n's being equal to zero. Conveying only a single piece of information does not represent much information. In this case, the average randomness is equal to zero. The larger the number of pieces of information being conveyed, the more information that is conveyed. The more the information that is conveyed, the smaller the difference between the P_n's and the larger the average randomness $\mathbb{H}$.

The average randomness is a *macroscopic parameter* of a stochastic system.

Consider two sets of events $S_A\{A_1, A_2, A_3, ..., A_n, ..., A_N\}$ and $S_B\{B_1, B_2, B_3, ..., B_m, ..., B_M\}$. One can define two random variables m and n. The random variables enumerate the events A_n and B_m, respectively. The joint average randomness of the two discrete random variables is

$$\mathbb{H}_{A\cap B} = \frac{1}{\ln Q}\sum_{m=1}^{m=M}\sum_{n=1}^{n=N} P_{m\cap n} \ln P_{m\cap n} \tag{7.16}$$

where Q is the probability of the system to which the sets of events are being compared. It is first assumed that the events of the two sets S_A and S_B are statistically independent. The joint average randomness of two statistically independent random variables is

$$\begin{aligned} &\text{a)}\quad \mathbb{H}_{A\cap B} = \frac{1}{\ln Q}\sum_{m=1}^{m=M}\sum_{n=1}^{n=N}(P_m P_n)\ln(P_m P_n) \quad \text{or} \\ &\text{b)}\quad \mathbb{H}_{A\cap B} = \frac{1}{\ln Q}\sum_{m=1}^{m=M}\sum_{n=1}^{n=N}(P_m P_n)\ln P_m + \frac{1}{\ln Q}\sum_{m=1}^{m=M}\sum_{n=1}^{n=N}(P_m P_n)\ln P_n \end{aligned} \tag{7.17}$$

Because the sum over the individual probabilities is equal to one, Equation 7.17 reduces to:

$$\text{a)}\quad \mathbb{H}_{A\cap B} = \frac{1}{\ln Q}\sum_{m=1}^{m=M} P_m \ln P_m + \frac{1}{\ln Q}\sum_{n=1}^{n=N} P_n \ln P_n \quad \text{or} \quad \text{b)}\quad \mathbb{H}_{A\cap B} = \mathbb{H}_B + \mathbb{H}_A \tag{7.18}$$

Thus, the joint average randomness $\mathbb{H}_{A\cap B}$ of two statistically independent sets of events is equal to the sum of the individual average randomnesses of the individual sets of events $\mathbb{H}_A$ and $\mathbb{H}_B$.

Assume next that the two sets S_A and S_B are not statistically independent. In this more general case, the joint average randomness $\mathbb{H}_{A\cap B}$ of two statistically dependent discrete sets of events is

$$\begin{aligned} &\text{a)}\quad \mathbb{H}_{A\cap B} = \frac{1}{\ln Q}\sum_{m=1}^{m=M}\sum_{n=1}^{n=N} P(m|n)P_n \ln(m|n)P_n \quad \text{or} \\ &\text{b)}\quad \mathbb{H}_{A\cap B} = \frac{1}{\ln Q}\sum_{m=1}^{m=M}\sum_{n=1}^{n=N} P(m|n)P_n \ln P(m|n) + \frac{1}{\ln Q}\sum_{m=1}^{m=M}\sum_{n=1}^{n=N} P(m|n)P_n \ln P_n \end{aligned} \tag{7.19}$$

Because the sum over m of the individual conditional probabilities $P(m|n)$ is equal to one:

$$\text{a)}\quad \mathbb{H}_{A\cap B} = \frac{1}{\ln Q}\sum_{n=1}^{n=N}\sum_{m=1}^{m=M} P(m|n)\ln P(m|n)P_n + \frac{1}{\ln Q}\sum_{n=1}^{n=N} P_n \ln P_n \quad \text{or}$$
$$\text{b)}\quad \mathbb{H}_{A\cap B} = \sum_{n=1}^{n=N} \mathbb{H}(B|n)P_n + \mathbb{H}_A \tag{7.20}$$

where $\mathbb{H}(B|n)$ is the conditional average randomness of the set of events S_B provided event A_n has occurred.

$$\mathbb{H}(B|n) \equiv \frac{1}{\ln Q}\sum_{n=1}^{n=N} P(m|n)\ln P(m|n) \tag{7.21}$$

The conditional average randomness $\mathbb{H}(A|m)$ of the set S_A provided an event B_m has occurred can, similarly, be derived.

One can extend this concept to multiple sets S_A, S_B, S_C, S_D, ..., etc., of statistically dependent random variables.

$$\mathbb{H}_{A\cap B\cap C..\cap D} = \frac{1}{\ln Q}\sum_{k=1}^{k=K}\sum_{l=1}^{l=L}\sum_{m=1}^{m=M}\cdots\sum_{n=1}^{n=N}\left[P(k|l)P(l|m)P(m|s)\ldots P(t|n)P_n\right] \times \ln P(k|l)P(l|m)P(m|s)\ldots P(t|n)P_n\Big] \tag{7.22}$$

This can be rewritten by expanding the logarithms, as was done in the case with two sets of events.

$$\mathbb{H}_{A\cap B\cap C..\cap D} = \frac{1}{\ln Q}\left[\sum_{k=1}^{k=K}\sum_{l=1}^{l=L} P(k|l)\ln P(k|l)P_l + \sum_{l=1}^{l=L}\sum_{m=1}^{m=M} P(l|m)\ln P(l|m)P_m + \ldots\right] \tag{7.23}$$

where:

$$P_j = \sum_{k=1}^{k=K}\sum_{l=1}^{l=L}\sum_{m=1}^{m=M} P(j|k)P(k|l)P(l|m)\ldots P_t \tag{7.24}$$

Equation 7.23 can be written in terms of conditional average randomnesses.

$$\mathbb{H}_{A\cap B\cap C\cap..\cap D} = \sum_{l=1}^{l=L}\mathbb{H}(A|l)P_l + \sum_{m=1}^{m=M}\mathbb{H}(B|m)P_m + \ldots \tag{7.25}$$

7.4 THE AVERAGE RANDOMNESS OF CONTINUOUS RANDOM VARIABLES

Now that the randomness of a continuous random variable has been defined, the average value of the randomness of a continuous random variable x can be derived. The average value of the randomness of a continuous random variable x is written as follows:

$$\mathbb{H} = \frac{1}{\ln Q}\int_0^a p(x)\ln\left[p(x)\lambda\right]dx \tag{7.26}$$

where $Q < 1$ is the probability of the system to which the random variable is being compared. Here, $0 < x \le a$. The average value of the randomness of a continuous random variable is also known as the *Boltzmann entropy.* Ludwig Boltzmann (1844–1906) lived in Austria and developed many of the fundamental concepts of probability theory. He was the first one to propose quantized description of physical phenomena.

The above-average randomness of a continuous random variable is only specified to within a constant. By noting that the logarithm of a product is equal to the sum of the logarithms, Equation 7.26 can be rewritten as follows:

$$\mathbb{H} = \frac{1}{\ln Q}\int_0^a p(x)\ln p(x)dx + \frac{\ln\lambda}{\ln Q}\int_0^a p(x)dx \tag{7.27}$$

Because the integral in the second term of Equation 7.27 is equal to one, Equation 7.27 can be written as:

$$\mathbb{H} = \frac{1}{\ln Q}\int_0^a p(x)\ln p(x)dx + \mathbb{H}_0 \tag{7.28}$$

where the constant $\mathbb{H}_0$ has to be chosen in such a way that the average randomness $\mathbb{H}$ is positive.

Ludwig Boltzmann

EXAMPLE 7.3

Consider again the Gaussian probability density p(x):

$$p(x) = \frac{1}{\sqrt{2\pi\sigma^2}} \exp\left[-\frac{(x-a)^2}{2\sigma^2}\right]$$

where $-\infty < x \leq \infty$. Next, the randomness r(x) of a continuous random variable x is calculated by using Equation 7.8:

$$\frac{1}{\ln Q} \ln\left[\lambda p(x)\right] = \frac{1}{\ln Q}\left[\ln \frac{\lambda}{\sqrt{2\pi\sigma^2}} - \frac{(x-a)^2}{2\sigma^2}\right]$$

Substitute the previous equation into Equation 7.26 to obtain an expression for the average randomness of the Gaussian random variable:

$$\mathbb{H}(\sigma) = \left[\frac{1}{\ln Q} \ln \frac{\lambda}{\sqrt{2\pi\sigma^2}}\right] \frac{1}{\sqrt{2\pi\sigma^2}} \int_{\infty}^{\infty} \exp\left[-\frac{(x-a)^2}{2\sigma^2}\right] dx -$$

$$\frac{1}{\ln Q} \frac{1}{\sqrt{2\pi\sigma^2}} \int_{\infty}^{\infty} \frac{(x-a)^2}{2\sigma^2} \exp\left[-\frac{(x-a)^2}{2\sigma^2}\right] dx$$

The first integral times the fraction immediately preceding is equal to one. The following transformation of variables is made to integrate the second integral:

$$\text{a)}\quad u \equiv \frac{x-a}{\sigma} \quad \text{then} \quad \text{b)}\quad dx = \sigma du$$

By substituting the preceding expression into the equation, the average randomness ℍ(s) is obtained:

$$\mathbb{H}(\sigma) = \left[\frac{1}{\ln Q} \ln \frac{\lambda}{\sqrt{2\pi\sigma^2}}\right] - \frac{1}{2 \ln Q} \frac{1}{(2\pi)^{\frac{1}{2}}} \int_{-\infty}^{\infty} u^2 \exp\left[-\frac{u^2}{2}\right] du$$

The integral times the fraction immediately preceding is again equal to one.

$$\mathbb{H}(\sigma) = \frac{1}{\ln Q}\left[\frac{1}{2} \ln \frac{\lambda^2}{2\pi} = \frac{1}{2} - \ln \sigma\right]$$

where $\lambda^2 < 2\pi\sigma^2/e$ in order for the average randomness $\mathbb{H}(\sigma)$ to be positive. Here, σ is a parameter. One can compare two different values of the average randomness for different values of σ.

$$\mathbb{H}(\sigma_2) - \mathbb{H}(\sigma_1) = \frac{\ln \sigma_1}{\ln Q} - \frac{\ln \sigma_2}{\ln Q}$$

This eliminates λ as well as other constants.

One can consider two sets of continuous random variables $S_A\{0 < x \le a\}$ and $S_B\{0 < y \le b\}$. The joint average randomness $\mathbb{H}(A \cap B)$ of two continuous random variables is:

$$\mathbb{H}(A \cap B) = \frac{1}{\ln Q} \int_0^a dx \int_0^b dy p(x \cap y) \ln\left[p(x \cap y)\lambda\right] \tag{7.29}$$

As was the case with the average randomness of a single continuous random variable, the joint average randomness of two continuous random variables is also only defined to within a constant.

$$\mathbb{H}(A \cap B) = \frac{1}{\ln Q} \int_0^a dx \int_0^b dy p(x \cap y) \ln p(x \cap y) + \frac{\ln \lambda}{\ln Q} \int_0^a dx \int_0^b dy p(x \cap y) \tag{7.30}$$

Because the double integral in the second term of Equation 7.30 is equal to one, Equation 7.30 can be written as:

$$\mathbb{H}(A \cap B) = \frac{1}{\ln Q} \int_0^a dx \int_0^a dy p(x \cap y) \ln p(x \cap y) + \mathbb{H}_o \tag{7.31}$$

where the constant $\mathbb{H}_o$ has to be chosen in such a way that the average randomness $\mathbb{H}(A \cap B)$ is positive.

First, assume that the random variables x and y are statistically independent. The joint average randomness $\mathbb{H}(A \cap B)$ of the two sets of statistically independent continuous random variables x and y is:

$$\mathbb{H}(A \cap B) = \frac{1}{\ln Q} \int_0^a dx \int_0^b dy p_A(x) p_b(y) \ln\left[p_A(x)\lambda_A p_B(x)\lambda_B\right] \tag{7.32}$$

Again, the joint average randomness $\mathbb{H}(A \cap B)$ of two sets of statistically independent random variables is equal to the sum of the random variables of the average randomnesses $\mathbb{H}_A$ and $\mathbb{H}_B$ of the individual random variables. Here, λ_A and λ_B are characteristic lengths. Similar to the discrete case, this can be written as:

$$\text{a)}\quad \mathbb{H}(A\cap B)=\frac{1}{\ln Q}\int_0^a dx p_A(x)\ln\left[p_A(x)\lambda_A\right]+\frac{1}{\ln Q}\int_0^b dy p_B(x)\ln\left[p_B(x)\lambda_B\right]$$

or (7.33)

$$\text{b)}\quad \mathbb{H}(A\cap B)=\mathbb{H}_A+\mathbb{H}_B$$

Of course, the average randomnesses $\mathbb{H}_A$ and $\mathbb{H}_B$ are only defined to within a constant. For the case where the two sets $S_A\{0 < x \le a\}$ and $S_B\{0 < y \le b\}$ of continuous random variables are not statistically independent, the joint average randomness $\mathbb{H}(A \cap B)$ is

$$\mathbb{H}(A\cap B)=\frac{1}{\ln Q}\int_0^a dx\int_0^b dy p(x|y)p(y)\ln\left[p(x|y)\lambda_1 p(y)\lambda_2\right] \tag{7.34}$$

This can be expanded to:

$$\mathbb{H}(A\cap B)=\frac{1}{\ln Q}\int_0^a dx\int_0^b dy p(x|y)p(y)\ln\left[p(x|y)\lambda_1\right]+\frac{1}{\ln Q}\int_0^b dy p(y)\ln\left[p(y)\lambda_2\right] \tag{7.35}$$

$$\mathbb{H}(A\cap B)=\int_0^b \mathbb{H}(A|y)p(y)\,dy+\mathbb{H}_B$$

where the conditional average randomness $\mathbb{H}(A|y)$ of two sets of continuous random variables is:

$$\mathbb{H}(A|y)\equiv\frac{1}{\ln Q}\int_0^a dx p(x|y)\ln\left[p(x|y)\lambda_1\right] \tag{7.36}$$

The conditional average randomness $\mathbb{H}(B|x)$ of two continuous random variables can similarly be derived. The conditional average randomnesses $\mathbb{H}(A|y)$ and $\mathbb{H}(B|x)$, similar to the other average randomnesses of continuous random variables, are only defined to within a constant.

7.5 THE AVERAGE RANDOMNESS OF RANDOM VARIABLES WITH VALUES THAT HAVE THE SAME PROBABILITY

The average randomness of sets of events where some of the events have equal probabilities of occurring are analyzed here. Consider the average value of the randomness of a set of events where the events in various groups have the same probability of occurring. For example, events A_1 and A_2 have a probability P_1 of occurring; events

A_3, A_4, and A_5 have a probability P_2 of occurring; events A_6, A_7, A_8, and A_9 have a probability P_3 of occurring, etc. The average randomness for this set of events is

$$\mathbb{H} = \frac{1}{\ln Q}\Big[P_1 \ln P_1 + P_1 \ln P_1 + P_2 \ln P_2 + P_2 \ln P_2 + P_2 \ln P_2 + P_3 \ln P_3 + P_3 \ln P_3 + P_3 \ln P_3 + P_3 \ln P_3 + P_4 \ln P_4 + P_4 \ln P_4 + P_4 \ln P_4 + \ldots\Big] \tag{7.37}$$

Equation 7.37 can be written as follows:

$$\mathbb{H} = \frac{1}{\ln Q}\left[2P_1 \ln P_1 + 3P_2 \ln P_2 + 4P_3 \ln P_3 + 3P_4 \ln P_4 + \ldots\right] \tag{7.38}$$

As has been shown before, the quantities 2, 3, 4, 3, etc., that multiply the $P_n \ln P_n$ are the number of values of the random variable with the same probability. In general, the average value of the randomness, where a number G_n of values of the random variable have the same probability is

$$\mathbb{H} = \frac{1}{\ln Q}\sum_{n=1}^{n=N} G_n P_n \ln P_n \tag{7.39}$$

where G_n is the density of events with the same probability.

EXAMPLE 7.4

Consider N = 5 spheres. Each sphere has a small flat area. The probability P_F of a sphere landing on its flat area is equal to 0.2, and the probability P_R that a sphere will not land on its flat area is equal to 0.8.

Note that $G_K = N!/K!(N-K)!$ numbers of the random variables K that have the same probability. The average randomness in binary bits is

$$\mathbb{H} = -\frac{1}{\ln 2}\sum_{K=0}^{K=5} \frac{5!}{K!(5-K)!} P_F^K P_R^{N-K} \ln\left[P_F^K P_R^{N-K}\right]$$

Because the logarithm of a product of terms is equal to the sum of the logarithms of the individual terms:

$$\mathbb{H} = -\frac{\ln P_F}{\ln 2}\sum_{K=0}^{K=5}\frac{5!K}{K!(5-K)!}P_F^K P_R^{N-K} - \frac{\ln P_R}{\ln 2}\sum_{K=0}^{K=5}\frac{5!(5-K)}{K!(5-K)!}P_F^K P_R^{N-K}$$

The first term, the K = 0 term, in the first sum is equal to zero. Thus, the first sum is of the form:

$$\sum_{K=1}^{K=5}\frac{N!}{(K-1)!(N-K)!}P_F^K P_R^{N-K}$$

Let L = K – 1 or K = L + 1, at K = 1 or L = 0 and at K = N or L = N – 1. Thus, the sum takes the following form:

$$NP_F\sum_{L=0}^{L=N-1}\frac{(N-1)!}{L![(N-1)-L]!}P_F^L P_R^{(N-1)-L}$$

Each term in the preceding sum represents the probability that L out of N – 1 spheres will land on their flat spots. The probability that none or any number of spheres will land on their flat spots is equal to one. Therefore, the previous sum is equal to one. Substituting into the expression for the average randomness:

$$\mathbb{H} = -\frac{NP_F \ln P_F}{\ln 2} - \frac{\ln P_R}{\ln 2}\sum_{K=0}^{K=5}\frac{5!(5-K)}{K!(5-k)!}P_F^K P_R^{N-K}$$

Let J = N – K or K = N – J; at K = 0, J = N and at K = N, J = 0. Substituting for K:

$$\mathbb{H} = -\frac{NP_F \ln P_F}{\ln 2} - \frac{\ln P_R}{\ln 2}\sum_{J=0}^{J=N}\frac{N!}{(J-1)!(N-J)!}P_F^{N-J} P_R^J$$

Using a method similar to the one used to evaluate the first sum:

$$\mathbb{H} = -N/\ln 2\,[P_F \ln P_F + P_R \ln P_R]$$

$$\mathbb{H} = -5/\ln 2\,[0.3218875825 + 0.1785148411]$$

$$\mathbb{H} = 3.609640475 \text{ binary bits}$$

If five coins were used with equal probabilities, $P_H = ½$ and $P_T = ½$, of landing on their head (H) or tail (T) sides instead of the five spheres, one would have obtained:

$$\mathbb{H} = N$$

or

$$\mathbb{H} = 5 \text{ coins or binary bits}$$

Indeed, a system of five coins is as random as a system of five coins. The average randomness of the coins is larger than the average randomness of the spheres because more is actually known about the spheres than about the coins. It is known that it is more likely for a sphere not to land on its flat spot than for it to land on its flat spot. Thus, the system of spheres is less random.

Similarly, the average randomness of a continuous random variable x that has regions with the same probability density is

$$\mathbb{H} = \frac{1}{\ln Q} \int_0^a g(x) p(x) \ln\left[p(x)\lambda\right] dx \tag{7.40}$$

where g(x) is the density of events with the same probability density p(x). Here, Q is the probability of the system to which the random variable is being compared. This average randomness, like the other randomnesses of continuous random variables, is only defined to within a constant.

$$\mathbb{H} = \frac{1}{\ln Q} \int_0^a g(x) p(x) \ln p(x) dx + \frac{\ln \lambda}{\ln Q} \int_0^a g(x) p(x) dx \tag{7.41}$$

Because the integral in the second term of Equation 7.41 is equal to one, Equation 7.41 can be written as:

$$\mathbb{H} = \frac{1}{\ln Q} \int_0^a g(x) p(x) \ln p(x) dx + \mathbb{H}_0 \tag{7.42}$$

Again, the constant $\mathbb{H}_0$ has to be chosen in such a way that the average randomness $\mathbb{H}$ is positive.

7.6 THE ENTROPY OF REAL PHYSICAL SYSTEMS AND A VERY LARGE NUMBER

The entropy $\mathbb{S}$ of a stochastic physical system is proportional to its average randomness $\mathbb{H}$.

$$\mathbb{S} = k\left[\mathbb{H} \ln \frac{1}{Q}\right] \tag{7.43}$$

where $\mathbb{H}$ is the average randomness given by Equations 7.12, 7.26, 7.28, or 7.39, and $k = 1.380662 \times 10^{-23}$ J per °K is Boltzmann's constant. Recall that for coins, 1/Q is equal to 2, the number of sides of a coin. For dice, 1/Q is equal to 6, the number of sides of a dice, etc. Thus, one can interpret 1/Q as the number of equal sides of a solid that can be rolled like a die. Then, Q is the probability of any one of these sides, say, making contact with the surface on which it is rolled. Because for a real physical system:

$$\text{a)}\quad k = \frac{1}{\ln\frac{1}{Q}} \quad \text{or} \quad \text{b)}\quad \frac{1}{Q} = \exp\left(\frac{1}{k}\right) \tag{7.44}$$

one can interpret the entropy of a real physical system as an average randomness, where the physical system is being compared to a stochastic system of solids with $10^{3.14555251} \times 10^{22}$ *equal sides*, a large number indeed. If each of the surfaces of this solid consisted of a single atom spaced approximately a few Å from its neighbor, this solid would have a radius of $10^{1.5752776255} \times 10^{22}$ light-years. This is substantially larger than the radius of the known universe (as of 2007).

EXAMPLE 7.5

The probability per unit energy, the probability density p(E), of a classical particle, is the Boltzmann density function:

$$p(E) = \left(\frac{m}{2\pi kT}\right)^{\frac{3}{2}} \exp\left(-\frac{E}{kT}\right)$$

where the energy E can have values between zero and infinity ($0 < E \le \infty$). Here, $k = 1.380662 \times 10^{-23}$ J per °K is Boltzmann's constant, and T is the absolute temperature. The density of states in energy for classical particles is:

$$g(E) = 2\pi\left(\frac{2}{m}\right)^{\frac{3}{2}} E^{\frac{1}{2}}$$

Let $Q = e^{-1}$. By substituting the Boltzmann density function into Equation 7.36, the average randomness is

$$\mathbb{H} = \frac{3k}{2}\ln\left(\frac{m\lambda}{2\pi kT}\right)\int_0^\infty g(E)p(E)dE + \frac{1}{T}\int_0^\infty g(E)p(E)E\,dE$$

where $\lambda^{3/2}$ is the constant that ensures that the randomness is positive for all values of the random variable. The first integral is equal to one, and the second integral is equal to the average value $\langle E \rangle$ of the energy. Thus, the average randomness is

$$\mathbb{H} = \frac{3k}{2}\ln\left(\frac{m\lambda}{2\pi kT}\right) + \frac{\langle E \rangle}{T}$$

where $\langle E \rangle$ is the average value of the energy as expressed by the last term of the previous equation.

7.7 THE CEPSTRUM

The cepstrum $\mathbb{C}(t)$ is defined as the inverse Fourier transform of the logarithm of the power density spectrum $s_{ff}(\omega)$.

$$\mathbb{C}(t) \equiv \frac{1}{2\pi}\int_0^{\infty} \ln s_{ff}(\omega) e^{j\omega t} d\omega \tag{7.45}$$

The power density spectrum $s_{ff}(\omega)$ will be discussed in Section 11.2. The cepstrum gives information about delays in the signal. It is used in the analysis of speech and radar signals. The cepstrum is another macroscopic stochastic parameter.

7.8 STOCHASTIC TEMPERATURE AND THE LEGENDRE TRANSFORM

A random variable that is not very random has just a few values that are very likely to occur, whereas the rest of its values have very small probabilities of occurring. On the other hand, most values of a random variable have more or less equal probabilities of occurring of a random variable with a large degree of randomness. Obviously, the degree of randomness of a random variable will influence its macroscopic parameters $\mathbb{U}$, such as the average value $\langle n \rangle$ of the random variable n, the variance σ^2, its m-th moment $\langle n^m \rangle$, etc. The degree of randomness of a random variable is expressed by its average randomness $\mathbb{H}$. Because all macroscopic stochastic parameters are functions of the individual probabilities, the various macroscopic parameters $\mathbb{U}$ must be functions of each other; that is, they are parametrically related. In particular, the macroscopic parameters $\mathbb{U}$ must be functions of the average randomness $\mathbb{H}$ (see Figure 7.1).

$$\mathbb{U} = \mathbb{U}(\mathbb{H}) \tag{7.46}$$

The stochastic temperature $\mathbb{T}_o$ is the change of a specific macroscopic parameter $\mathbb{U}$ with respect to the average randomness of the random variable. The stochastic

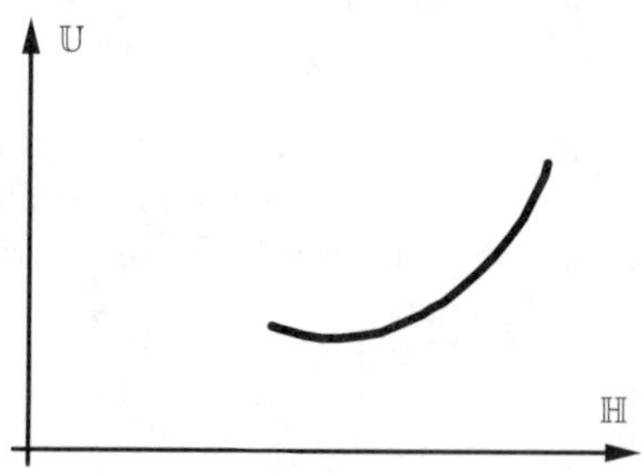

FIGURE 7.1 A plot of a macroscopic parameter of a stochastic system as a function of the average randomness of the same stochastic system.

temperature $\mathbb{T}_o$ can be expressed mathematically as the partial derivative of the macroscopic parameter $\mathbb{U}$ of a random variable n with respect to its average randomness $\mathbb{H}$.

$$\mathbb{T}_o \equiv \frac{\partial \mathbb{U}}{\partial \mathbb{H}} \tag{7.47}$$

The random variable could, for example, be the energy E of some physical system and the macroscopic parameter; its average value $\mathbb{U}(= \langle E \rangle)$. As was discussed in Section 7.6, the entropy $\mathbb{S} = (-k\mathbb{H}\ln Q)$ of a physical system is equal to its average randomness $\mathbb{H}$ times Boltzmann's constant $k = 1.380662 \times 10^{-23}$ Joules per °K. The physical temperature T is equal to the partial derivative of the average value $\langle E \rangle$ of the energy with respect to the entropy. Thus, the stochastic temperature $\mathbb{T}_o(= kT)$ is equal to the physical temperature T times Boltzmann's constant k. One of the first attempts to make a standard temperature scale occurred about AD 170, when Galen, in his medical writings, proposed a standard "neutral" temperature made up of equal quantities of boiling water and ice; on either side of this temperature were four degrees of heat and four degrees of cold, respectively.

It is often inconvenient to deal with a function of the average randomness. The average randomness is often difficult to control. This is especially true of the entropy of a physical system. It is often more convenient to express the macroscopic parameter as a function of the stochastic temperature $\mathbb{U}(\mathbb{T}_o)$. However, if one simply solves Equations 7.52 and 7.53 for the macroscopic parameter in $\mathbb{U}$ in terms of its derivative $\mathbb{T}_o$ with respect to its average randomness $\mathbb{H}$, some information is lost. If one tries to reconstruct the function $\mathbb{U}(\mathbb{H})$ again from a function of the derivative, this can only be done to within a constant of integration. If one tries to integrate Equation 7.47, a solution for $\mathbb{U}$ to within a constant of integration C is obtained.

$$\mathbb{U} = \int \mathbb{T}\, d\mathbb{H} + C \tag{7.48}$$

This is illustrated in Figure 7.2. All the curves in this illustration are solutions of Equation 7.48. The simplest way is, of course, to describe the macroscopic parameter $\mathbb{U}$ of Figure 7.2 as function of $\mathbb{H}$ is by specifying all values of $\mathbb{U}$ and the corresponding values of $\mathbb{H}$, as illustrated by the solid curve (see Figure 7.3). One can also describe the macroscopic parameter by specifying all the lines that are tangent to the curve $\mathbb{U}(\mathbb{H})$ (see Figure 7.4). Each tangent line can be specified by its slope and a point on the line. The slope of the curve $\mathbb{U}(\mathbb{H})$ at some point is, of course, equal to $\mathbb{T}_o$. Choosing the point on the line as the intersection of the tangent line with the line $\mathbb{H} = 0$, one can label this intersection point $\mathbb{F}$. Every tangent line has a different intersection point $\mathbb{F}$. The point $\mathbb{F}$, therefore, is a function of the slope $\mathbb{T}_o$. Indeed, one can use the intercept points $\mathbb{F}(\mathbb{T}_o)$ as the function of the stochastic temperature. This is the macroscopic parameter that is a function of the temperature.

Consider a single tangent line, as shown in Figure 7.4. From Figure 7.4, one can observe the following relation between the macroscopic parameter $\mathbb{U}$, the average randomness $\mathbb{H}$, and the intercept point $\mathbb{F}$.

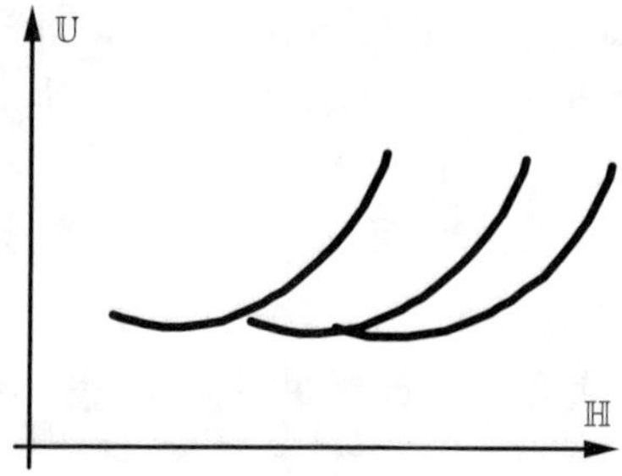

FIGURE 7.2 Reconstructed macroscopic stochastic parameter $\mathbb{U}$ from Equation 7.47. Each solution is for a different constant of integration.

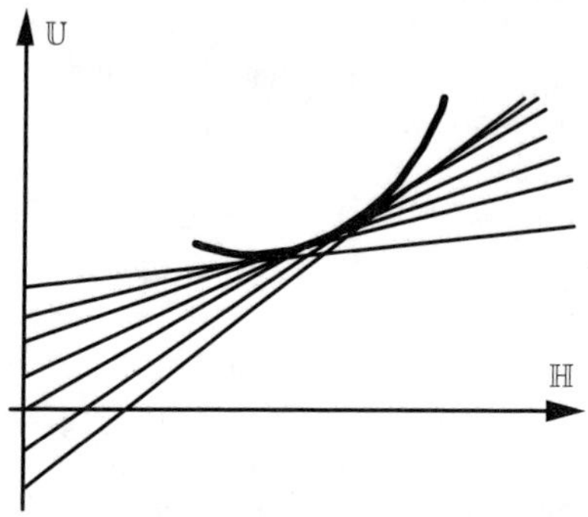

FIGURE 7.3 The function as described by lines tangent to the curve.

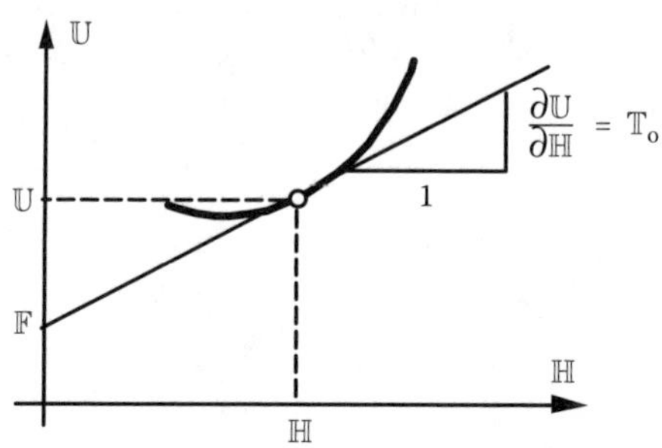

FIGURE 7.4 A single tangent line to the curve $\mathbb{U}(\mathbb{H})$, showing the slope $\mathbb{T}_o$ and the intercept point $\mathbb{F}$.

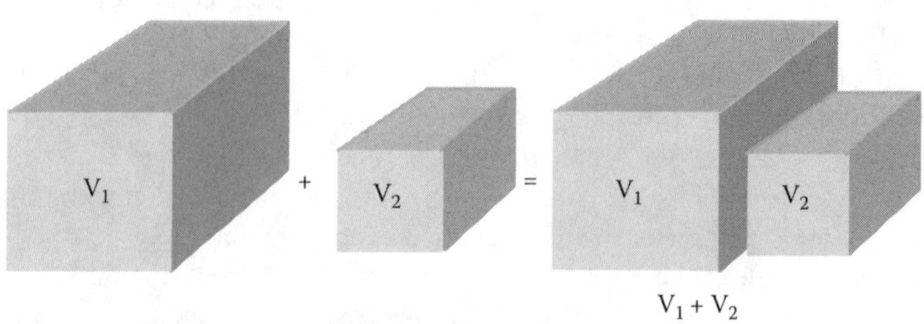

FIGURE 7.5 The volume v is an extensive parameter.

$$\frac{\partial \mathbb{U}}{\partial \mathbb{H}} = \frac{\mathbb{U} - \mathbb{F}}{\mathbb{H}} \tag{7.49}$$

By substituting Equation 7.47 into Equation 7.49, one obtains for the new macroscopic variable $\mathbb{F}$:

$$\mathbb{F} = \mathbb{U} - \mathbb{T}_o \mathbb{H} \tag{7.50}$$

This process is known as the *Legendre transform*.[3,4] In particular, $\mathbb{F}(\mathbb{T}_o)$ is the Legendre transform of the macroscopic variable $\mathbb{U}(\mathbb{H})$. The Legendre transform has many applications. The transform was developed by the mathematician Adrien-Marie Legendre (1752–1833).

Adrien-Marie Legendre

The quantity $\mathbb{F}$ is the stochastic free energy, the amount of the macroscopic parameter available "to do work." In a physical system with the random variable equal to the energy, the stochastic free energy $\mathbb{F}$ is equal to the physical free energy F. It is the amount of energy available to do work. Observe that the stochastic free energy $\mathbb{F}$, the quantity available to do work, is less than the macroscopic parameter $\mathbb{U}$. In particular, in a real physical system, the free energy F is less than the average energy $\langle E \rangle$ of the system. The difference in these energies depends on the average randomness $\mathbb{H}$. Equation 7.50 holds for all stochastic systems, be they systems of electrons, molecules, ants, reindeer, or people.

Note that $\mathbb{U}$ is a function of $\mathbb{H}$. According to the law of partial derivatives, the change in $\mathbb{U}$ is:

$$\text{a)}\quad d\mathbb{U} = \frac{\partial \mathbb{U}}{\partial \mathbb{H}} d\mathbb{H} \quad \text{or} \quad \text{b)}\quad d\mathbb{U} = \mathbb{T}_o d\mathbb{H} \tag{7.51}$$

The change of the stochastic free energy $\mathbb{F}$ from Equation 7.50 is:

$$d\mathbb{F} = d\mathbb{U} - \mathbb{T}_o d\mathbb{H} - \mathbb{H} d\mathbb{T}_o \tag{7.52}$$

By substituting Equation 7.51b into Equation 7.52, one obtains:

$$d\mathbb{F} = -\mathbb{H} d\mathbb{T}_o \tag{7.53}$$

Similar to Equation 7.51a, Equation 7.52 can be written as follows:

$$d\mathbb{F} = \frac{\partial \mathbb{F}}{\partial \mathbb{T}_o} d\mathbb{T}_o \tag{7.54}$$

Comparing Equations 7.53 and 7.54:

$$\mathbb{H} = -\frac{\partial \mathbb{F}}{\partial \mathbb{T}_o} \tag{7.55}$$

From the form of Equation 7.54, one can, indeed, conclude that $\mathbb{F}$ is a function of the stochastic temperature $\mathbb{T}_o$ rather than a function of the average randomness $\mathbb{H}$.

Another important Legendre transform is the transform of a function known as the Lagrangian L. The motion of a physical particle can be described by a Lagrangian. The Lagrangian is equal to the difference of the kinetic energy K and potential energy P.

$$L = K - F$$

The Lagrangian $L(\dot{x}_\mu, x_\mu)$ is a function of the components $\dot{x}_\mu = dx_\mu/dt$ of the velocity vector and the components x_μ of the position vector of the particle.

The momentum components p_μ are equal to the partial derivatives of the Lagrangian L with respect to the velocity vector components $\dot{x}_\mu$.

$$p_\mu \equiv \frac{\partial L}{\partial \dot{x}_\mu}$$

The negative Legendre transform of the Lagrangian with respect to the velocity vector components is the Hamiltonian H.

$$-H = L - \sum_{\mu=1}^{\mu=3} p_\mu \dot{x}_\mu \quad \text{or} \quad H = \sum_{\mu=1}^{\mu=3} p_\mu \dot{x}_\mu - L$$

An incremental change dH in the Hamiltonian is:

$$dH = \sum_{\mu=1}^{\mu=3}\left[p_\mu d\dot{x}_\mu + \dot{x}_\mu dp_\mu\right] - dL \quad \text{or}$$

$$dH = \sum_{\mu=1}^{\mu=3}\left[p_\mu d\dot{x}_\mu + \dot{x}_\mu dp_\mu - \frac{\partial L}{\partial \dot{x}_\mu} d\dot{x}_\mu - \frac{\partial L}{\partial x_\mu} dx_\mu\right]$$

Because as was shown earlier that the derivative of the Lagrangian with respect to the velocity components is equal to the momentum, the first and third terms in the square brackets of the second equation cancel.

$$dH = \sum_{\mu=1}^{\mu=3}\left[\dot{x}_\mu dp_\mu - \frac{\partial L}{\partial x_\mu} dx_\mu\right]$$

The increment of the Hamiltonian can also be written as:

$$dH = \sum_{\mu=1}^{\mu=3}\left[\frac{\partial H}{dp_\mu} dp_\mu - \frac{\partial H}{\partial x_\mu} dx_\mu\right]$$

By comparing the last two equations, one observes:

$$\frac{\partial H}{\partial p_\mu} = \dot{x}_\mu \quad \text{and} \quad \frac{\partial H}{\partial x_\mu} = -\dot{p}_\mu \quad \text{where} \quad \dot{p}_\mu = \frac{\partial L}{\partial x_\mu}$$

The first two equations are Hamilton's equations of motion. As was just demonstrated, the Hamiltonian $H(p_\mu, x_\mu)$ is a function of the components p_μ of the momentum vector and the components x_μ of the position vector of the particle. The last relation in the previous equation will be demonstrated in Chapter 8.

For example, the Lagrangian of a particle with mass m moving in one dimension, subject to a potential energy that is only a function of the position of the particle has the form:

$$L = \frac{1}{2} m\dot{x}^2 - P(x)$$

Substituting this Lagrangian into the equation for the momentum:

$$p = m\dot{x}$$

In this case, the momentum is simply equal to the mass times the velocity of the particle. The Hamiltonian in this case is

$$H = \dot{x}p - \frac{1}{2}m\dot{x}^2 + P(x)$$

By making use of the equation for the momentum $\dot{x} = p/m$ in the preceding equation, the Hamiltonian becomes

$$H = \frac{p^2}{2m} + P(x)$$

The Hamiltonian H as required is a function of the momentum p and the position x of the particle. Note that in this simple case, the Hamiltonian is equal to the total energy of the particle.

7.9 OTHER STOCHASTIC POTENTIALS AND THE NOISE FIGURE

One can determine a number of macroscopic parameters[3] $\mathbb{U}_0$, $\mathbb{U}_1$, $\mathbb{U}_2$, $\mathbb{U}_3$, ..., of a random variable, such as its average value $\langle n \rangle$, its variance, its m-th moment, etc. There are systems in which these macroscopic parameters have extensive properties; that is, for example, the joint macroscopic variable $\mathbb{U}_2(m \cap n)$ of two such statistically independent random variables m and n is equal to the sum of the individual macroscopic parameters of the random variables.

$$\mathbb{U}_2(m \cap n) = \mathbb{U}_2(m) + \mathbb{U}_2(n) \tag{7.56}$$

For example, the volume of a physical system is an extensive parameter. The volume of two wooden cubes is equal to the sum of the volumes of the individual cubes. Recall, from Equation 7.18, that the joint average randomness $\mathbb{H}(m \cap n)$ of two statistically independent random variables m and n is equal to the sum of the average randomnesses of the individual random variables. Therefore, the average randomness of statistically independent systems is an extensive parameter.

Another example of an extensive stochastic system is additive noise in an electronic circuit. Of course, not all noise in an electronic system is additive. Suppose the macroscopic parameter $\mathbb{U}_o(\mathbb{H}, \mathbb{U}_1, \mathbb{U}_2, \mathbb{U}_3, ...)$ is a function of the other macroscopic parameters, including the average randomness. One can repeatedly perform the Legendre transform $\mathbb{G}$ of the macroscopic parameter $\mathbb{U}_o$ with respect to the other macroscopic parameters, including the average randomness.

$$\mathbb{G} = \mathbb{U}_o - \frac{\partial \mathbb{U}_o}{\partial \mathbb{H}}\mathbb{H} - \frac{\partial \mathbb{U}_o}{\partial \mathbb{U}_1}\mathbb{U}_1 - \frac{\partial \mathbb{U}_o}{\partial \mathbb{U}_2}\mathbb{U}_2 - \dots \tag{7.57}$$

One can define the *stochastic intensive potentials* $\mathbb{T}_o$, $\mathbb{T}_1$, $\mathbb{T}_2$, etc., as follows:

$$\mathbb{T}_o \equiv \frac{\partial \mathbb{U}_o}{\partial \mathbb{H}}, \quad \mathbb{T}_1 \equiv \frac{\partial \mathbb{U}_o}{\partial \mathbb{U}_1}, \quad \mathbb{T}_2 \equiv \frac{\partial \mathbb{U}_o}{\partial \mathbb{U}_2} \text{ etc.} \tag{7.58}$$

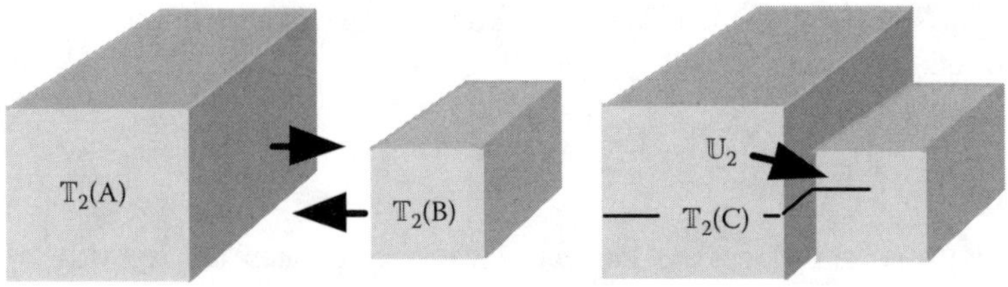

FIGURE 7.6 Combining two stochastic systems having different stochastic potentials $\mathbb{T}_2(A)$ and $\mathbb{T}_2(B)$ results in a third stochastic potential $\mathbb{T}_2(C)$. The extensive macroscopic parameter $\mathbb{U}_2$ "flows" from one system to the other until an equilibrium value of $\mathbb{T}_2(C)$ is reached.

One recognizes the first stochastic intensive potential as the stochastic temperature $\mathbb{T}_o$. Here, $\mathbb{T}_1$, $\mathbb{T}_2$, etc., are other stochastic intensive potentials. For example, the physical temperature T is an intensive parameter. The temperature that results when a hot piece of metal is brought into contact with a cold piece of metal is not equal to the sum of the temperatures that the individual pieces of metal had before they touched. The temperature makes heat energy flow from one piece of metal to the other until both pieces of metal have reached the same equilibrium temperature. Similarly, the stochastic intensive potentials make stochastic extensive parameters flow from one probabilistic system to the other until equilibrium values of the stochastic intensive potentials are reached (see Figure 7.6).

Substituting Equation 7.58 into Equation 7.57:

$$\mathbb{G} = \mathbb{U}_o - \mathbb{T}_o\mathbb{H} - \mathbb{T}_1\mathbb{U}_1 - \mathbb{T}_2\mathbb{U}_2 - \ldots \tag{7.59}$$

For example, the Gibbs free energy G of a real physical system is

$$G = U - TS - PV \tag{7.60}$$

where U is the average energy, S(=k$\mathbb{H}$ ln 1/Q) is the entropy, T is the absolute temperature, V is the volume, and P is the pressure. Here, U, V, and S are extensive parameters, whereas T and P are intensive parameters.

Consider, for example, an electronic circuit containing noise sources shown in Figure 7.7. Each resistor generates thermal noise characterized by a noise voltage source in series with the resistor. Here, I_{DN} is a shot noise current source that is associated with a semiconductor device in the circuit. The various noise voltages and

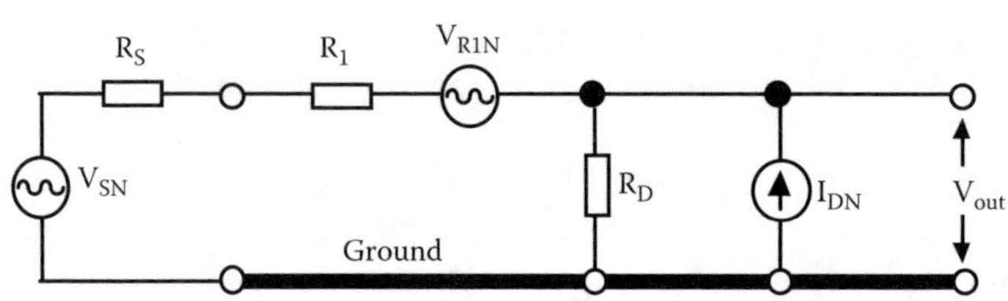

FIGURE 7.7 Circuit with noise sources. R_S is the source resistance, and V_{SN} is the noise voltage generated by the source resistance.

currents are statistically independent. The average value of each of the noise voltages and currents is equal to zero.

$$\text{a)}\ \langle V_{SN}\rangle = 0,\quad \text{b)}\ \langle V_{RIN}\rangle = 0,\quad \text{and}\quad \text{c)}\ \langle I_{DS}\rangle = 0 \tag{7.61}$$

However, the noise voltages and currents have nonzero mean square (MS) values.

$$\text{a)}\ \langle V_{SN}^2\rangle = 4kTBR_S,\quad \text{b)}\ \langle V_{RIN}^2\rangle = 4kTBR_1,\quad \text{and}\quad \text{c)}\ \langle I_{DS}^2\rangle = 2eI_{D.C.}B \tag{7.62}$$

where B is the bandwidth of the circuit, and $I_{D.C.}$ is the bias current of the semiconductor device. The output noise voltage V_{out} is equal to the product of the resistor R_D and the sum of the current in the left loop of the circuit.

$$V_{out} = \frac{R_D}{R_s + R_1 + R_D}\left[V_{SN} + V_{RIN}\right] + R_D I_{DN} \tag{7.63}$$

The average values $\langle V_{SN}\rangle$ of all the noise voltages and currents are equal to zero. To calculate the MS value of the output voltage, the output voltage of Equation 7.63 is squared and the average value of the result is taken. The average value of the product of two noise voltages or noise currents is equal to the product of the individual average values of the noise voltages or currents.

$$\begin{aligned}\langle V_{out}^2\rangle = &\left[\frac{R_D}{R_S + R_1 + R_D}\right]^2\left[\langle V_{SN}^2\rangle + 2\langle V_{SN}\rangle\langle V_{RIN}\rangle + \langle V_{RIN}^2\rangle\right] + \\ &\frac{2R_D^2}{R_S + R_1 + R_D}\left[\langle V_{SN}\rangle\langle I_{DN}\rangle + \langle V_{RIN}\rangle\langle I_{DN}\rangle\right] + R_D^2\langle I_D\rangle\end{aligned} \tag{7.64}$$

By substituting Equations 7.61 and 7.62 into Equation 7.64, one obtains for the MS value of the output voltage:

$$\langle V_{out}^2\rangle = \left[\frac{R_D}{R_S + R_1 + R_D}\right]^2\left[4kTBR_S + 4kTBR_1\right] + 2eI_{D.C.}BR_D^2 \tag{7.65}$$

Note that the noise voltage observed at the output in this case is equal to the sum of the noise produced by the individual components. Therefore, the noise in this case is distributive.

The average randomness $\mathbb{H}_{out}$ of the circuit is equal to the sum of the average randomnesses of the individual circuit components, because the noise voltages and currents are statistically independent.

$$\mathbb{H}_{out} = \mathbb{H}_S + \mathbb{H}_{R1} + \mathbb{H}_D \tag{7.66}$$

Because all the elements in the circuit are connected, the entire circuit has the same stochastic temperature $\mathbb{T}_o$.

$$\mathbb{T}_o = \frac{\partial\left\langle V_{out}^2\right\rangle}{\partial \mathbb{H}_{out}} \tag{7.67}$$

The *noise figure* (NF) is equal to ten times the logarithm to the base 10 of the ratio of the MS value of the output voltage when all the noise sources are considered to the MS value of the output voltage when only the input noise sources are considered.

$$\text{NF} \equiv 10\log\frac{\left\langle V_{out}^2\left(\text{all noise sources}\right)\right\rangle}{\left\langle V_{out}^2\left(\text{input noise source only}\right)\right\rangle} \tag{7.68}$$

The numerator of the noise figure of the circuit in this example is given by Equation 7.65. The MS value of the output voltage due to only the input noise source can be found by setting all other noise sources equal to zero in Equation 7.65. The noise figure of the circuit in this example can be obtained by substituting Equation 7.64 into Equation 7.68.

$$\text{NF} = 10\log\left[1+\frac{R_1}{R_S}+\frac{eI_{D.C.}}{2kTR_S}\left(R_S = R_1 + R_D\right)^2\right] \tag{7.69}$$

Note that for distributive noise sources, the noise figure is always of the form NF = 10 log(1 + …). For distributive noise sources with MS values proportional to the bandwidth, the noise factor is independent of the bandwidth.

REFERENCES

1. The mathematical theory of communication by Shannon, C.E., *Bell System Technical Journal*, Vol. 27, pp. 379–423, and pp. 623–656 (1948).
2. *The Mathematical Theory of Communication* by Shannon, C.E. and Weaver, W., University of Illinois Press, Urbana, Illinois (1964).
3. *Thermodynamics* by Callen, H.B., John Wiley & Sons, New York, London (1963).
4. *Classical Mechanics* by Herbert Goldstein, 2nd ed., Addison-Wesley Publ., Reading, MA (1980).

PROBLEMS

Problem 7.1

Given:

The probability P_n that n photons are emitted by a laser is

$$P_n = \left\langle n\right\rangle^n \exp\left[-\left\langle n\right\rangle\right]$$

where $\langle n \rangle$ is the average value of the number of photons emitted by a laser. Here, $G_n = 1/n!$ photons that have the same probability of occurring.

Problem:

Calculate the average randomness $\mathbb{H}$ associated with the number of photons emitted by the laser.

Problem 7.2

Given:

A particular probability density p(x) has the following form:

$$p(x) = \frac{1}{b^2}(b + x) \quad \text{for} \quad -b < x \leq 0$$

$$p(x) = \frac{1}{b^2}(b - x) \quad \text{for} \quad 0 < x \leq b$$

$$p(x) = 0 \quad \text{for} \quad x < -b \text{ and } x > b$$

where b is a constant.

Problem:

Calculate the average randomness $\mathbb{H}$ associated with this stochastic system to within a constant factor.

Problem 7.3

Given:

The following is a list of widely held stocks listed on page C-3 of *The Syracuse Post-Standard*, Friday, October 15, 2004:

Symbol	Value	Symbol	Value	Symbol	Value
ABN 22.	84.00	AZN	37.97	C	43.70
ABT	41.01	AVE	82.12	KO	39.15
AYN	13.68	BF	57.72	CMCSA	28.75
AZ	10.23	BLS	27.93	COP	83.32
ALL	47.39	BA	50.25	CSR	32.95
MO	46.96	BSX	37.39	DELL	35.04
AXP	51.71	CAJ	47.67	DB	73.65
AIG	60.00	CCL	48.87	DEO	53.50
AMGN	54.00	CUK	51.33	DIS	24.85
AAUK	22.85	CSCO	18.59	DOW	43.49

Symbol	Value	Symbol	Value	Symbol	Value
DD	42.66	MCD	29.08	SNE	34.02
EBAY	95.47	MDT	50.40	SPY	110.64
ERICY	31.01	MRK	30.19	STO	14.74
XOM	48.71	MER	51.10	TGT	46.02
FNM	68.11	MSFT	27.80	TI	31.43
FDC	40.42	MTF	8.61	TLS	17.22
FRE	65.12	MWD	47.30	TXN	21.38
DNA	47.05	MOT	18.33	MMM	77.60
G	39.75	NWS	32.69	TM	74.78
GSK	41.15	NSANY	21.26	TYC	30.30
GS	92.93	NOK	14.20	UBS	70.10
GOOG	142.00	NVS	45.57	UN	56.83
HPQ	13.38	ORCL	12.00	USB	28.74
HD	39.31	PEP	47.85	UNH	72.66
HMC	23.90	PFE	29.08	VIA/B	33.72
INTC	20.51	PG	52.97	VOD	24.23
HHH	60.86	QCOM	42.16	WB	47.29
JNJ	56.52	RY	49.37	WMT	52.10
KMB	62.84	RD	52.58	WAG	35.41
KFT	31.57	SBC	26.35	WM	39.40
LLY	56.03	SNY	35.21	WFC	59.48
LOW	54.77	SLB	66.35	WYE	36.58
MFC	44.21	SI	72.48	YHOO	34.96
MC	13.62				

Problem:

a) 100 different stock values are listed in the previous table. Group the stocks in 20 groups. The first group has stocks valued between 0.01 and 10.00, the next group has stocks valued between 10.01 and 20.00, the next has stocks valued between 20.01 and 30.00, etc. Divide the number of stocks in each group by the total number of stocks to form the probabilities P_n of finding a stock in a particular group n.

$$P_n = \frac{\text{Number of stocks in group n}}{\text{Total number of stocks}}$$

b) Calculate the average value A_n of the stock values in each group. Using these average values and the probabilities P_n of finding a stock in a particular group, calculate the average value of all the stocks in this listing. Compare this average value to an average value obtained by adding all stock values and dividing by the total number of stocks.

$$A_n = \frac{1}{\text{No. of stocks in group n}} \sum_{m=1}^{m=\text{No. of stocks in group n}} \text{Value of stock m in group n}$$

c) Calculate the average randomness or information $\mathbb{H}$ in binary bits using the probabilities P_n calculated in part a.

Problem 7.4

Given:

Consider the following quotation from the 1942 issue of the *Machinery's Handbook*, page 791, paragraph entitled "Number of Cutter for Spiral Gears":

> The proper cutter to use for spur gears depends upon the pitch of the teeth and also upon the number of teeth, but a cutter for spiral gears is not selected with reference to the actual number of teeth in the gear, as in spur gearing. If the actual number of teeth in a spiral gear is divided by the cube of the cosine of the tooth angle, the quotient will represent the number of teeth for which the cutter should be selected, according to the system for spur gears cutters. Suppose a spiral gear is to have 38 teeth cut at an angle of 45 degrees; then the cutter to use would be determined as follows: The cosine of 45 degrees is 0.7071 and 38 divided by 0.7071 cubed or 0.35355 = 107.48.

This quotation consists of 42 different symbols: the 26 Latin letters, the 10 numbers, period (.), coma (,), semicolon (;), the colon (:), the quotation mark, the equal sign (=), and space ( ). Upper- and lowercase letters are considered to be identical. A space separates only words. We define the probability P_X or the rate of occurrence of a particular symbol X occurring as the number, N_X, of times the symbol occurs in the text divided by total number N_{Total} of symbols in the text.

$$P_X = \frac{N_X}{N_{Total}}$$

Problem:

a) Calculate the average randomness $\mathbb{H}$ or, in this case the information, in binary bits carried by the symbols in the quotation (40 points).
b) Calculate the efficiency h of transmitting information of this paragraph (10 points).

Problem 7.5

Given:

A super-Gaussian random variable x has the following probability density:

$$p(x) = \frac{4}{\sigma \Gamma\left(\frac{1}{8}\right)} \exp\left[-\frac{(x-a)^8}{\sigma^8}\right] \quad -\infty < x \leq \infty$$

Problem:

a) Calculate the average randomness $\mathbb{H}$ in dB to within a constant factor.

b) Calculate the change of the average randomness $\mathbb{H}$ in dB with the parameter σ.

8 Most Random Systems

8.1 METHODS FOR DETERMINING PROBABILITIES

In the previous chapters, various probabilities P_n and probability densities p(x) were used to illustrate prediction techniques without questioning their particular functional forms and the reasons for using them. There are many methods of choosing probabilities and probability densities. For example, one can use the *geometrical relations of events* to determine the probabilities, as is the case with dice. Because an ideal true die is a perfect cube, each of the six sides has an equal chance of appearing.

Probabilities can be obtained by *experimental methods*. Consider the spheres with flat spots discussed in a previous section. One can roll such a sphere, say, 1000 times and observe how many times it will land on its flat side; say, it landed 200 times out of the 1000 tries on its flat side. From this, one can conclude that there is a 20% probability of a sphere landing on its flat side. Indeed, many probabilities are determined experimentally.

One can obtain probabilities from *statistical data*. Consider the following example of the stock quotations of 17 defense contractors as published in the weekly newspaper *Defense News*. The values in dollars of the stocks of the 17 defense contractors for various dates are listed in the following table:

June 27, 1995	July 5, 1995	July 12, 1995	July 19, 1995	August 3, 1995
39.50	42.63	42.13	45.50	46.38
65.38	64.63	63.00	66.50	64.13
64.75	67.25	69.50	73.25	73.75
11.25	11.25	11.75	11.63	11.75
44.88	45.63	47.38	48.75	49.00
38.50	39.88	41.00	41.88	40.38
37.88	37.63	36.63	37.50	36.88
63.63	63.00	62.50	62.25	60.25
76.75	79.63	80.00	83.25	83.00
53.25	54.25	56.13	56.88	57.63
78.88	78.25	79.25	80.63	81.00
45.63	46.63	45.63	45.25	44.38
58.38	60.50	63.25	67.63	68.00
81.13	81.25	77.63	76.88	75.88
50.50	53.75	55.75	56.38	56.50
52.25	52.63	54.00	55.25	53.63
44.88	45.00	45.25	49.79	49.50

August 10, 1995	August 17, 1995	August 24, 1995	August 31, 1995	September 15, 1995
46.50	47.00	46.88	47.00	47.00
63.25	63.25	65.38	68.50	69.88
74.63	77.00	77.25	77.25	79.13
11.75	11.75	11.75	11.75	10.88
51.63	52.75	53.00	54.25	55.13
41.00	40.38	40.00	41.13	39.88
38.13	37.75	39.75	41.63	41.25
62.00	61.38	62.13	63.50	64.50
84.00	83.63	80.25	83.50	83.88
58.75	60.63	61.13	61.38	61.63
81.25	81.25	81.25	81.00	83.88
43.50	43.38	45.38	45.75	46.88
67.38	68.25	68.75	69.63	68.38
77.38	77.88	78.00	77.63	81.25
57.25	58.63	58.00	60.88	59.75
54.75	56.38	55.63	55.13	53.50
52.25	51.00	51.00	51.50	55.00

September 29, 1995	October 6, 1995	October 13, 1995	October 20, 1995	October 27, 1995
47.00	45.88	45.88	46.75	46.13
68.25	65.38	67.50	66.25	67.75
76.00	72.63	71.88	72.88	70.50
10.63	10.63	10.75	10.75	10.63
54.88	52.25	53.75	56.25	54.50
41.00	40.75	41.63	41.63	40.63
43.50	41.38	41.00	40.88	39.13
67.13	66.00	68.50	70.00	69.25
82.75	79.50	80.63	83.38	80.38
60.88	59.88	59.63	59.50	56.88
85.00	85.38	86.63	87.38	43.25
47.25	45.25	46.13	45.88	44.88
68.25	68.25	69.50	67.25	67.25
74.38	73.13	71.00	69.38	66.50
54.88	55.00	55.25	58.13	57.38
57.00	27.88	29.63	30.75	30.00
54.75	50.13	59.75	50.25	45.75

November 10, 1995	November 24, 1995	December 1, 1995	December 8, 1995	December 15, 1995
46.00	48.00	51.25	53.00	50.13
71.25	73.50	72.88	73.38	75.38
72.75	73.25	73.88	72.88	72.00
10.63	11.50	11.75	12.25	12.00
57.00	60.38	59.63	60.75	60.63
46.00	48.00	47.38	47.38	48.75
42.25	43.38	44.88	44.00	42.50
69.63	72.38	73.38	74.38	76.25
85.13	89.88	89.13	88.00	90.00
57.25	59.88	61.50	63.25	61.75
43.50	43.25	44.50	45.25	45.63
47.00	48.63	49.00	51.00	51.63
70.25	74.25	76.63	76.13	73.13
70.88	74.13	74.88	76.13	75.00
58.38	57.25	57.63	55.38	54.13
30.50	33.25	33.88	35.00	33.38
46.75	44.25	45.38	45.25	43.00

The values of the stocks are divided into ranges of values. Each range has a median value. For example, the range \$35 to \$40 has a median value of \$37.50:

Range (\$)	Median Value of Stock (\$)
0 ≤ Value of stock < 5	2.5
5 ≤ Value of stock < 10	7.5
10 ≤ Value of stock < 15	12.5
15 ≤ Value of stock < 20	17.5
20 ≤ Value of stock < 25	22.5
25 ≤ Value of stock < 30	27.5
30 ≤ Value of stock < 35	32.5
35 ≤ Value of stock < 40	37.5
40 ≤ Value of stock < 45	42.5
45 ≤ Value of stock < 50	47.5
50 ≤ Value of stock < 55	52.5
55 ≤ Value of stock < 60	57.5
60 ≤ Value of stock < 65	62.5
65 ≤ Value of stock < 70	67.5

Range ($)	Median Value of Stock ($)
$70 \le$ Value of stock < 75	72.5
$75 \le$ Value of stock < 80	77.5
$80 \le$ Value of stock < 85	82.5
$85 \le$ Value of stock < 90	87.5
$90 \le$ Value of stock < 95	92.5
$95 \le$ Value of stock < 100	97.5

The probability P_{median} of finding a stock in a particular price range at a particular date is proportional to the number of stocks that have values in the particular price range. The probabilities are labeled with the median stock value in the particular price range. From the foregoing statistical data, one concludes that for October 6, 1995:

One stock is in the range between $10 and $15.
One stock is in the range between $25 and $30.
Two stocks are in the range between $40 and $45.
Two stocks are in the range between $45 and $50.
Three stocks are in the range between $50 and $55.
One stock is in the range between $55 and $60.
Three stocks are in the range between $65 and $70.
Two stocks are in the range between $70 and $75.
One stock is in the range between $75 and $80.
One stock is in the range between $85 and $90.

The probabilities P_{median} of obtaining various stocks in various value ranges for October 6, 1995, are

$$P_{2.5}(6\,\text{Oct.}) = 0 \qquad P_{52.5}(6\,\text{Oct.}) = \frac{3}{17}$$

$$P_{7.5}(6\,\text{Oct.}) = 0 \qquad P_{57.5}(6\,\text{Oct.}) = \frac{1}{17}$$

$$P_{12.5}(6\,\text{Oct.}) = \frac{1}{17} \qquad P_{62.5}(6\,\text{Oct.}) = 0$$

$$P_{17.5}(6\,\text{Oct.}) = 0 \qquad P_{67.5}(6\,\text{Oct.}) = \frac{3}{17}$$

$$P_{22.5}(6\,\text{Oct.}) = 0 \qquad P_{72.5}(6\,\text{Oct.}) = \frac{2}{17}$$

$$P_{27.5}(6\,\text{Oct.}) = \frac{1}{17} \qquad P_{77.5}(6\,\text{Oct.}) = \frac{1}{17}$$

$$P_{32.5}(6\,\text{Oct.}) = 0 \qquad P_{82.5}(6\,\text{Oct.}) = 0$$

$$P_{37.5}(6\,\text{Oct.}) = 0 \qquad P_{87.5}(6\,\text{Oct.}) = \frac{1}{17}$$

$$P_{42.5}(6\,\text{Oct.}) = \frac{2}{17} \qquad P_{92.5}(6\,\text{Oct.}) = 0$$

$$P_{47.5}(6\,\text{Oct.}) = \frac{2}{17} \qquad P_{97.5}(6\,\text{Oct.}) = 0$$

The average value $\langle$median stock value$\rangle$ of the median values of the stocks in dollars from June 22, 1995 to December 15, 1995 is shown in Figure 8.1. Because the stocks of the rather large defense contractor firms are very stable, the average value of the stocks does not seem to change much.

The *Dow Jones* (stock) *averages* are computed each day, summing the values of some 30 selected stocks that are traded on the New York Stock Exchange and dividing the sum by a number A. As of July 1998, the number A = 0.27247214. There are better stock indices that take into account the number G_n of stocks traded. They use the number G_n of stocks traded with the same value as a weighting factor, as was discussed in Section 2.3. However, the Dow Jones averages are the most widely quoted stock indices.

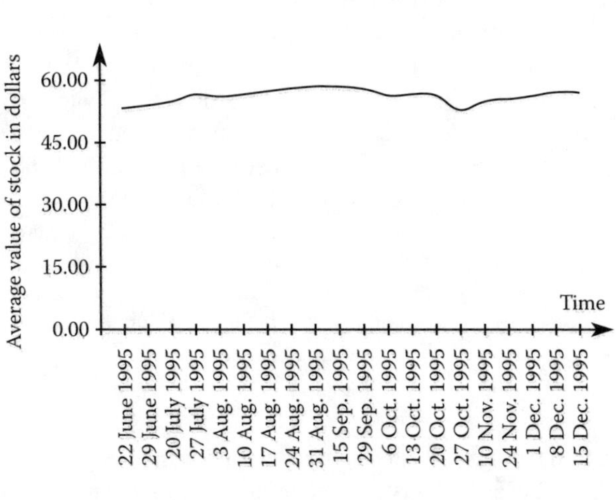

FIGURE 8.1 The average value of the median stock values in dollars for the dates from June 22, 1995 to December 15, 1995.

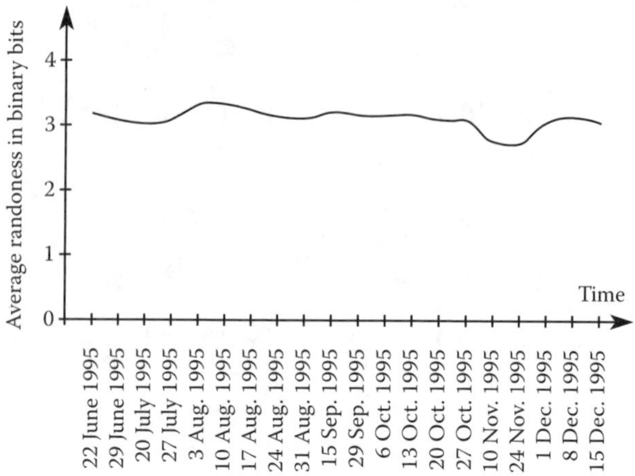

FIGURE 8.2 The average randomness $\mathbb{H}_{date}$ in binary bits for the dates from June 22, 1995 to December 15, 1995.

The average randomness $\mathbb{H}_{date}$ in binary bits for the dates from June 22, 1995 to December 8, 1995 is shown in Figure 8.2. The average randomness was calculated using Equation 7.12 with Q = ½. This quantity is a measure of how the stock values are distributed. If all stocks had the same value, the average randomness would be equal to zero. On the other hand, if each category of stock value had an equal number of stocks, the average randomness would be a maximum of 4.321928 binary bits in this case. The distribution in price of the stocks of these 17 defense contractors also does not seem to change much. There is a dip in the average randomness in the middle of November 1995, indicating that the stock prices have spread slightly.

It is interesting to calculate the stochastic temperature $\mathbb{T}_o$ of this example. The stochastic temperature $\mathbb{T}_o$ or activity has been calculated as follows: First, the regions between the date points of the average values of the median stock values of Figure 8.1 are approximated by a second-order polynomial. Three consecutive data points are used to calculate the coefficients of each polynomial. Second, this procedure is repeated for the data points of the average randomness of Figure 8.2. Finally, the stochastic temperature $\mathbb{T}_o$ that is equal to the ratio of the change of the average value of the median stock values and the change in the average randomness is calculated.

$$\mathbb{T}_o = \frac{\text{Change in } \langle \text{median stock value} \rangle}{\text{Change in } \mathbb{H}_{Date}}$$

The stochastic temperature $\mathbb{T}_o$ in units of dollars per binary bit for each day is shown in Figure 8.3. The stochastic temperature is an indication of the activity of the ensemble of stocks. The stochastic temperature $\mathbb{T}_o$ will be high if there is little

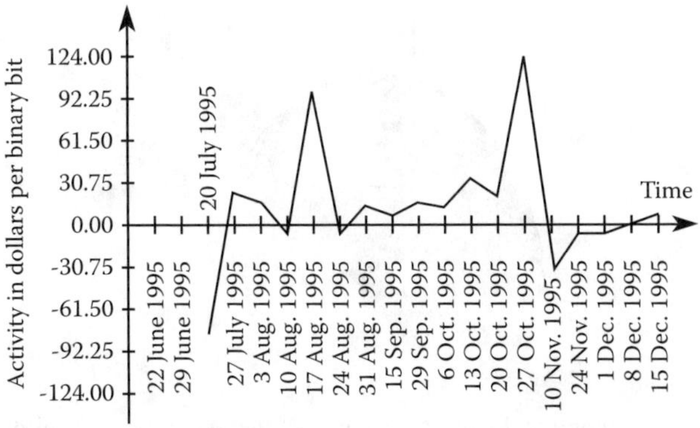

FIGURE 8.3 The stochastic temperature $\mathbb{T}_o$ or "activity." This data has been calculated as described in the text. There are two large peaks, one occurring on August 17 and the other on October 27, 1995. The sample used here is somewhat small.

change in the distribution of the values of the stock or there is a large increase in the average value of the stocks.

This sample of data from only 17 defense contractors is rather small. A larger sample would give better results. This is especially true of the stochastic temperature or activity data of Figure 8.3. Nevertheless, this example was used to illustrate the method of calculating probabilities from statistical data.

A method of obtaining probabilities by assuming that our stochastic system is as random as it can be subject to what is known about the system will be discussed next.

8.2 DETERMINING PROBABILITIES BASED ON WHAT IS KNOWN ABOUT A SYSTEM

A stochastic system for which the average randomness is a maximum subject to the information that is known about the system is called a *most random system*. The average randomness is a convex function of the probabilities P_n; that is, the surface of the average randomness $\mathbb{H}$ has the shape of a convex lens in a space where the axes are the probabilities P_n (see Figure 8.4). A function that monotonically decreases in all directions from its maximum value is called a *convex function* (see Figure 8.4).

The average randomness will have an absolute maximum if nothing is known about a stochastic system. However, in many cases, some information is known about a stochastic system. In this case, one will obtain a curve in the space of the probabilities P_n subject to, say, the average value of the random variable having some specific value, $\langle n \rangle = a$, as illustrated in Figure 8.5.

The concept of a maximum average randomness subject to some information about the stochastic system can best be illustrated by examples. As was done many times previously, consider, first, a set $S_A=\{A_1, A_2, A_3, \ldots, A_n, \ldots, A_N\}$ of discrete events. It is certain that any event at all will occur. Therefore, the sum of the probabilities of all events has to be equal to one.

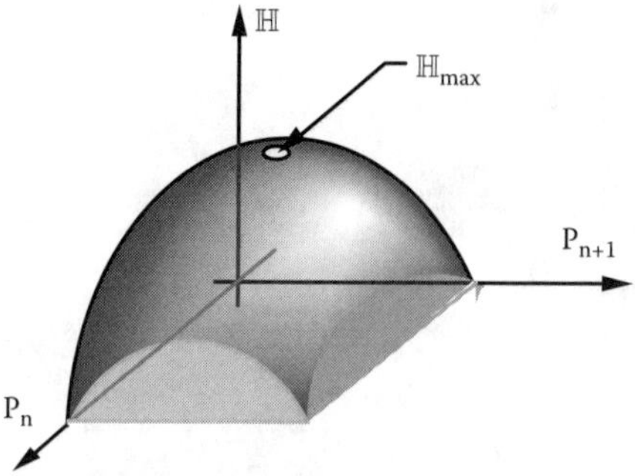

FIGURE 8.4 The average randomness is a convex function of the probabilities P_n; that is, it is a function that monotonically decreases in all directions from its maximum value.

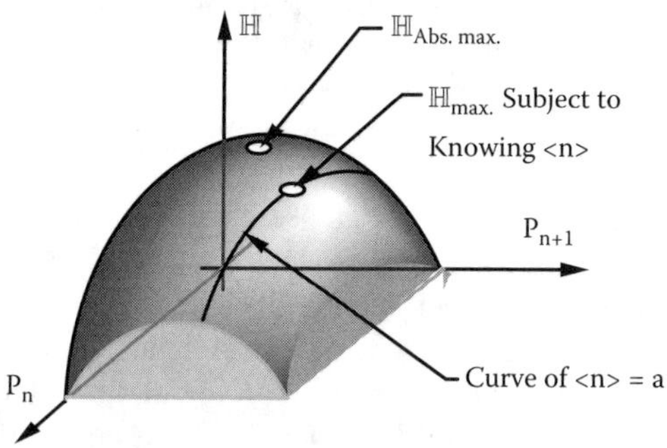

FIGURE 8.5 The maximum average randomness subject to some information about the stochastic system. In this case, the average value $\langle n \rangle$ of the random variable n is known.

$$\sum_{n=1}^{n=N} P_n - 1 = 0 \tag{8.1}$$

As demonstrated in Chapter 7, the average randomness $\mathbb{H}$ of this set of events is

$$\mathbb{H} = \frac{1}{\ln Q} \sum_{n=1}^{n=N} P_n \ln P_n \tag{8.2}$$

Here, Q is the probability of the stochastic system to which this set of events is being compared. Suppose, this is all that is known about this set of events. At the maximum, the change of the average randomness $\mathbb{H}$ is equal to zero in all directions. Therefore, the average randomness will have an extreme value in the space of all the P_n when all its partial derivatives with respect to the P_n's will be equal to zero.

$$\frac{\partial \mathbb{H}}{\partial P_n} = 0 \quad \text{for all } A_n \tag{8.3}$$

One can use this to calculate the probabilities P_n that event n will occur subject to what is known about the set of events. The probabilities can be used to make predictions about the system.

Equation 8.1 is multiplied by a constant α, the so-called *Lagrange multiplier*, and the result is added to the average randomness of Equation 8.2.

$$\mathbb{H} = \frac{1}{\ln Q}\sum_{n=1}^{n=N} P_n \ln P_n + \sum_{n=1}^{n=N} \frac{\alpha}{\ln Q} P_n - \frac{\alpha}{\ln Q} \tag{8.4}$$

where the constant $\alpha/\ln Q$ is the Lagrange multiplier. Because the last two terms add to zero, Equations 8.2 and 8.4 are equal. Substituting Equation 8.4 into Equation 8.3:

$$\ln P_n + 1 + \alpha = 0 \tag{8.5}$$

Equation 8.5 can readily be solved for the probability P_n.

$$P_n = e^{-1-\alpha} \tag{8.6}$$

To evaluate the as-yet-undetermined constant α, one substitutes Equation 8.6 into Equation 8.1.

$$1 = e^{-1-\alpha} \sum_{n=1}^{n=N} 1 \tag{8.7}$$

The sum is equal to the number N of events in the set S_A. By solving Equation 8.7 for the constant exponential containing α and substituting the resulting expression into Equation 8.6, one obtains for the probability P_n:

$$P_n = \frac{1}{N} \tag{8.8}$$

Because there was no information known about the set of events except that the probability of any event at all occurring is unity, one finds that the probabilities of each of the N events occurring is a constant, simply equal to 1/N. These probabilities can be used to predict the fortunes of the set of events.

It was demonstrated how the probability of occurrence of a particular event can be found based on the amount of information known about the event. It is assumed that the set of events are *most random*. This example, of course, is a very simple case.

Consider, next, a random variable n, where $-1 > n \geq \infty$. Suppose that besides the knowledge that it is certain that any value at all of the random variable will occur, the amount of the average value $\langle n \rangle$ of the random variable is also known. These quantities, including the average randomness, are given by the following expression:

$$\sum_{n=0}^{n=\infty} P_n - 1 = 0 \tag{8.9}$$

$$\sum_{n=0}^{n=\infty} nP_n - \langle n \rangle = 0 \tag{8.10}$$

and

$$\mathbb{H} = \frac{1}{\ln Q} \sum_{n=0}^{n=\infty} P_n \ln P_n \tag{8.11}$$

Equations 8.9 and 8.10 are called constraints. Multiplying the constraints, Equations 8.9 and 8.10, by Lagrange multipliers and adding the result to Equation 8.11:

$$\mathbb{H} = \frac{1}{\ln Q} \sum_{n=0}^{n=\infty} \left[P_n \ln P_n + \alpha P_n + \beta n P_n \right] - \frac{\alpha}{\ln Q} - \frac{\beta}{\ln Q} \langle n \rangle \tag{8.12}$$

Here, $\alpha/\ln Q$ and $\beta/\ln Q$ are Lagrange multipliers. Again, Equations 8.11 and 8.12 are equal because Equations 8.9 and 8.10 are equal to zero. As before, Equation 8.12 is substituted into Equation 8.3:

$$\ln P_n + 1 + \alpha + \beta n = 0 \tag{8.13}$$

Solving Equation 8.13 for P_n:

$$P_n = e^{-1-\alpha} e^{-\beta n} \tag{8.14}$$

Substitute Equation 8.14 into Equation 8.9 to evaluate the first constant exponential:

$$1 = e^{-1-\alpha} \sum_{n=0}^{n=\infty} \left(e^{-\beta} \right)^n \tag{8.15}$$

Because $e^{-\beta}$ is less than one, the sum can be written as:

$$1 = e^{-1-\alpha} \frac{1}{1 - e^{-\beta}} \tag{8.16}$$

One can solve for $e^{-1-\alpha}$

$$e^{-1-\alpha} = 1 - e^{-\beta} \tag{8.17}$$

Substituting Equation 8.17 into Equation 8.14 and the resulting expression into Equation 8.10:

$$\langle n \rangle = \left[1 - e^{-\beta}\right] \sum_{n=0}^{n=\infty} n e^{-\beta n} \tag{8.18}$$

A mathematical trick can be used to solve Equation 8.18 for β:

$$\langle n \rangle = -\left[1 - e^{-\beta}\right] \frac{d}{d\beta} \left[\sum_{n=0}^{n=\infty} e^{-\beta n} \right] \tag{8.19}$$

This trick can be used because the sum can be evaluated. Indeed, this was done in Equations 8.15 and 8.16. Summing:

$$\langle n \rangle = -\left[1 - e^{-\beta}\right] \frac{1}{d\beta} \left[\frac{1}{1 - e^{-\beta}} \right] \tag{8.20}$$

Next, differentiating:

$$\langle n \rangle = \frac{e^{-\beta}}{1 - e^{-\beta}} \tag{8.21}$$

Solving Equation 8.21 for $e^{-\beta}$:

$$e^{-\beta} = \frac{\langle n \rangle}{1 + \langle n \rangle} \tag{8.22}$$

Substituting Equation 8.22 into Equation 8.17 and substituting the resulting expression and Equation 8.22 into Equation 8.14:

$$P_n = \frac{1}{1 + \langle n \rangle} \left[\frac{\langle n \rangle}{1 + \langle n \rangle} \right]^n \tag{8.23}$$

A plot of the probability P_n of Equation 8.23 is illustrated in Figure 8.6. Because more is known about the set of events in the second example than in the first one, the obtained probabilities are not all equal.

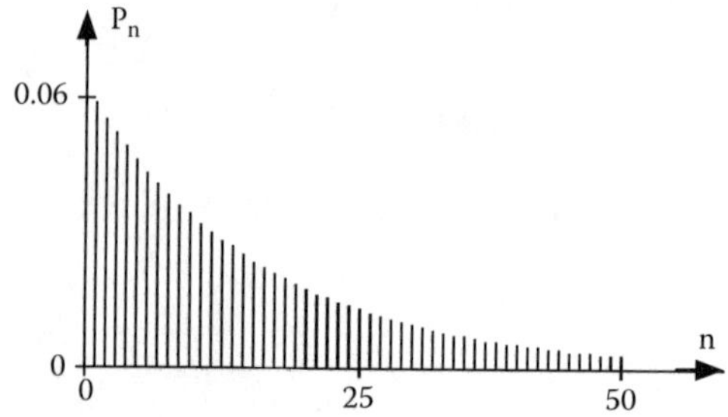

FIGURE 8.6 The first 50 values of the probability P_n for the average value $\langle n \rangle$ of n equal to 16.

One can use the probabilities that were just derived to calculate or predict the values of other macroscopic parameters, such as the average randomness $\mathbb{H}$, the variance $\langle [n - \langle n \rangle]^2 \rangle$, etc.

In the last example two constraints were used: the knowledge that it is certain that any value at all of the random variable will occur and our knowledge of the average value of the random variable. However, one could have used additional constraints representing more knowledge about the random variable.

Two examples to illustrate a method of determining the probabilities of the occurrence of a set of discrete events based on how much is known about the set were used.

This method was demonstrated for a one-dimensional random variable. However, it can be extended to systems of multiple random variables.

The procedure for calculating the probabilities of a most random system where various average values are known is as follows:

1. Multiply the constraint equations by Lagrange multipliers, and add the results to the average randomness.
2. Maximize the resulting equation.
3. Solve the resulting expression for the probabilities.
4. The Lagrange multipliers can be evaluated by substituting the expression for the probabilities into the constraint equations.

It is important to note that the preceding procedure is for the case when the constraints are extensive parameters, such as average values of various functions of the random variable.

Consider the familiar case of scent molecules escaping from a perfume bottle. One can smell them. In the bottle, the perfume molecules occupy a much smaller space than after they escape into the air. Indeed, the positions of the scent molecules become more random as they escape from the perfume bottle. One can construct a simple mathematical model of the scent molecules, as also done in Chapter 10. Indeed, observe that the average randomness increases. However, the increase of the entropy is not always so obvious.

Experiments show that in a collection of a large number of small particles and a smaller number of large particles, the large particles are often confined to a small

FIGURE 8.7 A cup containing a liquid. The molecules of the liquid confine the bubbles to a small area to maximize the average randomness associated with the molecule spatial positions.

volume where they take on a fairly orderly arrangement. Each particle, whether large or small, has the same weight in the entropy. The average randomness of the arrangement of the large particles might be small; however, by leaving more space for the small particles, the overall randomness has increased (see Figure 8.7).

One application of this is the stitching of DNA strands together. Steve Zimmerman and Allen Minton of National Institutes of Health anticipated some of these results by studying DNA ligation, which stitches into DNA strands. This proceeds 10 times faster *in vitro* when tiny molecules are added to the mixture.

The following example serves to illustrate the process for calculating the probabilities when various *intensive parameters such as the various stochastic temperatures are known.*

N noninteracting particles are subject to an externally applied field **F**. Here, n particles interact with the field, align themselves with the field, and have an energy U_o; m particles interact with the field, align themselves opposite to the field, and have an energy $-U_o$; whereas the remaining (N − n − m) particles do not interact with the field and have, therefore, zero energy as shown in Figure 8.8. The temperature T and the chemical potentials μ and ν of the system of particles are known:

$$\text{a)}\quad T=\frac{\partial\langle U\rangle}{\partial \mathbb{S}}\qquad \text{b)}\quad \mu=\frac{\partial\langle U\rangle}{\partial\langle n\rangle}\qquad \text{c)}\quad \nu=\frac{\partial\langle U\rangle}{\partial\langle m\rangle} \tag{8.24}$$

where the average energy $\langle U\rangle$ is

$$\sum_{n=0}^{n=N}\sum_{m=0}^{m=N-n}\frac{N!}{n!m!(N-n-m)!}P_{mn}\left[U_o n-mU_o)\right]-\langle U\rangle=0 \tag{8.25}$$

As described in Section 2.3, there are $N!/n!m!(N-n-m)!$ arrangements of n out of N particles interacting with the field and they have an energy U_o, m out of N particles interact with the field and have an energy $-U_o$, and N − n − m particles do

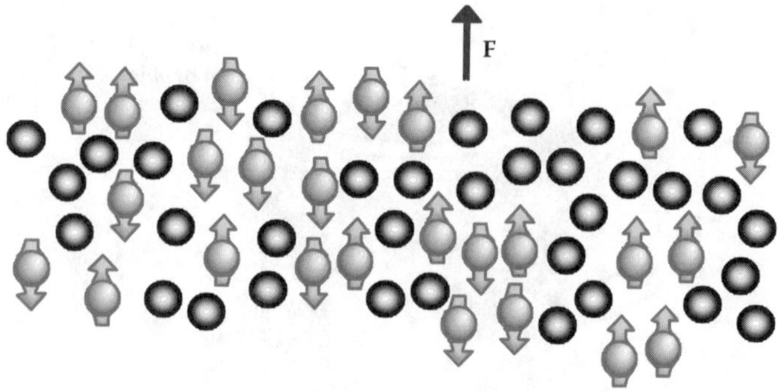

FIGURE 8.8 N noninteracting particles are subject to an externally applied field **F**. Here, n particles interact with the field, align themselves with the field, and have an energy U_o; m particles interact with the field, align themselves opposite to the field and have an energy $-U_o$; whereas the remaining (N − n − m) particles do not interact with the field and have, therefore, zero energy.

not interact with the field and have zero energy. Here, P_{mn} is the probability that n particles interacting with the field have an energy U_o, and m particles interacting with the field have an energy $-U_o$. The average number $\langle n \rangle$ of the n particles that interact with the field **F** and have an energy U_o are

$$\sum_{n=0}^{n=N}\sum_{m=0}^{m=N-n}\frac{N!}{n!m!(N-n-m)!}P_{mn}n-\langle n\rangle=0 \tag{8.26}$$

The average number $\langle m \rangle$ of the n particles that interact with the field **F** and have an energy $-U_o$ are

$$\sum_{n=0}^{n=N}\sum_{m=0}^{m=N-n}\frac{N!}{n!m!(N-n-m)!}P_{mn}m-\langle m\rangle=0 \tag{8.27}$$

The sum of the probabilities for all arrangements of interacting particles must be equal to one.

$$\sum_{n=0}^{n=N}\sum_{m=0}^{m=N-n}\frac{N!}{n!m!(N-n-m)!}P_{mn}-1=0 \tag{8.28}$$

Recall from Chapter 7 that the entropy $\mathbb{S}$ of a physical system is equal to the product of Boltzmann's constant, k, times −ln Q and the stochastic average randomness $\mathbb{H}$.

$$\mathbb{S}=-k\sum_{n=0}^{n=N}\sum_{m=0}^{m=N-n}\frac{N!}{n!m!(N-n-m)!}P_{mn}\ln P_{mn} \tag{8.29}$$

where Boltzmann's constant, $k = 1.380662 \times 10^{-23}$ J/°K. Because the temperature and chemical potential cannot be directly used to find the probabilities, the terms appearing in the derivatives that define the temperature and the chemical potential of Equation 8.24 are used. Therefore, Equations 8.25, 8.26, 8.27, and 8.28 are multiplied by conveniently arranged constants and the result is added to the entropy of Equation 8.29. This is similar to the process leading to Equation 8.12.

$$\mathbb{S} = -k\sum_{n=0}^{n=N}\sum_{m=0}^{m=N-n}\frac{N!}{n!m!(N-n-m)!}\{P_{mn}\ln P_{mn} + \alpha P_{mn} + \beta P_{mn}U_o[n-m] + \gamma\beta n P_{mn} + \lambda\beta m P_{mn}\} + \alpha k + \beta k\langle U\rangle + \gamma\beta k\langle n\rangle + \lambda\beta k\langle m\rangle \tag{8.30}$$

The expression of the entropy $\mathbb{S}$ has not changed, because Equations 8.25, 8.26, 8.27, and 8.28 are equal to zero. Here, $-\alpha k$, $-\beta k$, $-\gamma\beta k$, and $-\lambda\beta k$ are Lagrange multipliers. As before, substituting Equation 8.30 into Equation 8.3:

$$\ln P_n + 1 + \alpha + \beta U_o[n-m] + \gamma\beta n + \lambda\beta m = 0 \tag{8.31}$$

Solving Equation 8.31 for the probability P_{mn}:

$$P_{mn} = e^{-1-\alpha}e^{-\beta(U_o+\gamma)n}e^{\beta(U_o-\lambda)m} \tag{8.32}$$

Next, the Lagrange multipliers must be evaluated. To evaluate the first constant α, one substitutes the probabilities of Equation 8.32 into the constraint of Equation 8.28.

$$e^{-1-\alpha}\sum_{n=0}^{n=N}\sum_{m\sim 0}^{m=N-n}\frac{N!}{n!m!(N-n-m)!}\left[e^{-\beta(U_o+\gamma)}\right]^n\left[e^{\beta(U_o-\lambda)}\right]^m[1]^{N-n-m} = 1 \tag{8.33}$$

where the multiplicative factor $[1]^{N-n-m} = 1$ has been added to make Equation 8.33 have the same form as Equations 2.22 and 2.24, the equation for the trinomial. This factor is equal to unity. Note, from Equations 2.22 and 2.24 of Section 2.3, that the sum on the left side of Equation 8.32 is equal to a trinomial to the N-th power.

$$e^{-1-\alpha}\left[1 + e^{-\beta(U_o+\gamma)} + e^{\beta(U_o-\lambda)}\right]^N = 1 \tag{8.34}$$

By solving Equation 8.34 for $e^{-1-\alpha}$ and substituting the result into Equation 8.32, one obtains for the probabilities:

$$P_{mn} = \frac{e^{-\beta U_o(n-m)}e^{-\beta\gamma n}e^{-\beta\lambda m}}{\left[1 + e^{-\beta(U_o+\gamma)} + e^{\beta(U_o-\lambda)}\right]^N} \tag{8.35}$$

One can eliminate the first unknown constant α. To determine the other three constants β, γ, and λ, the logarithm of Equation 8.34 is taken:

$$\ln P_{mn} = -N\ln\left[1+e^{-\beta(U_o+\gamma)}+e^{\beta(U_o-\lambda)}\right]-\beta U_o(n-m)-\beta\gamma n-\beta\lambda m \qquad (8.36)$$

Multiplying Equation 8.36 by –k and taking the average value of the resulting expression:

$$\mathbb{S} = kN\ln\left[1+e^{-\beta(U_o+\gamma)}+e^{\beta(U_o-\lambda)}\right]+k\beta\langle U\rangle+k\beta\gamma\langle n\rangle+k\beta\lambda\langle m\rangle \qquad (8.37)$$

where use was made of Equations 8.25, 8.26, and 8.27 in the last three terms of Equation 8.37 and where one notes from Section 7.3 that the entropy, or average randomness, is equal to the average value of the randomness $\ln P_{mn}/\ln Q$. Here, $1/\ln Q = -k$. One next solves Equation 8.37 for the average energy $\langle U\rangle$ to be able to use Equation 8.24 to evaluate the remaining constants β, γ, and λ.

$$\langle U\rangle = \frac{\mathbb{S}}{\beta k}-\frac{N}{\beta}\ln\left[1+e^{-\beta(U_o+\gamma)}+e^{\beta(U_o-\lambda)}\right]-\gamma\langle n\rangle-\lambda\langle m\rangle \qquad (8.38)$$

The second constant β, the third γ, and fourth constant λ can be evaluated from the other pieces of information known about the system, its temperature T, and chemical potentials μ and ν. Substituting Equation 8.38 into Equation 8.24:

$$\begin{aligned}&\text{a)}\ \ T=\frac{1}{\beta k},\quad \text{b)}\ \ -\gamma-\mu,\quad \text{and}\quad \text{c)}\ \ -\lambda=\nu\quad \text{or}\\ &\text{d)}\ \ \beta=kT,\quad \text{e)}\ \ \gamma=-\mu,\quad \text{and}\quad \text{f)}\ \ \lambda=-\nu\end{aligned} \qquad (8.39)$$

The partial derivatives with respect to the entropy $\mathbb{S}$ of $\langle n\rangle$ and $\langle m\rangle$ were taken; that is, only the derivative with respect to $\mathbb{S}$ of the term containing $\mathbb{S}$ explicitly, with respect to $\langle n\rangle$ of the term containing $\langle n\rangle$ explicitly, and with respect to $\langle m\rangle$ of the term containing $\langle m\rangle$ explicitly was taken. This illustrates the use of more than one stochastic potential, the temperature and chemical potentials in this case. By substituting Equations 8.39d, 8.39e, and 8.39f into Equation 8.35, one finally obtains for the probability P_{mn} of a particular particle interacting with the applied field:

$$P_{mn} = \frac{\exp\left[-\dfrac{U_o-\mu}{kT}n\right]+\exp\left[\dfrac{U_o+\nu}{kT}m\right]}{\left[1+\exp\left(-\dfrac{U_o-\mu}{kT}\right)+\exp\left(\dfrac{U_o+\nu}{kT}\right)\right]^N} \qquad (8.40)$$

This is the desired result.

Next, the average energy and some of its properties will be investigated. To simplify the next step in the calculation, let

$$\text{a)}\quad x \equiv \exp\left[-\frac{U_o - \mu}{kT}\right] \quad \text{and} \quad \text{b)}\quad y \equiv \exp\left[\frac{U_o + \nu}{kT}\right] \tag{8.41}$$

Substituting Equation 8.41 into Equation 8.40 and substituting the resulting expression into Equation 8.25, one obtains for the average energy $\langle U \rangle$:

$$\langle U \rangle = \frac{U_o}{[1+x+y]^N} \sum_{n=1}^{n=N} \sum_{m=0}^{m=N-m} \frac{N!n}{n!m!(N-n-m)!} x^n y^m [1]^{N-n-m} - \frac{U_o}{[1+x+y]^N} \sum_{n=0}^{n=N} \sum_{m=1}^{m=N-m} \frac{N!m}{n!m!(N-n-m)!} x^n y^m [1]^{N-n-m} \tag{8.42}$$

where the first terms, n = 0 and m = 0, in Equation 8.25 are equal to zero. Therefore, one can start the appropriate terms in Equation 8.42 with n = 1 and m = 1, respectively. Note that:

$$\text{a)}\quad \frac{n}{n!} = \frac{1}{(n-1)!} \quad \text{and} \quad \text{b)}\quad \frac{m}{m!} = \frac{1}{(m-1)!} \tag{8.43}$$

One can make the following transformation of variables:

$$\begin{aligned} &\text{a)}\ k = n - 1, \text{ then at } n = 1, k = 0; \text{ at } n = N, k = N - 1, \text{ and } n = k + 1 \\ &\text{b)}\ 1 = m - 1, \text{ then at } m = 1, \ell = 0; \text{ at } m = N - 1, \ell = (N - 1) - n; m = \ell + 1 \end{aligned} \tag{8.44}$$

Also, note that because $1 \le \ell, n \le N - 1$ in the second sum of Equation 8.43. By substituting Equations 8.43 and 8.44 into Equation 8.42, one obtains for the average energy $\langle U \rangle$:

$$\langle U \rangle = \frac{NxU_o}{[1+x+y]^N} \sum_{k=0}^{k=(N-1)} \sum_{m=0}^{m=(N-1)-k} \frac{(N-1)!}{k!m![(N-1)-k-m]!} x^k y^m [1]^{(N-1)-k-m} - \frac{NyU_o}{[1+x+y]^N} \sum_{n=0}^{n=(N-1)} \sum_{l=0}^{l=(N-1)-n} \frac{(N-1)!}{n!l![(N-1)-n-l]!} x^n y^l [1]^{(N-1)-n-l} \tag{8.45}$$

The two sums in Equation 8.45 are equal to a trinomial to the (N – 1) power. Thus, Equation 8.45 reduces to:

$$\langle U \rangle = -NU_o \frac{\exp\left[\frac{U_o + \nu}{kT}\right] - \exp\left[-\frac{U_o - \mu}{kT}\right]}{1 + \exp\left[\frac{U_o + \nu}{kT}\right] + \exp\left[-\frac{U_o - \mu}{kT}\right]} \tag{8.46}$$

where use was made of Equation 8.41 in the previous result. The previous results assume that the temperature T as well as the two chemical potentials μ and ν are externally controlled.

An interesting case is when the energy U_o of each particle is equal to the magnetic moment M_o of each particle times the magnetic field H.

$$U_o = M_o H \tag{8.47}$$

The total magnetic moment M of the assembly of particles in this case is

$$M = \frac{\partial \langle U \rangle}{\partial H} \tag{8.48}$$

and the magnetic susceptibility x is

$$x = \frac{\partial M}{\partial H} \tag{8.49}$$

One can plot the magnetization M as a function of the applied magnetic field H in Figure 8.9 and the magnetic susceptibility as a function of the applied magnetic field H in Figure 8.10. Note that the assembly of particles undergoes a *phase transition* from a negative $M \leq 0$ magnetization state to a positive magnetization $M > 0$ state.

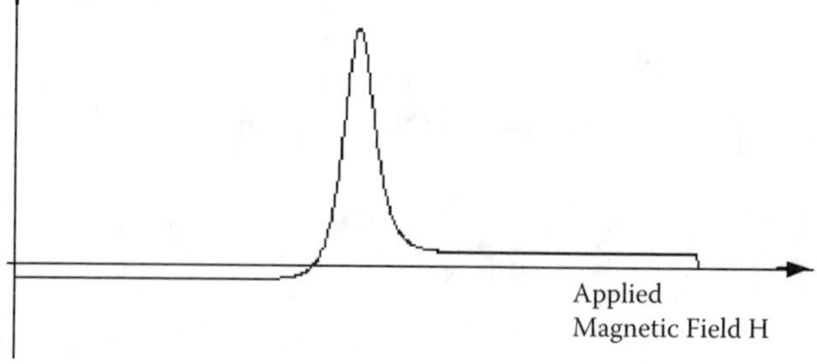

FIGURE 8.9 The magnetization M of the ensemble of particles as a function of the applied field.

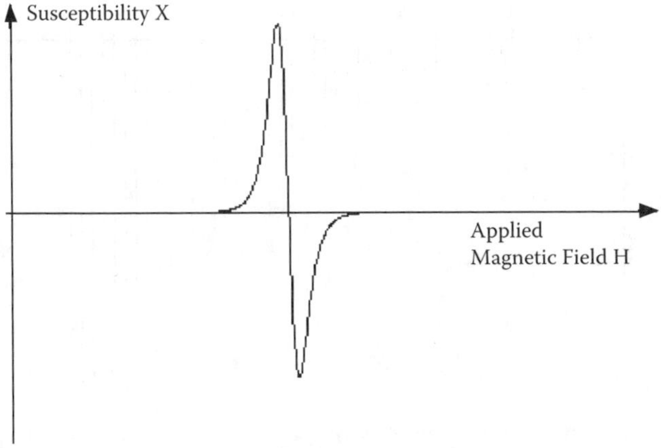

FIGURE 8.10 The magnetic susceptibility of the ensemble of particles of Figure 8.9 as a function of the applied field.

8.3 THE POISSON PROBABILITY AND ONE OF ITS APPLICATIONS

The Poisson probabilities are the probabilities of any number n of events out of a very large number N of events occurring, where the number of events with the same probability are equal to the coefficients of a binomial raised to a very large power $\left[\left(\lim/N \to \text{ very large}\right)\left(N!/n!\left(N-n!\right)\right)\right]$. The mitigating factor is that the probabilities decrease as the number of events occurring out of the very large number of possible events occurring increases $\left[\left(\lim/N \to \text{ very large}\right)P_n \approx 0\right]$. Stochastic systems in which the number of events with the same probability are equal to binomial coefficients have been discussed in Chapter 2.

For example, the probability P{10} of any 10 events out of 100 events occurring is 0.1304163, whereas the probability P{50} of any 50 out of the 100 events occurring is equal to 4.587985×10^{-23} (see Figure 8.11). The average value $\langle n \rangle$ of the number of events occurring was chosen to be equal to 10, which is much less than the total number of 100 events. Because the probabilities become negligibly small as the number of events in the arrangements increases, these probabilities can be approximated by Poisson probabilities.

The Poisson probability is best illustrated by an example. Assume that a monochromatic light source, such as a laser, can generate a maximum of N photons. Suppose one could identify each of the N individual photons. One could give each photon its own name, say photon Ab, photon Ac, photon Ad, etc., as shown in Figure 8.12.

Suppose that at some time the light source generates three photons. Photon Ac could be one of the three photons, whereas photon Ab is not among these three photons; or photon Av and photon Ap could be among the three photons. Indeed, there are N!/(3!(N – 3)!) different ways in which the N photons with individual names could be included among the three photons generated by the light source at a particular time; or more generally, there are N!/(n!(N – n)!) different ways n photons of the N photons with individual names can be included among n photons generated by the light source at any given time.

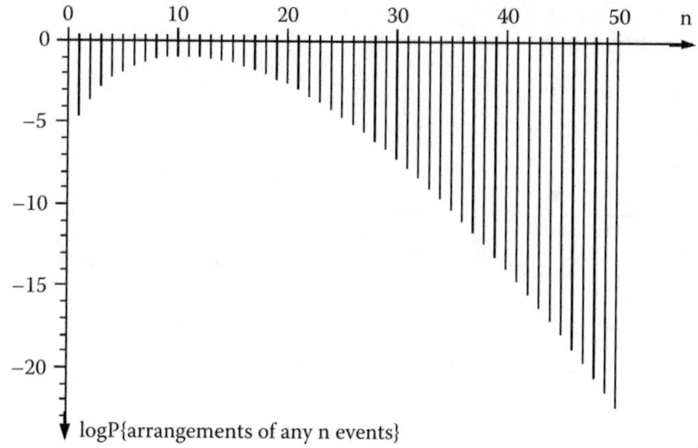

FIGURE 8.11 A plot of the logarithm of the probability P{arrangements of any n events} of any n events out of 100 events occurring. The average value $\langle n \rangle$ of n, in this case, was chosen to be equal to 10, which is much less than the 100 events. The probability P{10} of any 10 events out of 100 events occurring is 0.1304163, whereas the probability P{50} of any 50 out of the 100 events occurring is equal to 4.587985×10^{-23}.

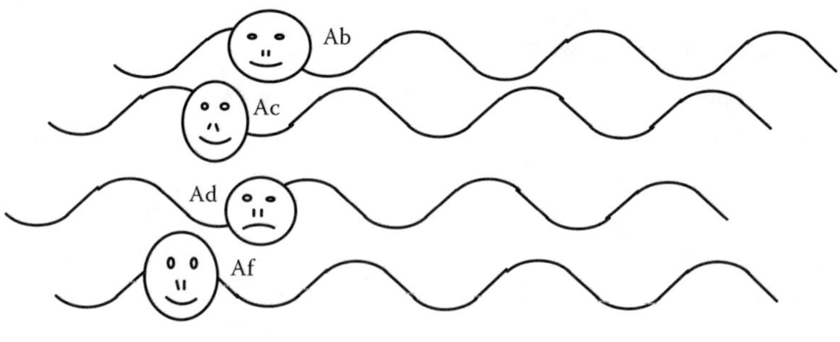

FIGURE 8.12 Individually identifiable photons Ab, Ac, Ad, Af, etc.

$$G_n = \frac{N!}{n!(N-n)!} \tag{8.50}$$

Because real photons are not individually identifiable, all arrangements of n photons have the same probability P_n of occurring. The quantity G_n is the number of arrangements of photons with the same probability. This can also be written as:

$$G_n = \frac{N(N-1)(N-2)\ldots(N+2-n)(N+1-n)}{n!} \tag{8.51}$$

One can approximate the number of arrangements of photons with the same probability G_n for very large values of the maximum number N of possible emitted photons compared to the actual number of photons n. This approximation can best be illustrated for a specific value of n. Suppose the light source emits four photons; n = 4. Then:

$$G_4 = \frac{N(N-1)(N-2)(N-3)}{4!} \tag{8.52}$$

For sufficiently large N, the number of arrangements of photons with the same probability G_4 can be approximated by:

$$G_4 \approx \frac{N^4}{4!} \tag{8.53}$$

Thus, for very large N, the numerator for all practical purposes becomes some power of N. Thus, one can approximate the numerator by a very large number, N^n.

$$G_n \approx \frac{N^n}{n!} \tag{8.54}$$

The probability P{emit any n photons} that the light source will emit any n of the N individually identifiable photons is

$$P\{\text{Emit any n photons}\} = G_n P_n \tag{8.55}$$

where P_n is the probability that the light source will emit a particular arrangement of n photons. By a particular arrangement of, say, three photons, it is meant that an arrangement consists of, for example, photon Av, photon Ab, and photon Ag. Note, of course, that all arrangements of three photons have the same probability P_n of occurring.

The average number $\langle n \rangle$ of photons emitted by the light source can be measured experimentally and is thus known. The average number $\langle n \rangle$ of photons for this case is approximately:

$$\sum_{n=0}^{n=\infty} \frac{N^n}{n!} nP_n - \langle n \rangle \approx 0 \tag{8.56}$$

where the sum was extended to infinity because it was assumed that the probabilities for large values of n become exceedingly small (see example of Figure 8.10). Similarly, the probability that all possible arrangements of photons will occur is approximately:

$$\sum_{n=0}^{n=\infty} \frac{N^n}{n!} P_n - 1 \approx 0 \tag{8.57}$$

The probability P_n associated with a particular arrangement of n photons can be calculated by finding a probability that maximizes the average randomness $\mathbb{H}$ or entropy appropriate for this case, subject to knowing the average number $\langle n \rangle$ of photons that has been experimentally measured. The average randomness for this case is

$$\mathbb{H} \approx \frac{1}{\ln Q} \sum_{n=0}^{n=\infty} \frac{N^n}{n!} P_n \ln P_n \tag{8.58}$$

To accomplish the maximizing process of the average randomness subject to what is known about the system, Equation 8.56 is multiplied by the constant $\alpha/\ln Q$ and Equation 8.57 is multiplied by the constant $\beta/\ln Q$. The resulting expressions are added to Equation 8.58. Because both Equations 8.56 and 8.57 are equal to zero, this will not change Equation 8.58.

$$\mathbb{H} = \frac{1}{\ln Q} \sum_{n=0}^{n=\infty} \frac{N^n}{n!} \left[P_n \ln P_n + \alpha P_n + \beta n P_n \right] - \frac{\alpha}{\ln Q} - \frac{\beta \langle n \rangle}{\ln Q} \tag{8.59}$$

The average randomness will be a maximum when all its partial derivatives with respect to each of the probabilities $P_1, P_2, \ldots, P_n, \ldots$, are equal to zero. Substituting Equation 8.59 into Equation 8.3:

$$\ln P_n + 1 + \alpha + \beta n = 0 \tag{8.60}$$

Solving for the probability P_n of a particular arrangement of n photons occurring:

$$P_n = e^{-1-\alpha} e^{-\beta n} \tag{8.61}$$

The constants a and b must now be evaluated. This is accomplished by successively substituting Equation 8.61 into the constraint equations, Equations 8.56 and 8.57. Substituting Equation 8.61 into Equation 8.57:

$$1 = e^{-1-\alpha} \sum_{n=0}^{n=\infty} \frac{(Nu)^n}{n!} \tag{8.62}$$

where:

$$u \equiv e^{-\beta} \tag{8.63}$$

Because the sum is the Taylor series expansion of the exponential e^{Nu}:

$$e^{-1-\alpha} = e^{-Nu} \tag{8.64}$$

By substituting Equations 8.63 and 8.64 into Equation 8.61, one obtains for the probability P_n associated with a particular arrangement of n photons:

$$P_n = e^{-Nu} u^n \tag{8.65}$$

Recall from Equation 8.63 that u is related to the second constant, which has yet to be evaluated. To evaluate the constant β or u, Equations 8.63 and 8.65 are substituted into the second constraint equation, Equation 8.56.

$$\text{a)} \quad \langle n \rangle = e^{-Nu} \sum_{n=1}^{n=\infty} \frac{n (Nu)^n}{n!} \quad \text{or} \quad \text{b)} \quad \langle n \rangle = e^{-Nu} \sum_{n=1}^{n=\infty} \frac{(Nu)^n}{(n-1)!} \tag{8.66}$$

The sum began at 1 instead of 0 because the term with n = 0 is equal to zero. To solve Equation 8.66b, the following transformation of variables is performed:

$$\text{a)} \quad m \equiv n-1 \quad \text{b)} \quad \text{at}\, n=1,\, m=0 \quad \text{and} \quad \text{c)} \quad n=m+1 \tag{8.67}$$

Substituting Equation 8.67 into Equation 8.66b:

$$\langle n \rangle = Nu\, e^{-Nu} \sum_{m=0}^{m=\infty} \frac{(Nu)^m}{m!} \tag{8.68}$$

As noted in the discussion following Equations 8.62 and 8.63, the sum in Equation 8.68, too, is equal to the exponential e^{Nu}, which cancels the exponential e^{-Nu}. Therefore, Equation 8.68 reduces to:

$$\langle n \rangle = Nu \tag{8.69}$$

Solving Equation 8.69 for u and substituting the resulting expression into the equation for the probability P_n associated with a particular arrangement of n photons, Equation 8.65:

$$P_n = e^{-\langle n \rangle} \frac{\langle n \rangle^n}{N^n} \tag{8.70}$$

We note that P_n contains the not-well-defined very large number N^n. By substituting Equations 8.54 and 8.70 into Equation 8.55, one obtains for the probability P{emit any n photons} that the light source will emit any n of the N photons:

$$P\{\text{Emit any n photons}\} = e^{-\langle n \rangle} \frac{\langle n \rangle^n}{n!} \tag{8.71}$$

Note that the not-well-defined very large number N^n was eliminated. Equation 8.71 is the *Poisson probability*. Of course, there are many other applications of the Poisson probability besides describing the number of light energy quanta, photons, emitted by a monochromatic light source such as a laser.

8.4 CONTINUOUS MOST RANDOM SYSTEMS

Next, the maximum value of the average randomness of a continuous system by analogy with the discrete system will be derived. Once again, examples will be used to illustrate the concept of a most random continuous system.

Consider a continuous random variable x, where $0 < x \le a$. It is known that any value at all of the random variable between zero and a will occur. This requires that the integral of the probability density p(x) from zero to a be equal to one.

$$\int_0^a p(x)dx - 1 = 0 \tag{8.72}$$

The *average randomness of the continuous random variable* x is

$$\mathbb{H} = \frac{1}{\ln Q}\int_0^a dxp(x)\ln[p(x)\lambda] \tag{8.73}$$

Here, Q is the probability of the stochastic system to which the random variable is being compared, and λ is a characteristic volume, area, or length. As was done in the discrete case, Equation 8.72 is multiplied by a constant, a so-called Lagrange multiplier, and adds the resulting expression to Equation 8.73.

$$\mathbb{H} = \frac{1}{\ln Q}\int_0^a dxp(x)\ln[p(x)\lambda] + \frac{\beta}{\ln Q}\int_0^a dxp(x) - \frac{\beta}{\ln Q} \tag{8.74}$$

where β is the constant or Lagrange multiplier. Equations 8.73 and 8.74 are equal because the last two terms of Equation 8.74 add to zero. To maximize the average randomness, a dummy variable α is introduced.

$$\mathbb{H}(\alpha) = \frac{1}{\ln Q}\int_0^a dx[p(x,\alpha)\ln p(x,\alpha) + p(x,\alpha)\ln\lambda + \beta p(x,\alpha)] - \frac{\beta}{\ln Q} \tag{8.75}$$

Recall that in the discrete case the maximum value of the average randomness was found by varying the discrete probabilities P_n. One could have used a dummy variable in the discrete case too.

Recall Occam's razor, that the simplest approach is the best. Thus, in the discrete case, the entropy was maximized by simply varying the probabilities. However, one could have used another method to adjust the probabilities until the maximum value of the average randomness was obtained. Assume that the discrete probabilities $P_n(\alpha)$ are not only functions of the random variable n but also functions of a dummy variable α. The average randomness $\mathbb{H}$ of Equation 8.2 subject to some constraints is

$$\mathbb{H}(\alpha)=\frac{1}{\ln Q}\big[-P_1(\alpha)\ln P_1(\alpha)-\beta P_1(\alpha)-\theta P_1(\alpha)g_1-P_2(\alpha)\ln P_2(\alpha)-$$
$$\beta P_2(\alpha)-\theta P_2(\alpha)g_2-\ldots-P_N(\alpha)\ln P_N(\alpha)-\beta P_N(\alpha)-\theta P_N(\alpha)g_N\big]+ \qquad (8.76)$$
$$\frac{\beta}{\ln Q}+\frac{\theta}{\ln Q}\langle g_n\rangle$$

where it was assumed that the average value $\langle g_n\rangle$ of some function g_n of the random variable n was known. Here, $\beta/\ln Q$ and $\theta/\ln Q$ are Lagrange multipliers. As will be shown, varying the average randomness $\mathbb{H}(\alpha)$ with respect to the dummy variable α will, indeed, vary the probabilities until the maximum value of the average randomness is obtained. The average randomness will have an extremum value when its partial derivative with respect to the dummy variable a is equal to zero.

$$\frac{\partial\mathbb{H}(\alpha)}{\partial\alpha}=0 \qquad (8.77)$$

To perform the variation, Equation 8.76 is substituted into Equation 8.77.

$$\left[\ln P_1(\alpha)+1+\beta+\theta g_1\right]\frac{\partial P_1(\alpha)}{\partial\alpha}+\left[\ln P_2(\alpha)+\beta+\theta g_2\right]\frac{\partial P_2(\alpha)}{\partial\alpha}+\ldots+$$
$$\left[\ln P_N(\alpha)+\beta+\theta g_N\right]\frac{\partial P_N(\alpha)}{\partial\alpha}=0 \qquad (8.78)$$

Because the partial derivatives $\partial P_n(\alpha)/\partial\alpha$ are, in general, not equal to zero, the terms in the square brackets must be equal to zero. Therefore:

$$\ln P_n+1+\beta+\theta g_n=0 \qquad (8.79)$$

where one dispensed with the dummy variable α. Note that Equation 8.79 is similar to Equation 8.5. Thus, an extremum value of the average randomness was obtained by assuming the probabilities P_n are not only functions of the random variable n but are also functions of a dummy variable α and varying the average randomness with respect to this dummy variable.

In the continuous case, one could say that there is a probability density for each value of the continuous random variable. To be able to vary the probability densities, it similarly was assumed that the probability density $p(x,\alpha)$ in Equation 8.75 is both a function of the random variable x and a dummy variable α. As has been demonstrated earlier the dummy variable α allows the variation of the probability densities. The average randomness will have an extremum value when its derivative with respect to α is equal to zero.

This process is known as the *method of the calculus of variation.* Substituting Equation 8.75 into Equation 8.77:

$$\int_0^a dx\left[\ln p(x,\alpha)+1+\ln\lambda+\beta\right]\frac{\partial p(x,\alpha)}{\partial\alpha}=0 \tag{8.80}$$

The fundamental lemma of the calculus of variation states that because $(\partial p(x,\alpha))/\partial\alpha$, in general, is not equal to zero, the quantity in the square brackets must be equal to zero. Indeed, this is what was done with Equations 8.78 and 8.79. Therefore:

$$\ln p(x)+1+\ln\lambda+\beta=0 \tag{8.81}$$

where the dummy variable α was discarded. Solving Equation 8.81 for the probability density p(x):

$$p(x)=\frac{1}{\lambda}e^{-1-\beta} \tag{8.82}$$

Because λ has the dimension of length, as expected, the probability density has dimensions of reciprocal length. Substituting Equation 8.60 into the constraint Equation 8.72 to evaluate the exponent containing the Lagrange multiplier constant β:

$$\text{a)}\quad 1=\frac{1}{\lambda}e^{-1-\beta}\int_0^a dx \quad\text{or}\quad \text{b)}\quad \frac{1}{\lambda}e^{-1-\beta}=\frac{1}{a} \tag{8.83}$$

Finally, substituting Equation 8.83b into Equation 8.82 to obtain the probability density:

$$p(x)=\frac{1}{a} \tag{8.84}$$

This example is similar to the first example of the discrete case. As in the first discrete example, nothing is known about the system except that it is certain that the random variable will have values between zero and a. Therefore, the system will be most random if the probability density p(x) is constant and equal to 1/a. Note that the characteristic length λ does not appear in the resulting probability density.

EXAMPLE 8.1

Another important example of the calculus of variation is the derivation of the equation of motion of a physical particle. A physical particle can be described by a function $L(\dot{x}_\mu, x_\mu)$ of all the components $\dot{x}_\mu = dx_\mu/dt$ of its velocity vector and the components of its position vector x_μ. This function is known as the Lagrangian. The Lagrangian is equal to the difference between the kinetic energy K and potential energy P.

$$L = K - P$$

The components of the velocity and position vectors are functions of time. One can define an action integral I as follows:

$$I \equiv \int_{t_a}^{t_b} L\left[\dot{x}_\mu(t), x_\mu(t)\right] dt$$

As was done with the average randomness, it is desired to find an extremum of the action integral. The average randomness of a continuous random variable was only a function of one variable p(x). This variable was varied to maximize the average randomness. Here, the time t takes the place of the random variable x. In this case, there are two types of functions $\dot{x}_\mu(t)$ and $x_\mu(t)$ that can be varied to find an extremum of the action integral. The procedure used earlier is employed. A dummy variable α is introduced. Here, the two types of functions, $\dot{x}_\mu(t,\alpha)$ and $x_\mu(t,\alpha)$, are not only functions of time but also functions of the dummy variable α. The action integral will have an extremum if its derivative with respect to the dummy variable a is equal to zero.

$$\frac{d}{d\alpha} I(\alpha) 0 \quad \text{where}\, I(\alpha) \equiv \int_{t_a}^{t_b} L\left[\dot{x}_\mu(t,\alpha), x_\mu(t,\alpha)\right] dt$$

It is assumed that the variations $d\dot{x}_\mu / d\alpha$ and $dx_\mu / d\alpha$ of $\dot{x}_\mu$ and x_μ are equal to zero at the start and finish of the path of the particle at time t equal to t_a and t_b. Substituting the second equation into the first equation:

$$\int_{t_a}^{t_b} \left[\frac{\partial L}{\partial \dot{x}_\mu} \frac{d\dot{x}_\mu}{d\alpha} + \frac{\partial L}{\partial x_\mu} \frac{dx_\mu}{d\alpha} \right] dt = 0$$

It is assumed that the variation, or the derivatives with respect to the dummy variables, of $\dot{x}_\mu(t,\alpha)$ and $x_\mu(t,\alpha)$ is equal to zero at the start and finish times t_a and t_b. Unfortunately, one cannot factor out a derivative with respect to the dummy variable as was done when maximizing the average randomness. To be able to factor out a derivative with respect to the dummy variable, the first term in the square bracket is integrated by parts.

$$\int_{t_a}^{t_b} \frac{\partial L}{\partial \dot{x}_\mu} \frac{d^2 x_\mu}{dt d\alpha} dt = \left[\frac{\partial L}{\partial \dot{x}_\mu} \frac{dx_\mu}{d\alpha} \right]_{t_a}^{t_b} - \int_{t_a}^{t_b} \frac{d}{dt}\left(\frac{\partial L}{\partial \dot{x}_\mu} \right) \frac{dx_\mu}{d\alpha} dt$$

Because it was assumed that the derivatives with respect to the dummy variables are equal to zero at the start and finish times t_a and t_b, the first term on the left of the previous equation is equal to zero. Substituting this result back into the original equation of the variation of the action integral:

$$\int_{t_a}^{t_b}\left[-\frac{d}{dt}\left(\frac{\partial L}{\partial \dot{x}_\mu}\right)+\frac{\partial L}{\partial x_\mu}\right]\frac{dx_\mu}{d\alpha}dt=0$$

The fundamental lemma of the calculus of variation states that because the derivative with respect to the dummy variable is not equal to zero; in general, the term in the square bracket must be equal to zero.

$$-\frac{d}{dt}\left(\frac{\partial L}{\partial \dot{x}_\mu}\right)+\frac{\partial L}{\partial x_\mu}=0$$

This is the *Euler Lagrange equation of motion* in the μ direction of the path of a physical particle. This equation is only equal to zero if there are no forces acting on the particle that cannot be derived from a Lagrangian.

8.5 PROPERTIES OF GAUSSIAN STOCHASTIC SYSTEMS

A very important example is as follows: Consider a random variable x defined between plus and minus infinity, $-\infty < x \leq \infty$. It is, of course, known that the probability of any value at all of the random variable occurring is equal to one. In addition to this, it is known that the average value $\langle x \rangle$ of the random variable is equal to a, and the variance σ^2 is also known. These three pieces of information are expressed by the following three equations:

$$\int_{-\infty}^{\infty} dx p(x,\alpha)-1=0 \tag{8.85}$$

$$\int_{-\infty}^{\infty} xdxp(x,\alpha)-a=0 \tag{8.86}$$

and

$$\int_{-\infty}^{\infty} x^2 dxp(x,\alpha)-(\sigma^2+a^2)=0 \tag{8.87}$$

where it was assumed that the probability density $p(x,\alpha)$ is not only a function of the random variable x but also of a dummy variable α that will allow the variation of the probability density to maximize the average randomness. The average randomness is given by:

$$\mathbb{H}=\frac{1}{\ln Q}\int_{-\infty}^{\infty} p(x,\alpha)\ln\left[p(x,\alpha)\lambda\right]dx \tag{8.88}$$

where λ is the as-yet-undetermined characteristic length. Equations 8.85, 8.86, and 8.87 are multiplied by appropriate Lagrange multipliers and added to Equation 8.88.

$$\mathbb{H}=\frac{1}{\ln Q}\int_{-\infty}^{\infty}dx\left[p(x,\alpha)\ln p(x,\alpha)+p(x,\alpha)\ln\lambda+\theta p(x,\alpha)-2\beta\gamma xp(x,\alpha)+\right.$$
$$\left.\beta x^2p(x,\alpha)\right]-\frac{\theta}{\ln Q}+\frac{2\beta\gamma a}{\ln Q}-\frac{\beta}{\ln Q}\left(\sigma^2+a^2\right) \tag{8.89}$$

Substituting Equation 8.89 into Equation 8.77, to perform the variation of the probability density $p(x,\alpha)$:

$$\ln p(x)+1+\ln\lambda+\theta-2\beta\gamma x+\beta x^2=0 \tag{8.90}$$

where the dummy variable is dispensed within Equation 8.90. Solving Equation 8.90 for the probability density $p(x)$:

$$p(x)=\frac{1}{\lambda}e^{-1-\theta+\beta\gamma^2}e^{-\beta\left[x^2-2\gamma x+\gamma^2\right]} \tag{8.91}$$

where $\beta\gamma^2$ was added in the exponent of the first exponential and $\beta\gamma^2$ was subtracted in the exponent of the second exponential to complete the square in the exponent of the second exponential. One will need this to evaluate the integrals in the constraint equations.

To evaluate the constant exponential multiplied by $1/\lambda$, the probability density given by Equation 8.91 is substituted in the first constraint equation, Equation 8.85.

$$1=\frac{1}{\lambda}e^{-1-\theta+\beta\gamma^2}\int_{-\infty}^{\infty}dx\, e^{-\beta(x-\gamma)^2} \tag{8.92}$$

To evaluate this and the subsequent integrals, the following transformation of variables is made:

$$\text{a)}\ \frac{u^2}{2}\equiv\beta(x-\gamma)^2\quad \text{or: b)}\ u=(2\beta)^{\frac{1}{2}}(x-\gamma),\quad \text{c)}\ x=\frac{u}{(2\beta)^{\frac{1}{2}}}+\gamma,$$
$$\text{d)}\ dx=\frac{du}{(2\beta)^{\frac{1}{2}}},\quad \text{e)}\ x^2=\frac{u^2}{2\beta}+\sqrt{\frac{2}{\beta}}u\gamma+\gamma^2 \tag{8.93}$$

To evaluate the integral of Equation 8.92, Equations 8.93a and 8.93d are substituted into Equation 8.92.

$$1=\frac{1}{\lambda}e^{-1-\theta+\beta\gamma^2}\frac{1}{(2\beta)^{\frac{1}{2}}}\int_{-\infty}^{\infty}du\,\exp\left(-\frac{u^2}{2}\right) \tag{8.94}$$

It is not proved here, but the integral appearing in Equation 8.94 has the following value:

$$\int_{-\infty}^{\infty} du \exp\left(-\frac{u^2}{2}\right) = (2\pi)^{\frac{1}{2}} \tag{8.95}$$

One can evaluate the constant exponential multiplied by 1/λ by substituting Equation 8.95 into Equation 8.94.

$$\frac{1}{\lambda} e^{-1-\theta+\beta\gamma^2} = \sqrt{\frac{\beta}{\pi}} \tag{8.96}$$

Next, substituting Equation 8.96 into the equation for the probability density, Equation 8.91:

$$p(x) = \sqrt{\frac{\beta}{\pi}} e^{-\beta(x-\gamma)^2} \tag{8.97}$$

The remaining Lagrange multipliers, β and γ, must now be evaluated. To evaluate γ, substitute Equation 8.97 into Equation 8.86.

$$a = \sqrt{\frac{\beta}{\pi}} \int_{-\infty}^{\infty} dx\, x\, e^{-\beta(x-\gamma)^2} \tag{8.98}$$

To evaluate the preceding integral, the same transformation of variables as in Equation 8.89 is used. Substituting Equations 8.93c and 8.93d into Equation 8.98:

$$a = \frac{1}{2(\pi\beta)^{\frac{1}{2}}} \int_{-\infty}^{\infty} du\, u \exp\left(-\frac{u^2}{2}\right) + \frac{\gamma}{(2\pi)^{\frac{1}{2}}} \int_{-\infty}^{\infty} du \exp\left(-\frac{u^2}{2}\right) \tag{8.99}$$

Because u is an antisymmetric function and exp $(-u^2/2)$ is a symmetric function, the first integral is equal to zero. Equation 8.95 can be used to evaluate the second integral to obtain the second Lagrange multiplier:

$$\gamma = a \tag{8.100}$$

The Lagrange multiplier γ is substituted into the equation for the probability density, Equation 8.92.

$$p(x) = \sqrt{\frac{\beta}{\pi}} e^{-\beta(x-a)^2} \tag{8.101}$$

The last Lagrange multiplier β must now be evaluated. To accomplish this, Equation 8.101 is substituted into the last constraint equation, Equation 8.87. Use is made of the transformation of variables of Equations 8.93d and 8.93e.

$$\sigma^2 + a^2 = \frac{1}{2\beta}\frac{1}{(2\pi)^{\frac{1}{2}}}\int_{-\infty}^{\infty} u^2 \exp\left(-\frac{u^2}{2}\right)du + \frac{a}{(\beta\pi)^{\frac{1}{2}}}\int_{-\infty}^{\infty} u \exp\left(-\frac{u^2}{2}\right)du + \frac{a^2}{(2\pi)^{\frac{1}{2}}}\int_{-\infty}^{\infty} \exp\left(-\frac{u^2}{2}\right)du \tag{8.102}$$

It is not proved here, but the first integral appearing in Equation 8.102 has the following value:

$$(2\pi)^{\frac{1}{2}} = \int_{-\infty}^{\infty} u^2 \exp\left(-\frac{u^2}{2}\right)du \tag{8.103}$$

Again, because u is an antisymmetric function and exp $(-u^2/2)$ is a symmetric function, the second integral is equal to zero. Equations 8.95 and 8.103 are used to evaluate the first and second integrals. Thus, one obtains for the last Lagrange multiplier β:

$$\text{a)}\quad \sigma^2 + a^2 = \frac{1}{2\beta} + a^2 \quad \text{or} \quad \text{b)}\quad \sigma^2 = \frac{1}{2\beta} \tag{8.104}$$

Substituting Equation 8.104 into Equation 8.101 for β, one finally obtains:

$$p(x) = \frac{1}{\sqrt{2\pi\sigma^2}}\exp\left[-\frac{(x-a)^2}{2\sigma^2}\right] \tag{8.105}$$

Equation 8.105 is the *Gaussian probability density.* The probability density is named after the 19th century scientist Carl Friedrich Gauss (1755–1855). A typical plot of a Gaussian probability density is shown in Figure 8.13. A stochastic system where the average value and the variance of the random variable are known will be most random if the probability density has the form of a Gaussian. Thus, a Gaussian probability density is used for a stochastic system where the average value $\langle x \rangle$ and the variance σ^2 of the random variable are known. This is the final result.

The probability that the Gaussian random variable x has values less or equal to w, the *distribution functions F(w) of Gaussian stochastic systems*, is an often-used quantity. Therefore, some properties of the distribution function $F_1(w)$ are investigated. Observe that the distribution function has acquired a 1 as a subscript. This will be discussed later. The distribution function $F_1(w)$ usually has the following form:

$$F_1(w) = \frac{1}{\sqrt{2\pi\sigma^2}}\int_{-\infty}^{w} dx \exp\left[-\frac{(x-a)^2}{2\sigma^2}\right] \tag{8.106}$$

Carl Friedrich Gauss in 1803.

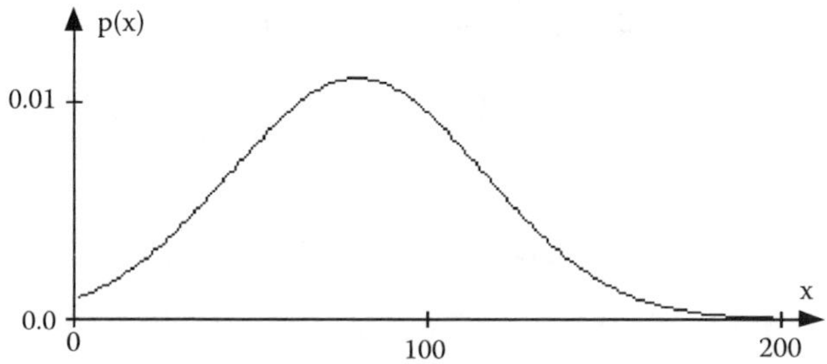

FIGURE 8.13 Plot of a section of a Gaussian probability density function having an average value a of the random variable of 80 and a standard deviation σ of 36.

where a is the average value of the Gaussian and σ^2 is its variance. No closed-form solution for this integral exists. At present, the value of this integral can readily be calculated numerically by the use of a computer. However, in the past, before computers were readily available, the values of the distribution function were tabulated. To simplify the tabulation, the following transformation of variables is made:

$$\text{a)}\quad u = \frac{x-a}{\sigma} \quad \text{b)}\quad du = \frac{1}{\sigma}dx \quad \text{c)}\quad \text{at } x = \infty \;\; u = -\infty$$
$$\text{d)}\quad \text{at } x = w \;\; u = \frac{w-a}{\sigma} \tag{8.107}$$

Substituting Equation 8.107 into Equation 8.106:

$$F_1(w) = \frac{1}{(2\pi)^{\frac{1}{2}}} \int_{-\infty}^{\frac{w-a}{\sigma}} du \exp\left[-\frac{u^2}{2}\right] \tag{8.108}$$

or

$$\text{a)}\quad F_1(\xi) \equiv \sqrt{\frac{1}{2\pi}} \int_{-\infty}^{\xi} du \exp\left[-\frac{u^2}{2}\right] \quad \text{and} \quad \text{b)}\quad \xi = \frac{w-a}{\sigma} \tag{8.109}$$

The Gaussian probability density is symmetric about u = 0, and the random variable has values between plus and minus infinity. The previous distribution function has the following properties:

1. $F(-\infty) = 0$
2. $F(\infty) = 1$
3. $F(0) = 0.5$
4. $F(-\xi) = 1 - F(\xi)$

Values of the distribution function are listed in Table 8.1.

Property 4 is illustrated in Figure 8.14. The area under the Gaussian probability density p(x) is equal to one. By symmetry of the Gaussian probability density function, one observes that $F_1(\xi) = 1 - F_1(-\xi)$. Distribution functions $F_1(-\xi)$ are listed as a function of ξ in Table 8.1.

Next, consider the problem of trying to evaluate the following integral related to a zero-mean Gaussian:

$$F_2(w) = \frac{1}{\sqrt{2\pi\sigma^2}} \int_{-\infty}^{w} dx\, x^2 \exp\left[-\frac{x^2}{2\sigma^2}\right] \tag{8.110}$$

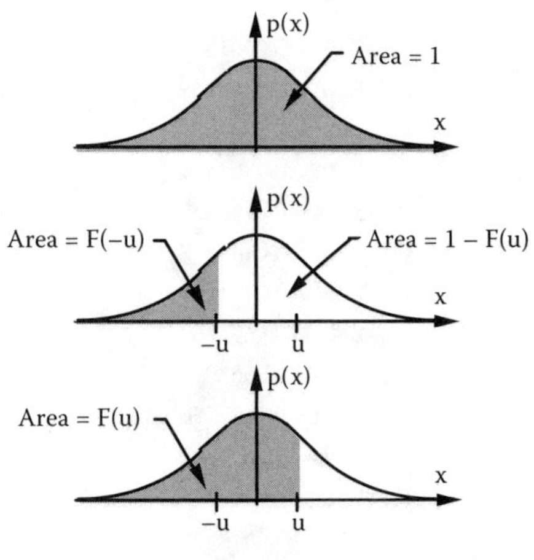

FIGURE 8.14 The area under the Gaussian probability density p(x) is equal to one. By symmetry of the Gaussian probability density function, observe that $F_1(-u) = 1 - F_1(u)$.

TABLE 8.1
Gaussian Distribution Functions

$F_1(-0.00) = 0.5$
$F_1(-0.02) = 0.492022$
$F_1(-0.04) = 0.484047$
$F_1(-0.06) = 0.476078$
$F_1(-0.08) = 0.468119$
$F_1(-0.10) = 0.460173$
$F_1(-0.12) = 0.452242$
$F_1(-0.14) = 0.44433$
$F_1(-0.16) = 0.436441$
$F_1(-0.18) = 0.428577$
$F_1(-0.20) = 0.420741$
$F_1(-0.22) = 0.412936$
$F_1(-0.24) = 0.405166$
$F_1(-0.26) = 0.397432$
$F_1(-0.28) = 0.389739$
$F_1(-0.30) = 0.382089$
$F_1(-0.32) = 0.374485$
$F_1(-0.34) = 0.366929$
$F_1(-0.36) = 0.359424$
$F_1(-0.38) = 0.351973$
$F_1(-0.40) = 0.344579$
$F_1(-0.42) = 0.337243$
$F_1(-0.44) = 0.329969$
$F_1(-0.46) = 0.322759$
$F_1(-0.48) = 0.315614$
$F_1(-0.50) = 0.308538$
$F_1(-0.52) = 0.301532$
$F_1(-0.54) = 0.294599$
$F_1(-0.56) = 0.28774$
$F_1(-0.58) = 0.280958$
$F_1(-0.60) = 0.274254$
$F_1(-0.62) = 0.267629$
$F_1(-0.64) = 0.261087$
$F_1(-0.66) = 0.254627$
$F_1(-0.68) = 0.248253$
$F_1(-0.70) = 0.241964$
$F_1(-0.72) = 0.235763$
$F_1(-0.74) = 0.229651$
$F_1(-0.76) = 0.223628$
$F_1(-0.78) = 0.217696$
$F_1(-0.80) = 0.211856$
$F_1(-0.82) = 0.206109$
$F_1(-0.84) = 0.200455$
$F_1(-0.86) = 0.194895$
$F_1(-0.88) = 0.18943$
$F_1(-0.90) = 0.184061$
$F_1(-0.92) = 0.178787$
$F_1(-0.94) = 0.173609$
$F_1(-0.96) = 0.168528$
$F_1(-0.98) = 0.163544$
$F_1(-1.00) = 0.158656$
$F_1(-1.02) = 0.153865$
$F_1(-1.04) = 0.14917$
$F_1(-1.06) = 0.144573$
$F_1(-1.08) = 0.140072$
$F_1(-1.10) = 0.135667$
$F_1(-1.12) = 0.131357$
$F_1(-1.14) = 0.127144$
$F_1(-1.16) = 0.123025$
$F_1(-1.18) = 0.119001$
$F_1(-1.20) = 0.11507$
$F_1(-1.22) = 0.111233$
$F_1(-1.24) = 0.107488$
$F_1(-1.26) = 0.103835$
$F_1(-1.28) = 0.100273$
$F_1(-1.30) = 0.0968006$
$F_1(-1.32) = 0.0934176$
$F_1(-1.34) = 0.0901227$
$F_1(-1.36) = 0.086915$
$F_1(-1.38) = 0.0837934$
$F_1(-1.40) = 0.0807567$
$F_1(-1.42) = 0.0778039$
$F_1(-1.44) = 0.0749338$
$F_1(-1.46) = 0.0721451$
$F_1(-1.48) = 0.0694367$
$F_1(-1.50) = 0.0668073$
$F_1(-1.52) = 0.0642556$
$F_1(-1.54) = 0.0617802$
$F_1(-1.56) = 0.05938$
$F_1(-1.58) = 0.0570535$
$F_1(-1.60) = 0.0547994$
$F_1(-1.62) = 0.0526162$
$F_1(-1.64) = 0.0505026$
$F_1(-1.66) = 0.0484573$
$F_1(-1.68) = 0.0464787$
$F_1(-1.70) = 0.0445655$
$F_1(-1.72) = 0.0427163$
$F_1(-1.74) = 0.0409296$
$F_1(-1.76) = 0.039204$
$F_1(-1.78) = 0.037538$
$F_1(-1.80) = 0.0359304$
$F_1(-1.82) = 0.0343796$
$F_1(-1.84) = 0.0328842$
$F_1(-1.86) = 0.0314428$
$F_1(-1.88) = 0.0300541$
$F_1(-1.90) = 0.0287166$
$F_1(-1.92) = 0.027429$
$F_1(-1.94) = 0.0261899$
$F_1(-1.96) = 0.024998$
$F_1(-1.98) = 0.0238518$
$F_1(-2.00) = 0.0227502$
$F_1(-2.02) = 0.0216918$
$F_1(-2.04) = 0.0206752$
$F_1(-2.06) = 0.0196993$
$F_1(-2.08) = 0.0187628$
$F_1(-2.10) = 0.0178645$
$F_1(-2.12) = 0.0170031$
$F_1(-2.14) = 0.0161774$
$F_1(-2.16) = 0.0153864$
$F_1(-2.18) = 0.0146288$
$F_1(-2.20) = 0.0139035$
$F_1(-2.22) = 0.0132094$
$F_1(-2.24) = 0.0125455$
$F_1(-2.26) = 0.0119107$
$F_1(-2.28) = 0.0113039$
$F_1(-2.30) = 0.0107242$
$F_1(-2.32) = 0.0101705$
$F_1(-2.34) = 9.64188E\text{-}03$
$F_1(-2.36) = 9.13748E\text{-}03$
$F_1(-2.38) = 8.65633E\text{-}03$
$F_1(-2.40) = 8.19755E\text{-}03$
$F_1(-2.42) = 7.76026E\text{-}03$
$F_1(-2.44) = 7.34364E\text{-}03$

TABLE 8.1 (continued)
Gaussian Distribution Functions

$F_1(-2.46)$ = 6.94686E-03	$F_1(-2.98)$ = 1.44125E-03	$F_1(-3.50)$ = 2.3263E-04
$F_1(-2.48)$ = 6.56913E-03	$F_1(-3.00)$ = 1.3499E-03	$F_1(-3.52)$ = 2.15774E-04
$F_1(-2.50)$ = 6.20967E-03	$F_1(-3.02)$ = 1.26388E-03	$F_1(-3.54)$ = 2.00064E-04
$F_1(-2.52)$ = 5.86775E-03	$F_1(-3.04)$ = 1.1829E-03	$F_1(-3.56)$ = 1.85428E-04
$F_1(-2.54)$ = 5.54263E-03	$F_1(-3.06)$ = 1.10669E-03	$F_1(-3.58)$ = 1.71798E-04
$F_1(-2.56)$ = 5.23362E-03	$F_1(-3.08)$ = 1.03501E-03	$F_1(-3.60)$ = 1.59109E-04
$F_1(-2.58)$ = 4.94002E-03	$F_1(-3.10)$ = 9.67605E-04	$F_1(-3.62)$ = 1.47302E-04
$F_1(-2.60)$ = 4.6612E-03	$F_1(-3.12)$ = 9.04256E-04	$F_1(-3.64)$ = 1.3632E-04
$F_1(-2.62)$ = 4.3965E-03	$F_1(-3.14)$ = 8.4474E-04	$F_1(-3.66)$ = 1.26108E-04
$F_1(-2.64)$ = 4.14531E-03	$F_1(-3.16)$ = 7.88847E-04	$F_1(-3.68)$ = 1.16618E-04
$F_1(-2.66)$ = 3.90704E-03	$F_1(-3.18)$ = 7.36376E-04	$F_1(-3.70)$ = 1.078E-04
$F_1(-2.68)$ = 3.68112E-03	$F_1(-3.20)$ = 6.87139E-04	$F_1(-3.72)$ = 9.96116E-05
$F_1(-2.70)$ = 3.46698E-03	$F_1(-3.22)$ = 6.40954E-04	$F_1(-3.74)$ = 9.20103E-05
$F_1(-2.72)$ = 3.2641E-03	$F_1(-3.24)$ = 5.9765E-04	$F_1(-3.76)$ = 8.49568E-05
$F_1(-2.74)$ = 3.07197E-03	$F_1(-3.26)$ = 5.57062E-04	$F_1(-3.78)$ = 7.84143E-05
$F_1(-2.76)$ = 2.89007E-03	$F_1(-3.28)$ = 5.19036E-04	$F_1(-3.80)$ = 7.23482E-05
$F_1(-2.78)$ = 2.71795E-03	$F_1(-3.30)$ = 4.83425E-04	$F_1(-3.82)$ = 6.6726E-05
$F_1(-2.80)$ = 2.55514E-03	$F_1(-3.32)$ = 4.50088E-04	$F_1(-3.84)$ = 6.15173E-05
$F_1(-2.82)$ = 0 2.40119E-03	$F_1(-3.34)$ = 4.18893E-04	$F_1(-3.86)$ = 5.66936E-05
$F_1(-2.84)$ = 2.25568E-03	$F_1(-3.36)$ = 3.89713E-04	$F_1(-3.88)$ = 5.22284E-05
$F_1(-2.86)$ = 2.11821E-03	$F_1(-3.38)$ = 3.6243E-04	$F_1(-3.90)$ = 4.80965E-05
$F_1(-2.88)$ = 1.98838E-03	$F_1(-3.40)$ = 3.3693E-04	$F_1(-3.92)$ = 4.42746E-05
$F_1(-2.90)$ = 1.86582E-03	$F_1(-3.42)$ = 3.13106E-04	$F_1(-3.94)$ = 4.07409E-05
$F_1(-2.92)$ = 1.75016E-03	$F_1(-3.44)$ = 2.90858E-04	$F_1(-3.96)$ = 3.7475E-05
$F_1(-2.94)$ = 1.64107E-03	$F_1(-3.46)$ = 2.70088E-04	$F_1(-3.98)$ = 3.44577E-05
$F_1(-2.96)$ = 1.5382E-03	$F_1(-3.48)$ = 2.50708E-04	$F_1(-4.00)$ = 3.16713E-05

There are no closed-form solutions for this integral. The integral of Equation 8.110 can be reformulated by making use of the transformation of variables shown in Equation 8.107 with a = 0. By making this transformation of variables, the following distribution function $F_2(w)$ is obtained:

$$F_2(w) = \frac{\sigma^2}{(2\pi)^{\frac{1}{2}}} \int_{-\infty}^{\frac{w}{\sigma}} du\, u^2 \exp\left[-\frac{u^2}{2}\right] \tag{8.111}$$

The Gaussian distribution function $F_1(w)$ had a subscript 1 and the average value $F_2(w)$ of the random variable squared has a subscript 2. Similarly define a function $F_2(\xi)$ is defined.

$$F_2(\xi) \equiv \sqrt{\frac{1}{2\pi}} \int_{-\infty}^{\xi} u^2 du \exp\left[-\frac{u^2}{2}\right] \tag{8.112}$$

A list of values of $F_2(\xi)$ for different values of x is given in Table 8.2. Again, by symmetry of the Gaussian probability density function, it is observed that $F_2(\xi) = 1 - F_2(-\xi)$. The values of the function $F_2(\xi)$ are listed in Table 8.2.

TABLE 8.2
Zero Mean Gaussian Variances

$F_2(-0.00) = 0.5$	$F_2(-0.66) = 0.466397$	$F_2(-1.32) = 0.313775$
$F_2(-0.02) = 0.499999$	$F_2(-0.68) = 0.463536$	$F_2(-1.34) = 0.307947$
$F_2(-0.04) = 0.499992$	$F_2(-0.70) = 0.460542$	$F_2(-1.36) = 0.302101$
$F_2(-0.06) = 0.499972$	$F_2(-0.72) = 0.457416$	$F_2(-1.38) = 0.296242$
$F_2(-0.08) = 0.499933$	$F_2(-0.74) = 0.454159$	$F_2(-1.40) = 0.290376$
$F_2(-0.10) = 0.499868$	$F_2(-0.76) = 0.450771$	$F_2(-1.42) = 0.284505$
$F_2(-0.12) = 0.499772$	$F_2(-0.78) = 0.447254$	$F_2(-1.44) = 0.278637$
$F_2(-0.14) = 0.499638$	$F_2(-0.80) = 0.443609$	$F_2(-1.46) = 0.272774$
$F_2(-0.16) = 0.49946$	$F_2(-0.82) = 0.439838$	$F_2(-1.48) = 0.266921$
$F_2(-0.18) = 0.499232$	$F_2(-0.84) = 0.435943$	$F_2(-1.50) = 0.261084$
$F_2(-0.2) = 0.498949$	$F_2(-0.86) = 0.431927$	$F_2(-1.52) = 0.255266$
$F_2(-0.22) = 0.498605$	$F_2(-0.88) = 0.42779$	$F_2(-1.54) = 0.249472$
$F_2(-0.24) = 0.498194$	$F_2(-0.90) = 0.423537$	$F_2(-1.56) = 0.243706$
$F_2(-0.26) = 0.49771$	$F_2(-0.92) = 0.41917$	$F_2(-1.58) = 0.237972$
$F_2(-0.28) = 0.497149$	$F_2(-0.94) = 0.414692$	$F_2(-1.60) = 0.232273$
$F_2(-0.30) = 0.496505$	$F_2(-0.96) = 0.410107$	$F_2(-1.62) = 0.226614$
$F_2(-0.32) = 0.495774$	$F_2(-0.98) = 0.4054176$	$F_2(-1.64) = 0.220999$
$F_2(-0.34) = 0.494951$	$F_2(-1.00) = 0.40062$	$F_2(-1.66) = 0.215431$
$F_2(-0.36) = 0.494032$	$F_2(-1.02) = 0.395739$	$F_2(-1.68) = 0.209913$
$F_2(-0.38) = 0.493012$	$F_2(-1.04) = 0.390759$	$F_2(-1.70) = 0.204449$
$F_2(-0.40) = 0.491887$	$F_2(-1.06) = 0.385691$	$F_2(-1.72) = 0.199042$
$F_2(-0.42) = 0.490654$	$F_2(-1.08) = 0.380537$	$F_2(-1.74) = 0.193695$
$F_2(-0.44) = 0.489308$	$F_2(-1.10) = 0.375304$	$F_2(-1.76) = 0.188411$
$F_2(-0.46) = 0.487848$	$F_2(-1.12) = 0.369995$	$F_2(-1.78) = 0.183192$
$F_2(-0.48) = 0.48627$	$F_2(-1.14) = 0.364615$	$F_2(-1.80) = 0.178041$
$F_2(-0.50) = 0.484571$	$F_2(-1.16) = 0.359168$	$F_2(-1.82) = 0.172961$
$F_2(-0.52) = 0.482748$	$F_2(-1.18) = 0.353659$	$F_2(-1.84) = 0.167953$
$F_2(-0.54) = 0.480801$	$F_2(-1.20) = 0.348093$	$F_2(-1.86) = 0.16302$
$F_2(-0.56) = 0.478726$	$F_2(-1.22) = 0.342476$	$F_2(-1.88) = 0.158164$
$F_2(-0.58) = 0.476522$	$F_2(-1.24) = 0.33681$	$F_2(-1.90) = 0.153387$
$F_2(-0.60) = 0.474188$	$F_2(-1.26) = 0.331103$	$F_2(-1.92) = 0.14869$
$F_2(-0.62) = 0.471723$	$F_2(-1.28) = 0.325358$	$F_2(-1.94) = 0.144075$
$F_2(-0.64) = 0.469127$	$F_2(-1.30) = 0.31958$	$F_2(-1.96) = 0.139543$

TABLE 8.2 (continued)
Zero Mean Gaussian Variances

$F_2(-1.98) = 0.135095$	$F_2(-2.66) = 0.0347635$	$F_2(-3.34) = 5.45679E\text{-}03$
$F_2(-2.00) = 0.130733$	$F_2(-2.68) = 0.033153$	$F_2(-3.36) = 5.12934E\text{-}03$
$F_2(-2.02) = 0.126457$	$F_2(-2.70) = 0.0316036$	$F_2(-3.38) = 4.81951E\text{-}03$
$F_2(-2.04) = 0.122268$	$F_2(-2.72) = 0.0301137$	$F_2(-3.40) = 4.52648E\text{-}03$
$F_2(-2.06) = 0.118167$	$F_2(-2.74) = 0.0286818$	$F_2(-3.42) = 4.24948E\text{-}03$
$F_2(-2.08) = 0.114154$	$F_2(-2.76) = 0.0273063$	$F_2(-3.44) = 3.98774E\text{-}03$
$F_2(-2.10) = 0.11023$	$F_2(-2.78) = 0.0259857$	$F_2(-3.46) = 3.74055E\text{-}03$
$F_2(-2.12) = 0.106396$	$F_2(-2.80) = 0.0247185$	$F_2(-3.48) = 3.5072E\text{-}03$
$F_2(-2.14) = 0.10265$	$F_2(-2.82) = 0.0235029$	$F_2(-3.50) = 3.28703E\text{-}03$
$F_2(-2.16) = 0.0989932$	$F_2(-2.84) = 0.0223377$	$F_2(-3.52) = 3.07938E\text{-}03$
$F_2(-2.18) = 0.0954259$	$F_2(-2.86) = 0.0212211$	$F_2(-3.54) = 2.88363E\text{-}03$
$F_2(-2.20) = 0.0919476$	$F_2(-2.88) = 0.0201518$	$F_2(-3.56) = 2.69919E\text{-}03$
$F_2(-2.22) = 0.0885579$	$F_2(-2.90) = 0.0191282$	$F_2(-3.58) = 2.52548E\text{-}03$
$F_2(-2.24) = 0.0852565$	$F_2(-2.92) = 0.0181489$	$F_2(-3.60) = 2.36196E\text{-}03$
$F_2(-2.26) = 0.0820429$	$F_2(-2.94) = 0.0172124$	$F_2(-3.62) = 2.2081E\text{-}03$
$F_2(-2.28) = 0.0789164$	$F_2(-2.96) = 0.0163172$	$F_2(-3.64) = 2.0634E\text{-}03$
$F_2(-2.30) = 0.0758764$	$F_2(-2.98) = 0.0154621$	$F_2(-3.66) = 1.92736E\text{-}03$
$F_2(-2.32) = 0.0729221$	$F_2(-3.00) = 0.0146455$	$F_2(-3.68) = 1.79954E\text{-}03$
$F_2(-2.34) = 0.0700527$	$F_2(-3.02) = 0.0138662$	$F_2(-3.70) = 1.67949E\text{-}03$
$F_2(-2.36) = 0.0672673$	$F_2(-3.04) = 0.0131227$	$F_2(-3.72) = 1.5668E\text{-}03$
$F_2(-2.38) = 0.0645649$	$F_2(-3.06) = 0.0124139$	$F_2(-3.74) = 1.46105E\text{-}03$
$F_2(-2.40) = 0.0619445$	$F_2(-3.08) = 0.0117383$	$F_2(-3.76) = 1.36186E\text{-}03$
$F_2(-2.42) = 0.0594048$	$F_2(-3.10) = 0.0110948$	$F_2(-3.78) = 1.26888E\text{-}03$
$F_2(-2.44) = 0.0569449$	$F_2(-3.12) = 0.0104821$	$F_2(-3.80) = 1.18175E\text{-}03$
$F_2(-2.46) = 0.0545634$	$F_2(-3.14) = 9.89905E\text{-}03$	$F_2(-3.82) = 1.10015E\text{-}03$
$F_2(-2.48) = 0.052259$	$F_2(-3.16) = 9.34448E\text{-}03$	$F_2(-3.84) = 1.02374E\text{-}03$
$F_2(-2.50) = 0.0500305$	$F_2(-3.18) = 8.81724E\text{-}03$	$F_2(-3.86) = 9.52246E\text{-}04$
$F_2(-2.52) = 0.0478765$	$F_2(-3.20) = 8.31623E\text{-}03$	$F_2(-3.88) = 8.85374E\text{-}04$
$F_2(-2.54) = 0.0457955$	$F_2(-3.22) = 7.84037E\text{-}03$	$F_2(-3.90) = 8.22854E\text{-}04$
$F_2(-2.56) = 0.0437863$	$F_2(-3.24) = 7.3886E\text{-}03$	$F_2(-3.92) = 7.64429E\text{-}04$
$F_2(-2.58) = 0.0418473$	$F_2(-3.26) = 6.95993E\text{-}03$	$F_2(-3.94) = 7.09855E\text{-}04$
$F_2(-2.60) = 0.039977$	$F_2(-3.28) = 6.55335E\text{-}3$	$F_2(-3.96) = 6.58902E\text{-}04$
$F_2(-2.62) = 0.0381739$	$F_2(-3.30) = 6.16791E\text{-}03$	$F_2(-3.98) = 6.1135E\text{-}04$
$F_2(-2.64) = 0.0364366$	$F_2(-3.32) = 5.80269E\text{-}03$	$F_2(-4.00) = 5.66993E\text{-}04$

EXAMPLE 8.2

Consider a Gaussian random variable x having an average value $\langle x \rangle = a$ of 2 and a variance σ^2 of 0.25, or standard divination σ of 0.5. The probability $P\{x \le 2.5\}$ that the random variable has values equal to or less than 2.5 needs to be calculated. One must first calculate ξ from Equation 8.107.

$$\xi = \frac{2.5 - 2}{0.5}, \quad \text{or} \quad \xi = 1$$

Next, the distribution function of $\xi = 1.0$ has to be calculated. From distribution function property No. 4, as discussed in the paragraph following Equation 8.109, one obtains:

$$F(1) = 1 - F(-1)$$

From the above table $F(-1.0) = 0.15865$; therefore,

$$F(1.0) = 1 - 0.15865, \; F(1.0) = 0.84135$$

and the result is that $P\{x \le 2.5\} = 0.84135$ in this case.

Another useful Gaussian probability density is the *joint Gaussian* probability density $p(x_1 \cap x_2 \cap x_3 \cap \ldots)$ of the random variables $x_1, x_2, x_3, \ldots$. A simple version of a joint Gaussian probability density for just two random variables x and y is illustrated below:

$$p(x \cap y) = \frac{1}{2\pi\sigma_x\sigma_y\sqrt{1-\rho^2}} \exp\left[\frac{\frac{(x-a)^2}{\sigma_x^2} - \frac{2\rho(x-a)(y-b)}{\sigma_x\sigma_y} + \frac{(y-b)^2}{\sigma_y^2}}{2(1-\rho^2)}\right] \tag{8.113}$$

where ρ has values between zero and one, $0 \le \rho < 1$.

One observes from Equation 8.113 that for ρ equal to zero, the joint Gaussian probability density $p(x \cap y)$ is equal to the product of two simple Gaussian probability densities each similar to the probability density of Equation 8.106. In this case the two random variables x and y would be statistically independent. Note that the joint Gaussian probability density is symmetric in form in its two random variables x and y. The joint Gaussian probability density of Equation 8.113 can be derived if it is assumed that the values of the two variances $\langle[x - \langle x\rangle]^2\rangle$ and $\langle[y - \langle y\rangle]^2\rangle$, the two average values $\langle x\rangle$ and $\langle y\rangle$, and the average value $\langle xy\rangle$ of the product of the random variables are known.

The various moments of the joint Gaussian probability density are investigated next. It is useful to calculate the first and second moments of the two joint Gaussian random variables.

The first moment $\langle x\rangle$ of the random variable x is:

$$\langle x\rangle = \frac{1}{2\pi\sigma_x\sigma_y\sqrt{1-\rho^2}} \int_{-\infty}^{\infty} dy \int_{-\infty}^{\infty} dx\, x \exp\left[-\frac{\frac{(x-a)^2}{\sigma_x^2} - \frac{2\rho(x-a)(y-b)}{\sigma_x\sigma_y} + \frac{(y-b)^2}{\sigma_y^2}}{2(1-\rho^2)}\right] \tag{8.114}$$

One can make the following transformation of variables:

$$\text{a)}\quad u \equiv \frac{x-a}{\sigma_x\sqrt{1-\rho^2}} \quad \text{b)}\quad x = \sigma_x\sqrt{1-\rho^2}\,u + a \quad \text{c)}\quad dx = \sigma_x\sqrt{1-\rho^2}\,du$$

$$\text{d)}\quad v \equiv \frac{y-b}{\sigma_y\sqrt{1-\rho^2}} \quad \text{e)}\quad y = \sigma_y\sqrt{1-\rho^2}\,v + b \quad \text{f)}\quad dy - \sigma_y\sqrt{1-\rho^2}\,dv \qquad (8.115)$$

By substituting Equations 8.115 into Equation 8.114, one obtains:

$$\langle x \rangle = \frac{\sqrt{1-\rho^2}}{2\pi}\int_{-\infty}^{\infty} dv \int_{-\infty}^{\infty} du\left(\sigma_x\sqrt{1-\rho^2}\,u + a\right)\exp\left[-\frac{u^2 - 2u\rho v + \rho^2 v^2}{2}\right]\times \exp\left[-\frac{v^2\left(1-\rho^2\right)}{2}\right] \qquad (8.116)$$

One can perform a second transformation of variables:

$$\text{a)}\quad w \equiv u - \rho v \quad \text{b)}\quad u = w + s\frac{\rho}{\sqrt{1-\rho^2}} \quad \text{c)}\quad du = dw$$

$$\text{d)}\quad s \equiv v\sqrt{1-\rho^2} \quad \text{e)}\quad v = \frac{s}{\sqrt{1-\rho^2}} \quad \text{f)}\quad dv = \frac{ds}{\sqrt{1-\rho^2}} \qquad (8.117)$$

By substituting Equations 8.117 into Equation 8.116, one obtains:

$$\langle x \rangle = \frac{1}{2\pi}\int_{-\infty}^{\infty} ds \int_{-\infty}^{\infty} dw\left(wo_x\sqrt{1-\rho^2} + s\rho\sigma_x + a\right)\exp\left[-\frac{w^2}{2}\right]\exp\left[-\frac{s^2}{2}\right] \qquad (8.118)$$

The first two integrals are equal to zero since w and s are antisymmetric functions and the exponentials are symmetric functions of w and s, respectively. One obtains for the first moment $\langle x \rangle$:

$$\text{a)}\quad \langle x \rangle = a \quad \text{and by symmetry} \quad \langle y \rangle = b \qquad (8.119)$$

The variance $\langle (x-a)^2 \rangle$ of the random variable x is:

$$\left\langle (x-\sigma)^2 \right\rangle = \frac{\sigma_x^2\sqrt{1-\rho^2}}{2\pi\sigma_x\sigma_y}\int_{-\infty}^{\infty} dy \int_{-\infty}^{\infty} dx \frac{(x-a)^2}{\sigma_x^2\left(1-\rho^2\right)} \times \exp\left[-\frac{\dfrac{(x-a)^2}{\sigma_x^2} - \dfrac{2\rho(x-a)(y-b)}{\sigma_x\sigma_y} + \dfrac{(y-b)^2}{\sigma_y^2}}{2\left(1-\rho^2\right)}\right] \tag{8.120}$$

By substituting Equations 8.115 into Equation 8.120, one obtains:

$$\left\langle (x-a)^2 \right\rangle = \frac{\sigma_x^2\left(1-\rho^2\right)^{\frac{3}{2}}}{2\pi}\int_{-\infty}^{\infty} dv \int_{-\infty}^{\infty} du\, u^2 \exp\left[-\frac{u^2 - 2u\rho v - \rho^2 v^2}{2}\right] \exp\left[-\frac{v^2\left(1-\rho^2\right)}{2}\right] \tag{8.121}$$

By substituting the second transformation of variables, Equation 8.117 into Equation 8.121, one obtains:

$$\left\langle (x-a)^2 \right\rangle = \frac{\sigma_x^2\left(1-\rho^2\right)}{2\pi}\int_{-\infty}^{\infty} ds \int_{-\infty}^{\infty} dw \left[w^2 + 2ws\frac{\rho}{\sqrt{1-\rho^2}} + \frac{\rho^2}{1-\rho^2}s^2\right] \exp\left[-\frac{w^2}{2}\right] \exp\left[-\frac{s^2}{2}\right] \tag{8.122}$$

By integrating and collecting terms for the variances:

$$\text{a)}\quad \left\langle (x-a)^2 \right\rangle = \sigma_x^2 \quad \text{and by symmetry} \quad \left\langle (y-b)^2 \right\rangle = \sigma_y^2 \tag{8.123}$$

The average value $\langle xy \rangle$ of the product of the random variables is:

$$\langle xy \rangle = \frac{1}{2\pi\sigma_x\sigma_y\sqrt{1-\rho^2}}\int_{-\infty}^{\infty} dy \int_{-\infty}^{\infty} dx\, xy \exp\left[-\frac{\dfrac{(x-a)^2}{\sigma_x^2} - \dfrac{2\rho(x-a)(y-b)}{\sigma_x\sigma_y} + \dfrac{(y-b)^2}{\sigma_y^2}}{2\left(1-\rho^2\right)}\right] \tag{8.124}$$

By substituting the transformation of variables, Equations 8.115 into Equation 8.124, one obtains:

$$\langle xy\rangle = \frac{\left(1-\rho^2\right)^{\frac{1}{2}}}{2\pi}\int_{-\infty}^{\infty} dv\int_{-\infty}^{\infty} du\Big[uv\sigma_x\sigma_y\left(1-\rho^2\right)+u\sigma_x\sqrt{1-\rho^2}\,b+ v\sigma_y\sqrt{1-\rho^2}\,a+ab\Big]\exp\left[-\frac{u^2-2u\rho v-\rho^2v^2}{2}\right]\exp\left[-\frac{v^2\left(1-\rho^2\right)}{2}\right] \tag{8.125}$$

By substituting the second transformation of variables, Equation 8.117 into Equation 8.125, one obtains:

$$\langle xy\rangle = \frac{1}{2\pi}\int_{-\infty}^{\infty} ds\int_{-\infty}^{\infty} dw\Big[ws\sigma_x\sigma_y\sqrt{1-\rho^2}+s^2\sigma_x\sigma_y\rho+w\sigma_x\sqrt{1-\rho^2}\,b+s\sigma_x b+ s\rho\sigma_y a+ab\Big]\exp\left[-\frac{w^2}{2}\right]\exp\left[-\frac{s^2}{2}\right] \tag{8.126}$$

By integrating and collecting terms for the variances:

$$\langle xy\rangle = ab+\rho\sigma_x\sigma_y \tag{8.127}$$

For the case when $\rho = 0$, that is, the two random variables x and y are uncorrelated, their average value $\langle xy\rangle$ reduces to the product ab of their individual average value.

8.6 IMPORTANT EXAMPLES OF STOCHASTIC PHYSICAL SYSTEMS

Consider a physical system consisting of a mass M suspended by a spring with spring constant K, a so-called Simple Harmonic Oscillator (SHO) as shown in Figure 8.15. Such a system when a quantum mechanical model of nature is used to describe its motion can only have any one of an infinite set of discrete energies $\hbar\omega\left[n+(1/2)\right]$. Here n is the number of energy quanta n = 0, 1, 2, …, ∞. The angular frequency ω with which the mass oscillates is equal to $\sqrt{K/M}$. The number of energy quanta n is a discrete random variable. The energy U_n of a quantum mechanical SHO is:

$$U_n = \hbar\omega\left[n+\frac{1}{2}\right] \tag{8.128}$$

where $\hbar = 1.054588664 \times 10^{-34}$ Joule-seconds is Planck's constant divided by 2π. This problem has applications not only to the SHO shown here but also to vibrations propagating through a crystal lattice and a quantum mechanical description of electromagnetic waves. In the electromagnetic wave case, the energy quanta are called photons.

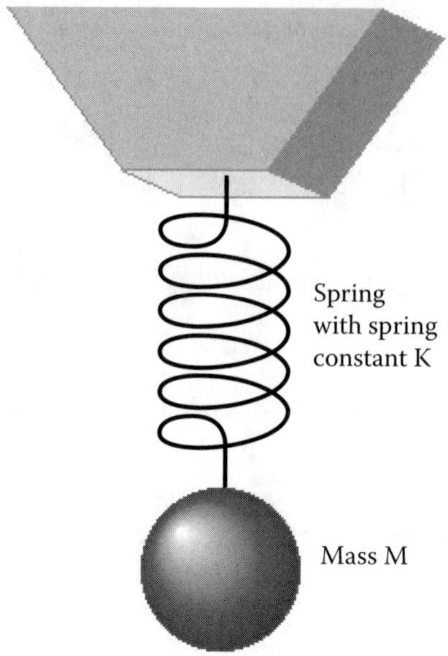

FIGURE 8.15 A simple harmonic oscillator consisting of a mass suspended by a spring.

One can deduce interesting thermal properties of the SHO and mechanisms that are analogous to it by maximizing its entropy $\mathbb{S}$ subject to what is known about it.

$$\mathbb{S} = -k\sum_{n=0}^{n=\infty} P_n \ln P_n \tag{8.129}$$

where k = 1.380662 × 10^{-23} Joules per °K is Boltzmann's constant. The sum of all probabilities P_n must be equal to one.

$$1-\sum_{n=0}^{n=\infty} P_n = 0 \tag{8.130}$$

In this case the temperature T is known. The temperature T is the change of the average energy $\langle U \rangle$ with entropy $\mathbb{S}$:

$$T \equiv \frac{\partial\langle U \rangle}{\partial \mathbb{S}} \tag{8.131}$$

When unable to use the equation for the temperature in maximizing the entropy, use the average value $\langle U \rangle$ of the energy instead. The expression for the average energy is:

$$\langle U \rangle - \sum_{n=0}^{n=\infty} n\hbar\omega P_n = 0 \tag{8.132}$$

The constant term $\frac{1}{2}\hbar\omega$ was absorbed in the average value. Equations 8.130 and 8.132 are multiplied by Lagrange multipliers and the results are added to the entropy $\mathbb{S}$ of Equation 8.129. Recall, again, from Chapter 7 that the entropy $\mathbb{S}$ of a physical system is equal to the product of Boltzmann's constant–k × ln Q and the stochastic average randomness.

$$\mathbb{S} = -k\sum_{n=0}^{n=\infty}\left[P_n \ln P_n + \lambda P_n + \beta n\hbar\omega P_n\right] + k\lambda + k\beta\langle U\rangle \tag{8.133}$$

where kλ and kβ are Lagrange multipliers. As before, by substituting Equation 8.133 into Equation 8.3:

$$\ln P_n + 1 + \lambda + \beta n\hbar\omega = 0 \tag{8.134}$$

By solving for the probability P_n:

$$P_n = e^{-1-\lambda} e^{-n\beta\hbar\omega} \tag{8.135}$$

One can substitute Equation 8.135 into the first constraint, Equation 8.130, to calculate the first Lagrange multiplier λk.

$$e^{-1-\lambda}\sum_{n=0}^{n=\infty}\left(e^{-\beta\hbar\omega}\right)^n = 1 \tag{8.136}$$

As was shown in Equation 8.6, since $e^{-\beta\hbar\omega}$ is less than one, the sum can be written as:

$$\text{a)}\quad e^{-1-\lambda}\frac{1}{1-e^{-\beta\hbar\omega}} = 1 \quad \text{or} \quad \text{b)}\quad e^{-1-\lambda} = 1 - e^{-\beta\hbar\omega} \tag{8.137}$$

The result is an equation for a function of the first Lagrange multiplier λ in terms as another function of the second Lagrange multiplier β. This allows the elimination of the first Lagrange multiplier λ. By substituting Equation 8.137b into Equation 8.135 for P_n:

$$P_n = e^{-n\beta\hbar\omega}\left(1 - e^{-\beta\hbar\omega}\right) \tag{8.138}$$

The probability P_n is only a function of the second Lagrange multiplier β at this stage. Since it is the temperature T rather than the average energy ⟨U⟩ that is known, the probability P_n of Equation 8.138 is substituted into the expression for the entropy $\mathbb{S}$, Equation 8.129.

$$\mathbb{S} = -k\ln\left(1 - e^{-\beta\hbar\omega}\right)\sum_{n=0}^{n=\infty} P_n + k\beta\sum_{n=0}^{n=\infty} n\hbar\omega P_n \tag{8.139}$$

The sum in the first term is equal to one, while the sum in the second term of Equation 8.139 is equal to the average energy ⟨U⟩.

$$\mathbb{S} = -k\ln\left(1 - e^{-\beta\hbar\omega}\right) + k\beta\left\langle U\right\rangle \tag{8.140}$$

Equation 8.140 can be solved for the average energy so that one will be able to take its derivative with respect to the entropy to obtain an expression for the physical temperature T.

$$\left\langle U\right\rangle = \frac{\mathbb{S}}{k\beta} + \frac{1}{\beta}\ln\left(1 - e^{-\beta\hbar\omega}\right) \tag{8.141}$$

The physical temperature T is equal to the partial derivative of the average energy ⟨U⟩ with respect to the entropy $\mathbb{S}$.

$$\text{a)}\quad T = \frac{1}{k\beta} \quad \text{or} \quad \text{b)}\quad \beta = \frac{1}{kT} \tag{8.142}$$

By substituting Equation 8.142b into the expression for the probability, one obtains the final result:

$$P_n = e^{-\frac{n\hbar\omega}{kT}}\left(1 - e^{-\frac{\hbar\omega}{kT}}\right) \tag{8.143}$$

In order to calculate the average energy ⟨U⟩, Equation 8.143 is used for the probability P_n in the equation for the average energy ⟨U⟩, Equation 8.132 to obtain:

$$\left\langle U\right\rangle = \left(1 - e^{-\beta\hbar w}\right)\sum_{n=0}^{n=\infty} n\hbar\omega e^{-n\beta\hbar\omega} \tag{8.144}$$

For convenience the constant β was retained in the above expression. As was done in Equation 8.19, one can use a mathematical trick to eliminate the sum in Equation 8.144.

$$\begin{aligned} &\text{a)}\quad \left\langle U\right\rangle = -\left(1 - e^{-\beta\hbar\omega}\right)\frac{d}{d\beta}\sum_{n=0}^{n=\infty}\left(e^{-\beta\hbar\omega}\right)^n \quad \text{or} \\ &\text{b)}\quad \left\langle U\right\rangle = -\left(1 - e^{-\beta\hbar\omega}\right)\frac{d}{d\beta}\left[\frac{1}{1 - e^{-\beta\hbar\omega}}\right] \end{aligned} \tag{8.145}$$

By performing the differentiation of Equation 8.145 one obtains:

$$\text{a)}\ \langle U\rangle = \frac{\hbar\omega e^{-\beta\hbar\omega}}{1-e^{-\beta\hbar\omega}} \quad \text{or} \quad \text{b)}\ \langle U\rangle = \frac{\hbar\omega}{e^{\beta\hbar\omega}-1} \tag{8.146}$$

By substituting Equation 8.142b into the expression for the average energy ⟨U⟩, Equation 8.146b becomes:

$$\langle U\rangle = \frac{\hbar\omega}{e^{\frac{\hbar\omega}{kT}}-1} \tag{8.147}$$

Equation 8.147 is the average energy of a physical system oscillating at a single frequency ω.

Since from Equation 8.128 the average energy $\langle U\rangle \approx \hbar\omega\langle n\rangle$, in this case, is proportional to the average number of energy quanta ⟨n⟩, the average number of energy quanta is:

$$\langle n\rangle = \frac{1}{e^{\frac{\hbar\omega}{kT}}-1} \tag{8.148}$$

Equation 8.148 is the *Bose-Einstein distribution function.* The result of Equation 8.148 is not a distribution function as it was defined in Chapter 3. It is the average value of energy quanta of a quantum mechanical SHO. Nevertheless, in physics jargon it is known as the Bose-Einstein distribution function. The average number of quanta ⟨n⟩ of all physical systems that behave like SHOs are given by this equation. For example, the average number of light quanta known as photons is given by Equation 8.148, where ω is the oscillating frequency of the light wave.

In order to obtain an interesting physical result for electromagnetic waves, the light emitted by a hot body will be investigated. The light propagates in all directions. An electromagnetic wave propagation in some particular direction is characterized by a wave vector **q**. The wave vector **q** is in the direction of propagation of the electromagnetic wave. The magnitude $q=\sqrt{q_x^2+q_y^2+q_z^2}$ of the wave vector is equal to $2\pi/\lambda$. Here λ is the wavelength of the electromagnetic radiation. If one considers light waves propagating in a closed room similar to the discussion in Chapter 6, it is required that there must be an integer number of half wavelength in each direction in the room.

$$\text{a)}\ q_x = \frac{\pi m_x}{L}, \quad \text{b)}\ q_y = \frac{\pi m_y}{L}, \quad \text{c)}\ q_z = \frac{\pi m_z}{L}, \tag{B.1}$$
$$\text{d)}\ q = \frac{\pi m}{L}, \quad \text{e)}\ m = \frac{Lq}{\pi}$$

Recall that each light wave quanta has an energy $\hbar\omega = \hbar cq$. Thus, all wavevectors with a magnitude q have the same energy. All wavevectors with a magnitude q terminate on the surface of a sphere in m space with radius m.

$$\text{a)}\quad 4\pi m^2 dm = \frac{4L^3}{\pi^2} q^2 dq \quad \text{or} \quad \text{b)}\quad 4\pi m^2 dm = \frac{4V}{\pi^2} q^2 dq \tag{B.2}$$

The magnitude of the wavevector q is related to the angular frequency ω

$$q = \frac{\omega}{c} \tag{B.3}$$

By substituting Equation B.3 into Equation B.2, one obtains:

$$\frac{4\pi m^2 dm}{V} = \frac{4}{\pi^2 c^3} \omega^2 d\omega \tag{B.4}$$

The change in the energy density per unit volume u can be obtained by multiplying Equation 8.147 by Equation B.4.

$$\text{a)}\quad du = \frac{4}{\pi^2 c^3} \frac{\hbar\omega^3 d\omega}{\exp\left[\frac{\hbar\omega}{kT}\right] + 1} \qquad \text{b)}\quad \frac{du}{d\omega} = \frac{4(kT)^3}{\pi^2 \hbar^2 c^3} \frac{\left(\frac{\hbar\omega}{kT}\right)^3}{\exp\left[\frac{\hbar\omega}{kT}\right] + 1} \tag{B.5}$$

This is the *black body radiation law.* The change of the energy density with frequency of electromagnetic waves radiated by a hot (black) body at a temperature T and frequency ω is given by Equation B.5b. A normalized black body radiation curve is shown in Figure 8.16.

The most famous "black body" is the universe itself. The Cosmic Background Radiation (CBR) was discovered by two engineers, Arno A. Penzias and Robert W. Wilson[3] in 1965, while measuring noise of telephone microwave antennas. They found that there was a background microwave signal that could not be eliminated no matter what they did. It has since been found that when the intensity of this signal is plotted as a function of its frequency ω, it exactly matches a black body radiation curve for a temperature of 2.725°K. It has a peak at a frequency of 159.95 GHz. The CBR had first been predicted by George Gamow in 1948 and later by Ralph Alpher and Robert Herman in 1950.

To calculate the total energy density U radiated by the black body, Equation B.5b is integrated over all frequencies.

$$U = \frac{4(kT)^4}{\pi^2 \hbar^3 c^3} \int_0^\infty \frac{x^3}{e^x - 1} dx \tag{B.6}$$

where

$$x = \frac{\hbar\omega}{kT} \tag{B.7}$$

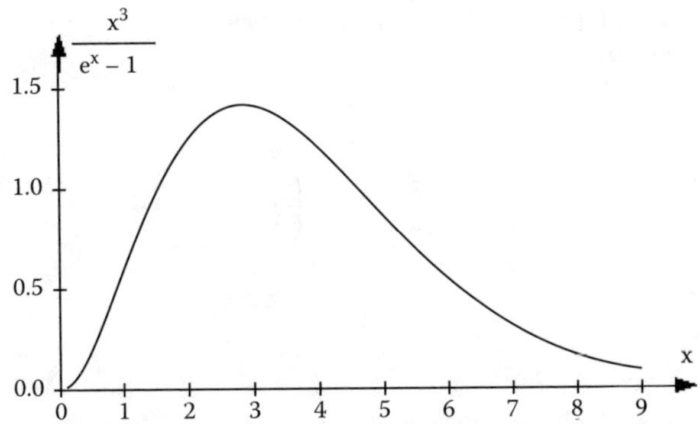

FIGURE 8.16 Normalized black body radiation curve. Here x = ħω/kT. The maximum occurs at x = 2.8169999.

Since the integral is equal to a constant, the total energy density U radiated by the black body is proportional to the temperature to the fourth power.

$$U = PT^4 \tag{B.8}$$

where P is a constant. Returning to Equation B.2, the wave vector q can be expressed in terms of the wavelength λ.

$$\text{a)}\quad q = \frac{2\pi}{\lambda} \qquad \text{b)}\quad dq = -\frac{2\pi}{\lambda^2}d\lambda \tag{B.9}$$

By substituting Equations B.9 into Equation B.2 one obtains:

$$\text{a)}\quad 4\pi m^2 dm = \frac{4V}{\pi^2}q^2 dq \qquad \text{b)}\quad \frac{4\pi m^2 dm}{V} = \frac{32\pi}{\lambda^4}d\lambda \tag{B.10}$$

One can express the average energy of Equation 8.147 in terms of the wavelength λ:

$$\langle U \rangle = \frac{\dfrac{hc}{\lambda}}{\exp\left[\dfrac{hc}{\lambda kT}\right] + 1} \tag{B.11}$$

The change in the energy density per unit volume can be obtained by multiplying Equation B.11 by Equation B.10b. By multiplying Equations B.10b and B.11, one obtains the following expression for the change of the energy density with wavelength radiated by a hot (black) body becomes:

$$\frac{du}{d\lambda} = -\frac{32\pi(kT)^5}{(hc)^4} \frac{\left(\frac{hc}{\lambda kT}\right)^5}{\exp\left[\frac{hc}{\lambda kT}\right]+1} \tag{B.12}$$

This, too, is a black body radiation law. It is the change of the energy density of electromagnetic waves with wavelength λ radiated by a hot (black) body at a temperature T and wavelength λ. A normalized black body radiation curve is shown in Figure 8.17.

Recall that one could have any number of vibration energy quanta n associated with an energy $\hbar\omega$ of a SHO. Only one particle can have a particular energy U in a system of interacting particles. Actually, this is not quite true. Each particle in a system of interacting particles must have a unique set of five quantum numbers. The quantum numbers are the values of the three components of the momentum vector, the value of the energy, and the value of the spin angular momentum.

The case where only one particle can have a particular energy U in a system of interacting particles can be illustrated by analyzing the properties of an ensemble of noninteracting systems. However, each system consists of interacting particles and can thus, only have a number of distinct energies. Since the systems do not

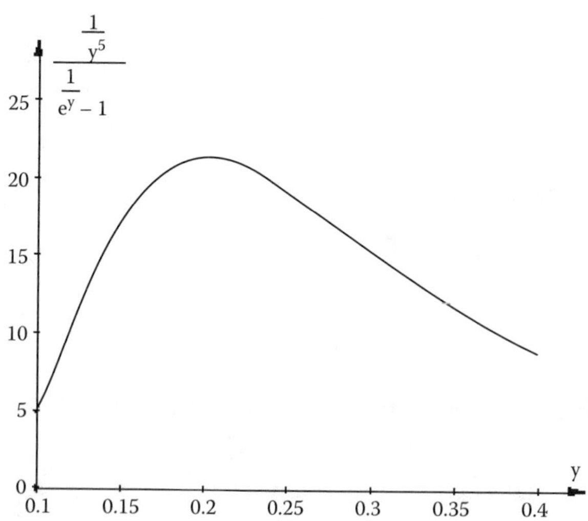

FIGURE 8.17 Normalized black body radiation curve. Here y = λkT/hc. The maximum occurs at y = 0.2014052366.

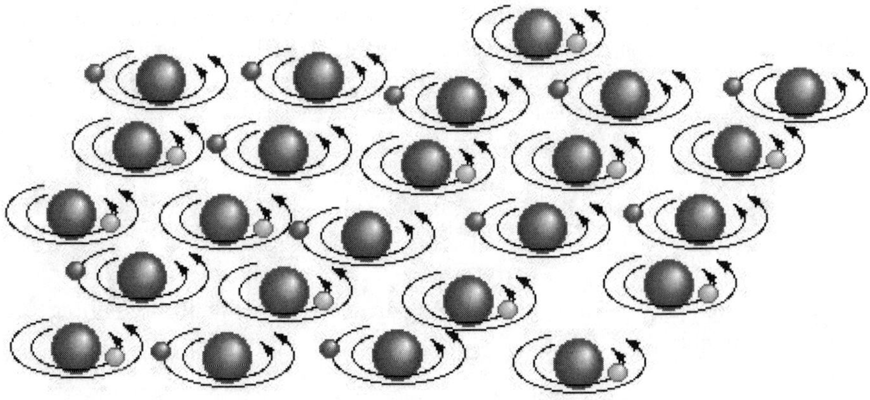

FIGURE 8.18 An ensemble of N noninteracting systems. The particles within each system interact. We assume for simplicity that each system can only have two energy levels: an energy U or zero energy. This is symbolically illustrated by systems with lightly shaded particles on the inner orbit having zero energy, while the systems with darkly shaded particles on the outer orbit have an energy U. In general the number of systems N(U) depends on the energy U.

interact, any number of systems can have the same energy. Indeed, the number of systems N(U) usually depends on the energy U. This is schematically illustrated in Figure 8.18.

Consider an ensemble of N noninteracting systems. Since the particles within each system interact, each system can only have discrete energy levels. It is assumed for simplicity that each system can only have two energy levels: an energy U or zero energy, as symbolically illustrated in Figure 8.18. Here n out of the N(U) systems has an energy U, while the remaining N – n systems have zero energy. The entropy $\mathbb{S}$ of such an assembly of systems is:

$$\mathbb{S} = -k\sum_{n=0}^{n=N} \frac{N!}{n!(N-n)!} P_n \ln P_n \tag{8.149}$$

The temperature T and the chemical potential μ are known. Recall that the temperature is the change of the average energy ⟨U⟩ with entropy $\mathbb{S}$. The chemical potential μ is a temperature-like function associated with a change of the average energy ⟨U⟩ with the average number ⟨n⟩ of systems. The temperature T and chemical potential μ are:

$$\text{a)}\quad T \equiv \frac{\partial\langle U\rangle}{\partial\mathbb{S}} \quad \text{and} \quad \text{b)}\quad \mu \equiv \frac{\partial\langle U\rangle}{\partial\langle n\rangle} \tag{8.150}$$

The sum of the probability of all arrangements of the systems must be equal to one.

$$\sum_{n=0}^{n=N} \frac{N!}{n!(N-n)!} P_n - 1 = 0 \tag{8.151}$$

Neither the temperature T nor the chemical potential can be directly used in maximizing the entropy. Therefore, the average energy ⟨U⟩ and average number ⟨n⟩ of n systems having an energy U, while the remaining systems have zero energy of the ensemble of systems, are used. The average energy ⟨U⟩ is:

$$\sum_{n=0}^{n=N} \frac{N!}{n!(N-n)!} UnP_n - \langle U \rangle = 0 \tag{8.152}$$

The average number ⟨n⟩ of n systems having an energy U while the remaining systems have zero energy is:

$$\sum_{n=0}^{n=N} \frac{N!}{n!(N-n)!} nP_n - \langle n \rangle = 0 \tag{8.153}$$

One multiplies Equations 8.151, 8.152, and 8.153 by Lagrange multipliers (and adds the results to the entropy $\mathbb{S}$ of the system, Equation 8.149.) Recall from Chapter 7 that the entropy of a physical system is equal to the product of Boltzmann's constant k and the stochastic average randomness.

$$\mathbb{S} = -k \sum_{n=0}^{n=N} \frac{N!}{n!(N-n)!} \left[P_n \ln P_n + \alpha P_n + \beta nUP_n + \beta\gamma nP_n \right] + \alpha k + \beta k \langle U \rangle + \beta\gamma k \langle n \rangle \tag{8.154}$$

where $k\alpha$, $k\beta$, and $k\beta\gamma$ are Lagrange multipliers. By substituting Equation 8.154 into Equation 8.3 one obtains:

$$\ln P_n + 1 + \alpha + \beta nU + \beta\gamma n = 0 \tag{8.155}$$

One can, next, solve for the probabilities

$$P_n = e^{-1-\alpha} e^{-\beta(U+\gamma)n} \tag{8.156}$$

One can substitute Equation 8.156 into the first constraint, Equation 8.151, in order to eliminate the first constant α.

$$e^{-1=\alpha} \sum_{n=0}^{n=N} \frac{N!}{n!(N-n)!} \left[e^{-\beta(U+\gamma)} \right]^n [1]^{N-n} \tag{8.157}$$

where the multiplicative factor $(1)^{N-n} = 1$ was added in order to make Equation 8.157 have the same form as Equation 2.17, the equation for the binomial. This factor is equal to unity. Note, from Equation 2.17 of Section 2C, that the sum on the left side of Equation 8.157 is equal to a binomial to the N-th power.

$$e^{-1-\alpha}\left[e^{-\beta(U+\gamma)}+1\right]^N=1 \tag{8.158}$$

One, next, solves Equation 8.158 for $e^{-1-\alpha}$ and substitutes the resulting expression into the expression for the probability P_n, Equation 8.156, to eliminate the first Lagrange multiplier α.

$$P_n=\frac{e^{-\beta(U+\gamma)n}}{\left[e^{-\beta(U+\gamma)}+1\right]^N} \tag{8.159}$$

The first step in determining the value of the remaining Lagrange multipliers is to take the logarithm of the probability.

$$\ln P_n=-N\ln\left[e^{-\beta(U+\gamma)}+1\right]-\beta Un-\beta\gamma n \tag{8.160}$$

As was done in Equation 8.36, one takes the average value of the terms in Equation 8.160 and multiplies the result by –k.

$$\mathbb{S}=kN\ln\left[e^{-\beta(U+\gamma)}+1\right]+\beta k\langle U\rangle+\beta\gamma k\langle n\rangle \tag{8.161}$$

where Equations 8.152 and 8.153 were used in the last two terms of Equation 8.161. One next solves Equation 8.161 for the average energy $\langle U\rangle$ in order to be able to use Equations 8.150 to evaluate the remaining constants β and γ.

$$\langle U\rangle=\frac{\mathbb{S}}{\beta k}-\frac{N}{\beta}\ln\left[e^{-\beta(U+\gamma)}+1\right]-\gamma\langle n\rangle \tag{8.162}$$

We substitute Equation 8.163 into Equations 8.150 to evaluate the constants β and γ.

$$\text{a)}\quad T=\frac{1}{k\beta}\quad\text{or}\quad\text{b)}\quad\beta=\frac{1}{kT}\quad\text{and}\quad\text{c)}\quad\mu=-\gamma \tag{8.163}$$

One finally substitutes Equations 8.163b and 8.163c into Equation 8.160 to obtain an expression of the probability P_n in terms of the temperature and chemical potential:

$$P_n=\frac{\exp\left(-\frac{U-\mu}{kT}n\right)}{\left[\exp\left(-\frac{U-\mu}{kT}\right)+1\right]^N} \tag{8.164}$$

It is interesting to calculate the average number ⟨n⟩ of systems having an energy U. By substituting Equation 8.164 into Equation 8.153 one obtains:

$$\langle n\rangle = \frac{1}{\left[\exp\left(-\frac{U-\mu}{kT}\right)+1\right]^{N}} \sum_{n=1}^{n=N} \frac{N!}{(n-1)!(N-n)!}\left[\exp\left(-\frac{U-\mu}{kT}\right)\right]^{n}[1]^{N-n} \tag{8.165}$$

One can make the following transformation of variables:

$$l = n - 1, \text{ therefore } n = l + 1, \text{ at } n = N, l = N - 1 \tag{8.166}$$

By substituting Equations 8.166 into Equation 8.165 one obtains:

$$\langle n\rangle = \frac{N\exp\left(-\frac{U-\mu}{kT}\right)}{\left[\exp\left(-\frac{U-\mu}{kT}\right)+1\right]^{N}} \sum_{l=0}^{l=N-1} \frac{(N-1)!}{l!\left[(N-1)-l\right]!}\left[\exp\left(-\frac{U-\mu}{kT}\right)\right]^{l}[1]^{(N-1)-l} \tag{8.167}$$

The sum in Equation 8.167 is equal to a binomial to the (N – 1)-th power. This cancels binomial terms to the (N – 1)-th power in the denominator.

$$\langle n\rangle = \frac{N\exp\left(-\frac{U-\mu}{kT}\right)}{\exp\left(-\frac{U-\mu}{kT}\right)+1} N(U) \tag{8.168}$$

In order to simplify Equation 8.168, the numerator and denominator of Equation 8.168 is multiplied by $\exp\left[(U-\mu)/(kT)\right]$. Note that the total number of systems N(U) is a function of the energy.

$$\langle n\rangle = \frac{1}{\exp\left[\frac{U-\mu}{kT}\right]+1} N(U) \tag{8.169}$$

Equation 8.169 is the product of the *Fermi-Dirac distribution function* and the number N(U) of systems with energy U.

$$f(U) = \frac{1}{\exp\left[\frac{U-\mu}{kT}\right]+1} \tag{8.170}$$

The Fermi-Dirac result of Equation 8.170 is not a distribution function as was defined in Chapter 3. It is the average value of the number of systems per total number of systems having an energy U and a chemical potential μ.

The average number ⟨n⟩ of particles in the assembly of systems is then

$$\langle n_T \rangle = \int_{U_1}^{U_2} f(U)N(U)du \tag{8.171}$$

By substituting Equations 8.163b and 8.163c into Equation 8.161 and rearranging the terms one obtains:

$$\mathbb{G} \equiv \langle U \rangle - T\mathbb{S} - \mu \langle n \rangle \tag{8.172}$$

where the free energy $\mathbb{G}$, the energy available to do work is:

$$\mathbb{G} \equiv -N(U)kT\ln\left[1 + \exp\frac{U-\mu}{kT}\right] \tag{8.173}$$

Here free energy $\mathbb{G}$ is the second Legendre transform of the average energy ⟨U⟩ with respect to both the temperature T and chemical potential μ.

8.7 THE LIMIT OF ZERO AND VERY LARGE TEMPERATURES

Consider a physical system that can have a set of energies $S_E\{U_1, U_2, ..., U_n, ..., U_N\}$ where $U_{n+1} > U_n$. The temperature T of the system is known. The temperature T is given by Equations 8.24a or 8.150a, where ⟨U⟩ is the average value of the energy and $\mathbb{S}$ is the entropy. The entropy is, again, given by:

$$\mathbb{S} = -k\sum_{n=0}^{n=\infty} P_n \ln P_n \tag{8.174}$$

The probability that the system will have any energy at all, of course, is:

$$\sum_{n=1}^{n=N} P_n - 1 = 0 \tag{8.175}$$

As before, the equation for the temperature, Equations 8.24a or 8.150a cannot directly be used in the maximization of the entropy. The expression for the average energy ⟨U⟩ has to be used.

$$\sum_{n=1}^{n=N} U_n P_n - \langle U \rangle = 0 \tag{8.176}$$

As was done a number of times before, Equations 8.175 and 8.176 are multiplied by Lagrange multiplier constants and the results added to the average randomness or, in this case, the entropy $\mathbb{S}$ of the system, Equation 8.174.

$$\mathbb{S} = -k\sum_{n=1}^{n=N}\left[P_n \ln P_n + \alpha P_n + \beta U_n P_n\right] + \alpha k + \beta k\langle U\rangle \tag{8.177}$$

where k is the previously discussed Boltzmann's constant. As was shown before, the entropy will have a maximum where its derivatives with respect to each probability P_n are equal to zero. By differentiating the entropy with respect to each of the probabilities, one obtains:

$$\ln P_n + 1 + \alpha + \beta U_n = 0 \tag{8.178}$$

One solves for the probability P_n:

$$P_n = e^{-\alpha-1}e^{-\beta U_n} \tag{8.179}$$

By substituting Equation 8.179 into Equation 8.175 for $e^{-\alpha-1}$:

$$e^{-\alpha-1} = \frac{1}{\sum_{n=1}^{n=N} e^{-\beta U_n}} \tag{8.180}$$

Since the values of the energies Un were not specified, one cannot evaluate the sum. By substituting Equation 8.180 into Equation 8.179, one obtains the following expression for the probabilities:

$$P_n = \frac{e^{-\beta U_n}}{\sum_{m=1}^{m=N} e^{-\beta U_m}} \tag{8.181}$$

By taking the logarithm of the probability of Equation 8.181, taking the average value of the resulting expression and multiplying it by –k,

$$\mathbb{S} = k\ln\left[\sum_{m=1}^{m=N} e^{-\beta U_m}\right] + \beta k\langle U\rangle \tag{8.182}$$

By solving Equation 8.182 for ⟨U⟩:

$$\langle U\rangle = \frac{\mathbb{S}}{\beta k} - \frac{1}{\beta}\ln\left[\sum_{m=1}^{m=N} e^{-\beta U_m}\right] \tag{8.183}$$

In order to evaluate the constant β, one substitutes Equation 8.183 into Equation 8.24a or 8.15a.

$$a)\ T = \frac{1}{\beta k} \quad \text{or} \quad b)\ \beta = \frac{1}{kT} \tag{8.184}$$

One substitutes Equation 8.184b into Equation 8.181 in order to obtain an expression for the probabilities P_n as a function of the temperature T.

$$P_n = \frac{\exp\left[-\dfrac{U_n}{kT}\right]}{\sum_{m=1}^{m=N} \exp\left[-\dfrac{U_m}{kT}\right]} \tag{8.185}$$

Since $U_{n+1} > U_n$ in the limit of very small temperatures T, the first term of the sum, $\exp\left[-U_1/kT\right]$, will, by far, be the largest term. By the same argument the ratio of all $\exp\left[-U_n/kT\right]/\exp\left[-U_1/kT\right]$ will be very small for very small temperatures, except for the case when n = 1. Thus, in the limit as the temperature goes to zero, the probability that the system has an energy U_1 is equal to one and all other probabilities are equal to zero.

$$a)\ \lim_{T\to 0} P_1 = 1 \quad \text{and} \quad b)\ \lim_{T\to 0} P_n = 0 \text{ for } n \neq 1 \tag{8.186}$$

Thus, in the limit as the temperature goes to zero, the average energy is equal to the lowest energy and the entropy is equal to zero.

$$a)\ \lim_{T\to 0} \langle U\rangle = U_1 \quad \text{and} \quad b)\ \lim_{T\to 0} \mathbb{S} = 0 \tag{8.187}$$

This result is often stated. However, like the arrow of time, it is a consequence of the system being probabilistic. Equation 8.187b is *Nernst's theorem*. For the case when there is more than one event having the lowest energy U_1, the entropy $\mathbb{S}$ in the limit as the temperature goes to zero is equal to the Boltzmann's constant times the logarithm of the number of events G_1 with energy U_1.

For very large temperatures all the exponentials $\exp\left[-U_n/kT\right]$ become approximately equal to one. Thus, the probabilities P_n in the limit as the temperature goes to infinity are all equal.

$$\lim_{T\to \infty} P_n = \frac{1}{N} \tag{8.188}$$

Therefore, the average energy $\langle U\rangle$ and the entropy $\mathbb{S}$ in the limit as the temperature goes to infinity are:

$$a)\ \lim_{T\to \infty} \langle U\rangle = \frac{1}{N}\sum_{n=1}^{n=N} U_n \quad \text{and} \quad b)\ \lim_{T\to \infty} \mathbb{S} = k \ln N \tag{8.189}$$

While most random systems here were thought, information will next be taught.

REFERENCES

1. "Gentle Force of Entropy Bridges Disciplines" by David Kestenbaum, *Science,* Vol. 279, 20, March 1998.
2. *Phys. Rev. Letters,* 12 Jan. Issue, by Peter Kaplan, Arjun Yodh, and Anthony Dinsmore.
3. A.A. Penzias and R.W. Wilson, *Astrophysics Journal,* Vol. 142, p. 537 (1965).

PROBLEMS

PROBLEM 8.1

Given:

Consider the random variable n = 1, 2, 3, …, 4N. The average value $\langle f_n \rangle$ of the function f_n is equal to ⅗.

$$f_n = \cos n\pi$$

Problem:

a) Calculate the probabilities P_n of the random variable n occurring provided the set of events enumerated by n are most random.
b) Calculate the average value $\langle g_n \rangle$ of the function $g_n = \sin(n\pi/2)$ (10 points).
c) Calculate the average randomness or entropy $\mathbb{H}$ in binary bits of this set of events for N = 1000.

PROBLEM 8.2

Given:

Consider a very large number N of balls each having a flat spot. The average number $\langle K \rangle$ of balls landing on their flat spot is known to be 10. Since the average number of balls landing on the flat spot is much smaller than the total number of balls

Ball with very small flat spot.

the probability of a large number of balls landing on their flat spot must be very small. Since there are a very large number of balls and the probability of a large number of balls landing on their flat spot is small, one can take the upper limit of the number of balls to be infinity for simplicity. In this case the number G_K with the same probability can be approximated by

$$G_K \approx \frac{N}{K!}$$

where N is a very large number.

Note:

$$e^u = \sum_{K=0}^{K=\infty} \frac{u^k}{K!}$$

Problem:

a) Calculate the probability P_K of any particular set of K balls landing on their flat spot assuming the system is most random.
b) Calculate the variance σ_{nn}^2 of numbers of balls landing on their flat spot.

Problem 8.3

Given:

The average value $\langle x^{2n} \rangle$ of the random variable x^{2n} is known.

$$\langle x^{2n} \rangle = b$$

Here $-\infty < x \leq \infty$ and n is an of the order of 10 to 100.

Problem:

a) Calculate the probability density p(x) per unit length assuming the system is most random.
b) Calculate the variance σ_{xx}^2 of the random variable x.

Problem 8.4

Given:

The energy of a particle is nu where the energy quanta u is a constant and n = 0, 1, 2, 3, … . There are N(μ)/n! particles with the same energy nu. The temperature T of the radiation is known.

$$T = \frac{\partial \langle nu \rangle}{\partial H}$$

where $\langle un \rangle$ is the average energy and $\mathbb{H}$ is the entropy. Boltzmann's constant k = $1.340661998 \times 10^{-23}$ Joules °K.

Problem:

a) Calculate the probabilities P_n that the system has an energy nu assuming the system is most random.
b) Calculate the average number $\langle n \rangle$ of energy quanta.

Problem 8.5

Given:

The average value $\langle x^8 \rangle$ of the 8-th power of the random variable x, $-\infty < x \leq \infty$, is known.

Problem:

a) Calculate the probability density p(x) assuming the system is most random.
b) Calculate the variance σ^2_{xx} of the system.

Problem 8.6

Given:

The average value $\langle x^8 \rangle$ of the 8-th power of a random variable x is known. Here $0 \leq x < \infty$.

$$\langle x^8 \rangle = b$$

Problem:

a) Calculate the probability density p(x) per unit length x assuming the system is most random.
b) Calculate the average value $\langle x \rangle$ of x.

Problem 8.7

Given:

The number G_n of events with the same probability is:

$$G_n = \frac{N!}{n!(N-n)!}$$

The temperature T is known:

$$T = \frac{\partial \langle Un \rangle}{\partial \mathbb{S}}$$

where $\langle Un \rangle$ is the average energy. Here U is a constant and n = 0, 1, 2, ..., N is a random variable.

Problem:

a) Calculate the probabilities P_n assuming the system is most random.
b) Calculate the average value $\langle Un \rangle$ of the energy Un.

PROBLEM 8.8

Given:

The kinetic energy U in the special relativistic model of mechanics has the following form:

$$U = \frac{m_o c^2}{\sqrt{1 - \frac{v^2}{c^2}}}$$

where $m_o = 1.673658468 \times 10^{-27}$ kg is the rest mass of a hydrogen molecule, c = 2.99792458×10^{8} m/sec is the speed of light in free space, and v is the velocity of a particle. The velocity v has values between zero and c. The temperature T is known.

$$T = \frac{\partial \langle U \rangle}{\partial \mathbb{H}}$$

Here $\langle U \rangle$ is the average energy, the average randomness $\mathbb{H}$ is measured in units of Boltzmann's constant k = $1.38066244 \times 10^{-23}$ Joules °K, and the temperature T has values between zero and 5×10^{13} °K.

Problem:

a) Assume the system is "most random." Derive an expression for the probability density p(v) dependent on the magnitude of the velocity v. You might not be able to evaluate all the integrals in this expression.
b) Calculate and plot the average energy $\langle U \rangle$ as a function of the temperature T for temperatures from zero to 5×10^{13} degrees K. The integrals will have to be evaluated numerically.

9 Information

9.1 INFORMATION CONCEPTS

Information can be encoded in many different ways. One popular method of encoding information is to use binary arithmetic. Binary arithmetic uses ones and zeros only. Binary arithmetic can be encoded by using a switch. An open switch would be a zero, and a closed switch would be a one. Present-day electronic computers consist of a very large number of electronic switches. Thus, it is natural for them to use binary arithmetic. It is interesting to speculate how quickly one can switch a binary bit. One form of the Heisenberg uncertainty principle discussed in Chapter 6 states that the product of the time t and the energy E has to be greater or equal to $h/2, \left(t \geq h/2E\right)$. Here, $h = 6.62617636 \times 10^{-34}$ J-s is Planck's constant; that is, the larger the energy, the faster the information can be changed. This is also known as the *Margolus–Levitin theorem.*[1] It is named after two scientists, Norman Margolus of the Massachusetts Institute of Technology and Lev Levitin of Boston University. Thus, a minimum energy of only 3.313088×10^{-22} J is required to switch a binary bit in 1 ps (1×10^{-12} s); 3.313088×10^{-22} J delivered in 1×10^{-12} s is equal to 3.313088×10^{-10} W. Information, of course, can be encoded in many other forms.

It is not necessary to have a computer to process information. Ordinary matter stores information in the thermal motion of its atoms.

As described in Chapter 7, the average randomness of Equation 7.12 can be used to describe both randomness and information.

$$\mathbb{H} = \frac{1}{\ln Q} \sum_{n=1}^{n=N} P_n \ln P_n \qquad (9.1)$$

Here, $0 < n \leq N$ and, in general, $Q < 1$ is the probability of a stochastic system to which the random variable n is being compared. For information measured in binary bits, $Q = ½$. The quantity $\mathbb{H}$ is also known as the *Shannon entropy.* In 1948 C.E. Shannon[2,3] suggested that this expression be used as a mathematical model for information.

Recall that when initially discussing the randomness in Chapter 7, various probabilities P_n were compared to systems of coins or binary bits. The randomness R_n is equal to the number of coins or binary bits that are as random as the probability P_n. Another interpretation is that the randomness R_n is equal to the number of binary bits necessary to convey some information. The average information is just the average value of the randomness R_n.

$$H = -\sum_{n=1}^{n=N} P_n \ln_2 P_n \qquad (9.2)$$

where the average randomness in this case is measured in binary bits.

An interesting example is to investigate how efficiently a language conveys information. From the preceding discussion, a language would convey the most information if it were to use, on average, all letters and symbols equally often. To obtain an estimate of how efficiently languages convey information, three paragraphs from three different types of publications *Scientific American*, a German physics book, and Shakespeare's play *A Midsummer Night's Dream* are analyzed. For example, in this analysis the 26 letters, the period, comma, semicolon, apostrophe, and space expressing the English language are counted as symbols. All numbers such as "nineteen hundred and forty-six" are spelled out. The number of occurrences of each symbol N_{symbol} is divided by the total number N_{total} of symbols in the paragraph, and this quantity is defined as the rate P_{symbol} at which the particular symbol occurs.

$$\text{a)}\quad P_{Symbol} = \frac{N_{Symbol}}{N_{Total}} \qquad \text{b)}\quad P_{Symbol\ Optimum} = \frac{1}{\text{Number of different symbols}} \tag{9.3}$$

Equation 9.2 is used to calculate the Shannon information in binary bits. Obviously, the language would be most efficient if it used each of the symbols an equal number of times. In this optimal case $P_{Symbol\ Optimum}$ would be given by Equation 9.3b.

The efficiency η of the language is obtained by dividing the Shannon information or entropy of the paragraph by the maximum Shannon information calculated from $P_{Symbol\ Optimum}$.

The paragraph from *Scientific American*, Volume 233, No. 6, December 1975, in the article entitled "The Arrow of Time" by David Layzer, pages 56 to 69, the first paragraph under the subheading "Information and Entropy" on page 60 reads:

> The processes that define the historical and the thermodynamic arrows of time generate information and entropy, respectively. As Claude E. Shannon of the Massachusetts Institute of Technology showed in nineteen hundred and forty six, information is a property of statistical description of physical systems. It is measured in bits, or binary digits; one bit is the quantity of information needed to decide between two equally likely possibilities. Information can also be regarded as a property of physical systems themselves, a measure of how highly organized they are. A fundamental theorem proved by Shannon shows that the information content of a system is the minimum number of bits needed to encode a complete statistical description of the system.

The paragraph has 753 symbols. The text contains 31 different symbols; the 26 letters, the period (.), the comma (,), the apostrophe ('), the semicolon (;), and the space ( ). No distinction between upper- and lowercase letters was made. An information value $H_1 = 4.14154$ binary bits is obtained. The average value of the conditional average randomnesses $H_2 = 2.94956$ binary bits and an efficiency η of 0.835966. H_1 is the information of Equation 9.2. The conditional average randomnesses H_2 are associated with the 31×31 conditional probabilities P(symbol X follows $|$ symbol Y) that symbol X will follow symbol Y. h_2 is the average value of the conditional average randomnesses. $h_2 = \mathbb{H}(S_{Follows} \mid \text{Symbol X})$. Here, S_{Follow} is the set of all possible symbols following another symbol, for example, in the word *THE*, "H" follows "T."

Of course, this set contains all symbols. H_2 is given by the first term on the left-hand side of Equation 7.20 of Chapter 7.

A German paragraph from *Optik und Atomphysik* by R.W. Pohl, Springer-Verlag, Berlin, 1958, page 5, is as follows:

> Die Physik ist und bleibt eine Erfahrungswissenschaft. Wie in den anderen Gebieten, haben auch in der Optik Beobachtung und Experiment den Ausgangspunkt zu liefern. Zweckmässigerweise beginnt man auch in der Optik mit einfachsten Erfahrungen des täglichen Lebens. Jeder Mensch kennt den unterschied von klarer und trüber Luft, von klarer und trüber Flüssigkeit. Trübe Luft enthält eine Unmenge winziger Schwebeteilchen, meist Qualm, Dunst oder Staub genannt. In gleicher weise werden Flüssigkeiten durch winzige Schwebeteilchen getrübt. Wir trüben z. B. klares Wasser durch eine Spur chinesischer Tusche, d .h. feinest verteilten Kohlenstaub, oder durch einige Tropfen Milch, d. h. eine Aufschwemmung von Fett und Käseteilchen von mikroskopisher Kleinheit.

The German paragraph has 756 symbols, which include the 26 conventional Latin letters a through z; the umlauts ä, ö, ü; the comma (,); the period (.); and the space ( ) for a total of 32 different symbols. No distinction between upper- and lowercase letters was made. An information $H_1 = 4.27134$ binary bits is obtained. The average value of the conditional average randomnesses $H_2 = 2.79688$ binary bits and an efficiency η of 0.854267.

The recitation by Robin at the end of the play *Midsummer Night's Dream*, Act V, Scene 1 is:

> If we shadows have offended,
> Think but this, and all is mended,
> That you have but slumb'red here
> While these visions did appear.
> And, as I am an honest Puck,
> If we have unearned luck
> Now to 'scape the serpent's tongue,
> We will make amends ere long;
> Else the Puck a liar call.
> So, good night unto you all.
> Give me your hands, if we be friends,
> And Robin shall restore amends.

The recitation by Robin at the end of the play *A Midsummer Night's Dream* from *The Complete Plays and Poems of William Shakespeare*, Riverside Press, Cambridge, 1942, page 114, has 490 symbols. The text contains 31 different symbols: the 26 letters, the period (.), the comma (,), the apostrophe ('), the semicolon (;), and the space ( ). No distinction between upper- and lowercase letters was made. An information $H_1 = 4.19785$ binary bits is obtained. The average value of the conditional average randomnesses $H_2 = 2.73065$ binary bits and an efficiency η of 0.847331.

This recitation was encrypted using a simple Enigma machine system. The German World War II encryption machine (Figure 9.1), known as the Enigma machine,[4]

FIGURE 9.1 German Enigma encryption machine. (Morton Swimmer, 1995. Used with permission.)

consisted of a keyboard and several encryption wheels. A three-wheel Enigma encryption machine, described in the following text, can generate

3,283,883,513,796,974,198,700,882,069,882,752,878,379,955,261,095,623,685,444, 055,315,226,006,433,616,627,409,666,933,182,371,154,802,769,920,000,000,000

or approximately 3.3×10^{114} different combinations for each letter. The signal from each type key went to a contact on the right side of a stationary wheel. The contacts on this stationary wheel made contact with the contacts on the left side of the first encryption wheel. The right side of this wheel also had the same number of contacts. However, the contacts on the left and right sides of the wheel were interconnected in a random pattern. The contacts on the right side of the first encryption wheel made contact with a set of contacts of the left side of the second wheel. After each letter was typed, the first encryption wheel was stepped so that each contact on the encryption wheel faced a different contact on the stationary wheel for the next letter. The second encryption wheel, too, had a set of contacts on its right side. Again, the contacts on the left and right sides of the second wheel were interconnected randomly. There were three to five such encryption wheels in each Enigma machine. After each letter was typed, each encryption wheel was stepped so that the next letter was encrypted differently. In the German machine the first wheel made a complete revolution before the second wheel was stepped once. The second wheel made a complete revolution before the next wheel was stepped. A fairly good description of the Enigma machine and its very interesting history can be found at the following Web site: www.codesandciphers.org.uk/enigma/.

Toward the end of World War II the U.S. also developed a wheel-based encryption machine called SIGABA. It had 15 wheels instead of the 3 wheels of the German machine. The wheels could be stepped at random by more than one letter, not like the speedometer-like stepping mechanism of the German machine. Usually at least one, but sometimes as many as four, rotors were stepped simultaneously.

For simplicity, here only a single encryption wheel is used. The encryption wheel has the following properties:

A	becomes	F
B	becomes	W
C	becomes	G
D	becomes	T
E	becomes	I
F	becomes	U
G	becomes	.
H	becomes	N
I	becomes	S
J	becomes	Q
K	becomes	,
L	becomes	Z
M	becomes	B
N	becomes	O
O	becomes	A
P	becomes	R
Q	becomes	C
R	becomes	V
S	becomes	D
T	becomes	Y
U	becomes	E
V	becomes	K
W	becomes	;
X	becomes	J
Y	becomes	L
Z	becomes	X
,	becomes	space

.	becomes	P
’	becomes	’
;	becomes	M
Space	becomes	H

As in the Enigma machine, this encoding wheel is stepped by one letter after each letter is encoded. The letters in, say, *THE*, are encoded as follows: From the preceding chart, T is encoded as a Y; the next letter, H, is encoded by a stepped wheel as an S, and E is encoded as a period. Thus, THE becomes YS. The encoded recitation from *A Midsummer Night’s Dream is*:

S.WXSILASBLGHBKR.KVGX UPMM;PV.SABGUXX.PCDMNBA’YCDHFEH,
J.HQFGGFK,ILUF;MQDBTDAV..DNWZIX;BFF’.ID.G BADBSMDCFDA.
XUPTTQ;PGM ’OAIWYHTUY,N’VYHBUYCPVYGLKFPUH CUTBU’
GVYUY; A,UKD YXFXY;ZZJXXHAGHBPWQSDXDBPU,RPO’
HUVULUSPFMSHL’NUF VLIKNXBLLNBUYRS HFETP.’; ’ZG FGKTOUNES,
LZE’’EI.DI,LGYLTWPETHSGLSI EBQPLCCO UY’DGIO;BSXLRFHRUW;
ZQUXB RXWZGRUWW ,DJCHY.,ZFATAQUNEOJAHBDWECMGGPE;
’UOFPV.IFVJ.RUNNEDZMB CRP’ADNIV;HSQHPZN.
SKGLDXKNGLFZOXBCLP,XKI’LY,VRMNWR.;
FUROSWRZADAT J GLQDJXSHPROWQRG;B.DYSPRHW’WECDFLTX,E

The first letter S was an I; the second symbol (the period) was encoded with a stepped wheel and was an F; the third symbol, W, was encoded with the twice-stepped wheel and was a space, etc. For the encoded recitation by Robin, an information H_1 = 4.89167 binary bits is obtained. The average value of the conditional average randomnesses H_2 = 3.5363 binary bits and an efficiency η of 0.987379. Note that the encoded text is much more efficient in the use of the symbols.

The notation of H_1 for the Shannon information, of h_2 for the conditional average randomness associated with one symbol being followed by another symbol, and of H_2 for the average value of the conditional randomnesses is the same as used by Dennis J. Stallings[5] in his analysis of the Voynich manuscript (Figure 9.2).

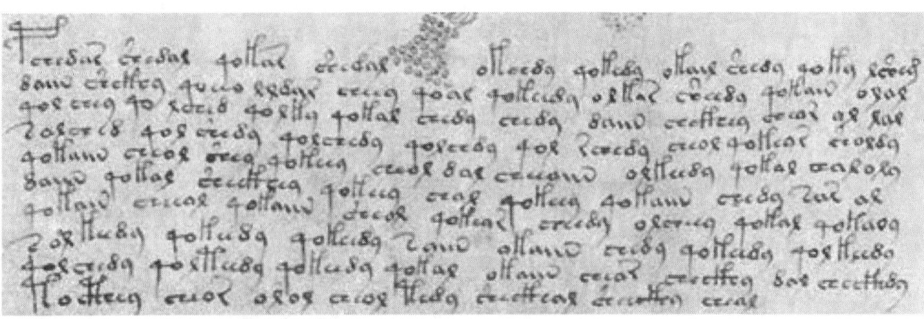

FIGURE 9.2 Typical section of Voynich text.

In 1912 a book collector named Wilfrid M. Voynich bought a 235-page-long medieval manuscript written in an unknown script and what appeared to be an unknown language or cipher. He purchased it from the Jesuit College at the Villa Mondragone Frascati near Rome, in Italy. However, despite the efforts of many well-known cryptologists and scholars, the book remains undeciphered. It is known from a letter by J. M. Marci written in 1665 or 1666 that the manuscript was bought by Emperor Rudolph II of Bohemia (1552–1612) for 600 ducats, about $983,000 in 2007 U.S. currency. The manuscript somehow passed to Jacobus de Tepenecz, the private physician and director of Rudolph's botanical gardens. His signature is present in folio 1r. It is speculated that this must have happened after 1608, when Jacobus Hocicki received his title "de Tepenecz." Since 1969 it has been at Yale University, at the Beinecke Rare Book Library with a catalog number MS 408. Thus, 1608 is the earliest definite date for the manuscript. In the 1940s and 1960s, the eminent cryptologist William F. Friedman made several valiant attempts at deciphering the Voynich manuscript, aided by groups of experts, but he also did not find any solution. William F. Friedman is famous for deciphering the Japanese "Purple" encoding machine in World War II. A copy of the text can be obtained from:

Robert Babcook
c/o Beinecke Rare Book Library
1603a Yale Station
New Haven, CT 06520

Now, almost 100 years later, the Voynich manuscript still stands as probably the most elusive puzzle in the world of cryptography. Not a single word of this most mysterious manuscript, written probably in the second half of the 15th century, can be understood.

In the courtyard of the American Central Intelligence Agency headquarters in Langley, Virginia, stands a sculpture known as Kryptos (Figure 9.3), Greek for hidden.[6] Ed Scheidt, a CIA master of encryption, created the coding system used in the sculpture, and sculptor James Sanborn created the physical sculpture. A large portion of it has been deciphered. However, another portion remains a mystery. The sculpture has the following encrypted message:

THE LEFT SIDE OF SCULPTURE:

EMUFPHZLRFAXYUSDJKZLDKRNSHGNFIVJ
YQTQUXQBQVYUVLLTREVJYQTMKYRDMFD
VFPJUDEEHZWETZYVGWHKKQETGFQJNCE
GGWHKK?DQMCPFQZDQMMIAGPFXHQRLG
TIMVMZJANQLVKQEDAGDVFRPJUNGEUNA
QZGZLECGYUXUEENJTBJLBQCRTBJDFHRR
YIZETKZEMVDUFKSJHKFWHKUWQLSZFTI
HHDDDUVH?DWKBFUFPWNTDFIYCUQZERE
EVLDKFEZMOQQJLTTUGSYQPFEUNLAVIDX

FIGURE 9.3 Kryptos sculpture in the courtyard of the Central Intelligence Agency headquarters in Langley, Virginia. (James Sanborn, 1990. Used with permission.)

FLGGTEZ?FKZBSFDQVGOGIPUFXHHDRKF
FHQNTGPUAECNUVPDJMQCLQUMUNEDFQ
ELZZVRRGKFFVOEEXBDMVPNFQXEZLGRE
DNQFMPNZGLFLPMRJQYALMGNUVPDXVKP
DQUMEBEDMHDAFMJGZNUPLGEWJLLAETG
ENDYAHROHNLSRHEOCPTEOIBIDYSHNAIA
CHTNREYULDSLLSLLNOHSNOSMRWXMNE
TPRNGATIHNRARPESLNNELEBLPIIACAE
WMTWNDITEENRAHCTENEUDRETNHAEOE
TFOLSEDTIWENHAEIOYTEYQHEENCTAYCR
EIFTBRSPAMHHEWENATAMATEGYEERLB
TEEFOASFIOTUETUAEOTOARMAEERTNRTI
BSEDDNIAAHTTMSTEWPIEROAGRIEWFEB
AECTDDHILCEIHSITEGOEAOSDDRYDLORIT
RKLMLEHAGTDHARDPNEOHMGFMFEUHE
ECDMRIPFEIMEHNLSSTTRTVDOHW?OBKR
UOXOGHULBSOLIFBBWFLRVQQPRNGKSSO
TWTQSJQSSEKZZWATJKLUDIAWINFBNYP
VTTMZFPKWGDKZXTJCDIGKUHUAUEKCAR

THE RIGHT SIDE OF SCULPTURE:

ABCDEFGHIJKLMNOPQRSTUVWXYZABCD
AKRYPTOSABCDEFGHIJLMNQUVWXZKRYP
BRYPTOSABCDEFGHIJLMNQUVWXZKRYPT
CYPTOSABCDEFGHIJLMNQUVWXZKRYPTO
DPTOSABCDEFGHIJLMNQUVWXZKRYPTOS
ETOSABCDEFGHIJLMNQUVWXZKRYPTOSA
FOSABCDEFGHIJLMNQUVWXZKRYPTOSAB
GSABCDEFGHIJLMNQUVWXZKRYPTOSABC
HABCDEFGHIJLMNQUVWXZKRYPTOSABCD
IBCDEFGHIJLMNQUVWXZKRYPTOSABCDE
JCDEFGHIJLMNQUVWXZKRYPTOSABCDEF
KDEFGHIJLMNQUVWXZKRYPTOSABCDEFG
LEFGHIJLMNQUVWXZKRYPTOSABCDEFGH
MFGHIJLMNQUVWXZKRYPTOSABCDEFGHI
NGHIJLMNQUVWXZKRYPTOSABCDEFGHIJ
OHIJLMNQUVWXZKRYPTOSABCDEFGHIJL
PIJLMNQUVWXZKRYPTOSABCDEFGHIJLM
QJLMNQUVWXZKRYPTOSABCDEFGHIJLMN
RLMNQUVWXZKRYPTOSABCDEFGHIJLMNQ
SMNQUVWXZKRYPTOSABCDEFGHIJLMNQU
TNQUVWXZKRYPTOSABCDEFGHIJLMNQUV
UQUVWXZKRYPTOSABCDEFGHIJLMNQUVW
VUVWXZKRYPTOSABCDEFGHIJLMNQUVWX
WVWXZKRYPTOSABCDEFGHIJLMNQUVWXZ
XWXZKRYPTOSABCDEFGHIJLMNQUVWXZK
YXZKRYPTOSABCDEFGHIJLMNQUVWXZKR
ZZKRYPTOSABCDEFGHIJLMNQUVWXZKRY

According to one of the people trying to break the code, there are multiple layers. The solution to some of it is another puzzle. (This is referred to as superencipherment.) In fact, the encryption system had stumped everyone who tried to decrypt it for years. However, now several people have come close to breaking the code.

For the left side of the Kryptos sculpture, an information H_1 = 4.60202 binary bits is obtained. The average value of the conditional average randomnesses is H_2 = 3.94931 binary bits and an efficiency η of 0.96785. Note that this encoded text, similar to the encoded recitation by Robin, is much more efficient in the use of the symbols. This text has only 27 symbols consisting of the 26 Latin letters and the question mark (?). It appears that the other punctuation marks such as period, comma, and most important, the space, are missing.

For the right side of the Kryptos sculpture, an information H_1 = 4.70035 binary bits is obtained. The average value of the conditional average randomnesses H_2 = 0.389225 binary bits and an efficiency η of 0.988531. Note that this encoded text, similar to the other encoded texts, is much more efficient in the use of the symbols. This text has only 26 symbols consisting of the 26 Latin letters. It appears that the other punctuation marks such as period, comma, and most important, the space, are missing.

An early information transmission system was the *Morse code system.* It was patented in 1854 by Samuel Morse. Samuel Morse was born in 1791 in Charleston, Massachusetts, and passed away in 1872 in New York City. Before he developed his code, he was a professor of sculpture and painting at New York University. The Morse code is used to this day, especially in emergency situations, because it requires very little bandwidth for transmission. A manually operated Morse code communication system requires a bandwidth of about 1 Hz.

The International Morse Code system is shown in the following chart:

```
A .-         N -.         0 -----
B -...       O ---        1 .----
C -.-.       P .--.       2 ..---
D -..        Q --.-       3 ...--
E .          R .-.        4 ....-
F ..-.       S ...        5 .....
G --.        T -          6 -....
H ....       U ..-        7 --...
I ..         V ...-       8 ---..
J .---       W .--        9 ----.
K -.-        X -..-       Fullstop (Period) .-.-.-
L .-..       Y -.--       Comma --..--
M --         Z --..       Query ..--..
```

The following additional symbols were used for encoding in Morse code the recitation by Robin at the end of the play *A Midsummer Night's Dream*, Act V, Scene 1:

```
Apostrophe -.... (the number 6)
Semicolon --... (the number 7)
Space ---.. (the number 8)
```

The recitation by Robin written in Morse code with the foregoing additional special symbols has 713 dashes and 838 dots. An information $H_1 = 0.9953096$ binary bits and an efficiency η also of 0.9953096 are obtained. Thus, not only does information transmission using Morse code require very little bandwidth, it is also very efficient.

From the foregoing limited data, it appears that English and German texts use their symbols with a remarkably high efficiency of about 85% to convey information. The space and the letter E seem to be the most commonly appearing symbols in both languages. In German, E is slightly more common than the space, whereas the opposite is true in English. The efficiency varies much less from sample to sample than the probabilities P_{symbol} of any particular symbol occurring. Note that the efficiency increased substantially after the text was encrypted. Encryption tends to randomize the symbols. Both encrypted samples demonstrate this here even though different schemes were used to encrypt each sample. This type of information is used in code breaking. Perhaps, incorporating or not incorporating certain symbols such as the space, etc., gives different efficiencies. This would tell the code breaker what symbols are included and excluded in the code. Of course, Morse code is a very efficient system for information transmission.

9.2 INFORMATION IN GENES

The genetic code is a long string of four nucleotide molecules, cytosine, guanine, adenine, and thymine, designated by the letters C, G, A, and T, respectively (Figure 9.4). Chargaff's rules state that a complete double-stranded DNA molecule from any cell of all organisms has a 1:1 ratio of pyrimidine and purine bases and, more specifically, that the amount of guanine, G, is equal to cytosine, C, and the amount of adenine, A, is equal to thymine, T. This was discovered by Austrian chemist Erwin Chargaff. This has implications for the genetic information H_1. The effect on the information of errors occurring in the string is also considered. For example, the probability that the DNA strings from two humans are completely alike is very small. Thus, it must differ from some nominal nucleotide arrangement. Indeed, the difference in the nucleotide sequence of the DNA of different humans is used for identification purposes.

Perhaps there is a minimum of information in its DNA string that an organism has to have to be alive?

In addition to calculating the information H_1 associated with the rate the four nucleotides appear in a gene information associated with the distance $X_{K,M}$ between like symbols. Here M denotes the symbol, and K denotes the distance in the number of symbols between the symbols M. The conditional rate $P(X_K|S_M)$ is the rate at which a particular space or distance $X_{K,M}$ between symbols of type M in a string occurs. $P(X_{K,M}|S_M)$ is the ratio of the number $N_{K,M}$ of types of the particular distances

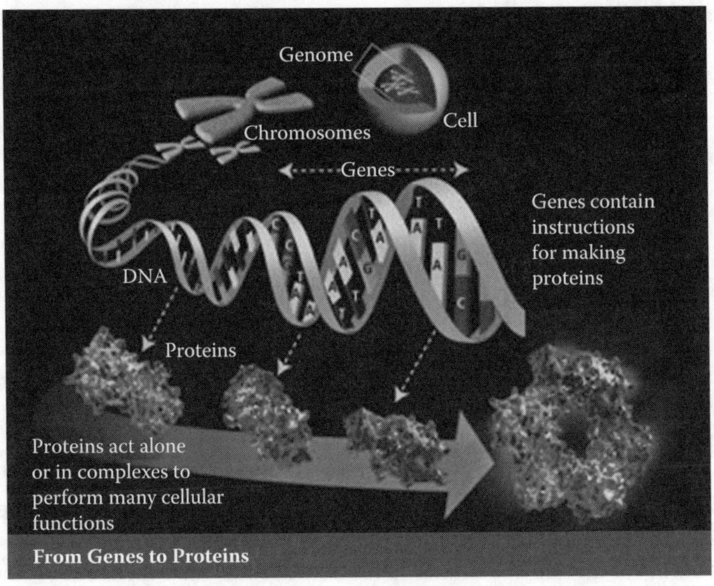

FIGURE 9.4 Cells, chromosomes, genes, and DNA.

$X_{K,M}$ between letters of type M occurs divided by the total number $N_{TD,M}$ of different distances between symbols of type M in the string.

$$P\left(X_{K,M}\middle|S_M\right)=\frac{N_{K,M}}{N_{TD,M}} \tag{9.4}$$

For example, in the word "BETWEEN," the distances between the E's are $X_{0E} = 0$ spaces, $X_{1E} = 1$ space, and $X_{3E} = 3$ spaces. All other spaces, in this case, are equal to zero. In this case, the number of each of these spaces is $N_{0E} = 1$, $N_{1E} = 1$ and $N_{3E} = 1$ for the symbol E. There are $N_{TD,E} = 5$ total number of spaces for symbol E. Thus, for the word "BETWEEN" $P(0|S_E) = 0.6$, $P(1|S_E) = 0.2$, and $P(3|S_E) = 0.2$. All other conditional distance rates are equal to 0. A similar argument as the one leading to Equation 7.21 can be used to formulate a conditional Shannon information $h_D(G_{X,M}|S_E)$ in binary bits, where $G_{x,m}$ is the set of all distances between symbols of type M.

$$h_D\left(G_{X,M}\middle|S_M\right)\equiv\frac{1}{\ln\frac{1}{2}}\sum_{K=1}^{K=N_{TD,M}} N_{K,M}P\left(X_{K,M}\middle|S_M\right)\ln P\left(X_{K,M}\middle|S_M\right) \tag{9.5}$$

where $G_{X,M}$ is the set of all distances between type M symbols. A weighting factor equal to the number of equal distances $N_{K,M}$ is included to make the conditional information dependent on the size of the sequence. For the word BETWEEN $h_D(G_{X,M}|S_E) = 1.58496$, and assuming no errors, all other conditional informations are equal to zero.

The average information H_D associated with distances between symbols is

$$H_D=\sum_{M=1}^{M=N_S}\sum_{J=1}^{J=N_S} h_D\left(G_{X,M}\middle|S_M\right)P\left(S_M\middle|S_J\right)P_J \tag{9.6}$$

For the word BETWEEN, $H_D = 0.67927$ binary bits assuming there are no errors. The net information I_D associated with the distances between like symbols is

$$I_D = H_1 + H_2 - H_D \tag{9.7}$$

A string that has a completely periodic arrangement of these letters does not convey much information; that is, a string where the distance between all X's is the same conveys little information.

For example, in the often-studied ϕX174 virus, the DNA molecule, which has only a single circular strand for part of its life cycle, consists of approximately 5375 nucleotides[3]. The nucleotides are grouped into nine known genes. These are responsible, in turn, for coding the amino acid sequences of nine different proteins. For example, gene B (shown in the following text) codes for a small protein that is part of the virus.

ATGGAACAACTCACTAAAAACCAAGCTGTCGCTACTTCCCAAGAAGCTGTTCAGA
ATCAGAATGAGCCGCAACTTCGGGATGAAAATGCTCACAATGACAAATCTGTCCA
CGGAGTGCTTAATCCAACTTACCAAGCTGGGTTACGGACGCGACGCCGTTCAACC
AGATATATTGAAGCAGAACGCAAAAAGAGAGATGAGATTGAGGCTGGGAAAAGTT
ACTGTAGCCGACGTTTTGGCGGCGCAACCTGTGACGACAAATCTGCTCAAATTTA
TGCGCCGCTTCGATAAAAATGATTGGCGTATCCAACCTGCAGAGTTTTATCGCTT
CCAATGACGCAGAAGTTAACACTTTCGGATATTTCTGA

Number of types of symbols	4	
Number of symbols	368	
Use of symbol information H_1	1.98442	binary bits
Efficiency η of use of symbols	0.992207	
Information H_2	1.94984	binary bits
Information of symbol distance H_D	35.1999	binary bits
Total information I_D	37.1843	binary bits

Efficiencies η of the use of symbols for text in English, Shakespearean English, German, Hebrew, and even the Voynich text, are typically on the order of 0.85 to 0.86. Encrypted texts tend to use symbols more efficiently than regular text. For example, the efficiency of the use of symbols in the encrypted text on James Sanborn's Kryptos sculpture in the courtyard of the CIA headquarters in Langley, Virginia, is equal to 0.96785. This is true for most encryption methods. The values for the use of symbol information H_1 and the average information H_2 associated with one type of symbol being followed by another type of symbol are also typical values for text. The informations H_1 and H_2 are independent of the text length for sufficiently long texts.

9.3 INFORMATION TRANSMISSION OF DISCRETE SYSTEMS

Recall from Chapter 3 that a transmission system can be characterized by conditional probabilities P(Received Signal|Transmitted signal) that a certain signal is received provided a specific signal was transmitted. Besides conveying less-than-perfect information, a transmission system introduces noise into the system.

To make a mathematical model of the information, two sets $S_A\{A_1, A_2, ..., A_n, ..., A_N\}$ and $S_B\{B_1, B_2, ..., B_m, ..., B_N\}$ of random events are considered. Event B_m is the transmission of information m. The arrival of message n at the other end of the transmission system is event A_n. The transmission system is characterized by conditional probabilities $P(A_n \text{ Received}|B_m \text{ Transmitted})$ that message n has been received, or that event A_n has occurred, provided message m has been transmitted or event B_m has occurred. The noise $\mathbb{N}$ introduced into the system is associated with these conditional probabilities. The noise $\mathbb{N}$ is given by the average value of the conditional average randomness $\mathbb{H}(A|B_m)$ of Equation 7.21.

$$\mathbb{N} = \sum_{m=1}^{m=N} \mathbb{H}\left(A\middle|B_m\right) Q_m \tag{9.8}$$

where Q_m is the probability that message m has been transmitted. Recall from the previous examples that the probability Q_m can be interpreted as the number of times the message m has been transmitted. For the noise $\mathbb{N}$ to be measured in binary bits, $Q = ½$.

In the examples of the various English paragraphs in Section 9.1, the probability P_{symbol} of a particular symbol occurring was determined by dividing the number of times the symbol occurred in the paragraph by the total number of symbols in the paragraph. Thus, this probability is equal to the rate at which the symbol occurred in the paragraph. The Shannon formulation of the transmitted information $\mathbb{H}_B$ is given by an expression similar to Equation 9.1:

$$\mathbb{H}_B = \frac{1}{\ln Q}\sum_{m=1}^{m-N} Q_m \ln Q_m \tag{9.9}$$

The information $\mathbb{I}$ received over the transmission system is the difference between the transmitted information $\mathbb{H}_B$ and the noise $\mathbb{N}$.

$$\mathbb{I} = \mathbb{H}_B - \mathbb{N} \tag{9.10}$$

As all the parameters in Equation 9.6 are real and positive, the received information $\mathbb{I}$ is less than the transmitted information $\mathbb{H}_B$.

Consider several examples: In one case the conditional probabilities $P(A_n|B_n)$ that the correct information has been received are large, whereas the conditional probabilities P(m Received|m Transmitted) that the incorrect information has been received are small.

$$\begin{array}{lll} P(A_1|B_1) = 0.998 & P(A_1|B_2) = 0.001 & P(A_1|B_3) = 0.01 \\ P(A_2|B_1) = 0.001 & P(A_2|B_2) = 0.998 & P(A_2|B_3) - 0.01 \\ P(A_3|B_1) = 0.001 & P(A_3|B_2) = 0.001 & P(A_3|B_3) = 0.98 \end{array}$$

The channels that transmit messages B_1 and B_2 are better than the third channel, which is more random. It is assumed that any of the three pieces of information are transmitted with equal probability; $Q_m = ⅓$. The noise $\mathbb{N}$ can be calculated by using Equation 9.4:

$$\mathbb{N} = \frac{1}{\ln\frac{1}{2}}\left\{\begin{array}{l} 2\times[0.998\ln 0.998 + 2\times 0.001\ln 0.001]\frac{1}{3} \\ +[2\times 0.01\ln 0.01 + 0.98\ln 0.98]\frac{1}{3} \end{array}\right\}$$

$$\mathbb{N} = 0.069022892 \text{ binary bits}$$

The transmitted information $\mathbb{H}_B$ from Equation 9.5 is

$$\mathbb{H}_B = \frac{1}{\ln\frac{1}{2}} \times 3 \times \left[\frac{1}{3}\ln\frac{1}{3}\right]$$

$$\mathbb{H}_B = 1.584962501 \text{ binary bits}$$

The received information $\mathbb{I}$ according to Equation 9.6 is

$$\mathbb{I} = 1.515939609 \text{ binary bits}$$

The probabilities of receiving the various pieces of information are

$$P_{1\ \text{Rec'd}} = 0.336333$$
$$P_{2\ \text{Rec'd}} = 0.336333$$
$$P_{3\ \text{Rec'd}} = 0.327333$$

Note that the better channels, channels 1 and 2, each transmit slightly more information than the third channel, which is more random.

An example where the maximum values do not occur along the diagonal of the conditional probability matrix is a binary counter. This is discussed in more detail in the March–April 2001 issue of *American Scientist*.[7] Consider a binary counter capable of counting up to 4. The probabilities that the counter was initially in various states are:

Q_{00} is the probability that the counter was initially in state 00
Q_{01} is the probability that the counter was initially in state 01
Q_{10} is the probability that the counter was initially in state 10
Q_{11} is the probability that the counter was initially in state 11

For simplicity, let the conditional probabilities P(XY|UV) that the counter will make a transition to state XY provided it was in state UV be:

$$\begin{array}{llll}
P(00|00) = 1 - 2q - p^2 & P(00|01) = p & P(00|10) = q & P(00|11) = q^2 \\
P(01|00) = q & P(01|01) = 1 - q - q^2 - p & P(01|10) = p^2 & P(01|11) = q \\
P(10|00) = q & P(10|01) = q^2 & P(10|10) = 1 - 2q - p^2 & P(10|11) = p \\
P(11|00) = p^2 & P(11|01) = q & P(11|10) = q & P(11|11) = 1 - q - q^2 - p
\end{array}$$

where p is the probability that a desired digit change will occur, and q is the probability that an undesired digit change will occur. For an ideal counter that steps by one, $P(11|00)$, $P(00|01)$, $P(01|10)$, and $P(10|11)$ would be equal to 1, and the rest of the conditional probabilities would be equal to zero; that is, for an ideal counter, $p = 1$ and $q = 0$. The digit change is either from a 1 to a 0 or from a 0 to a 1. For example, one digit is changed by going from state 00 to state 01; thus, $P(01|00) = q$, while two digits have to change in going from state 01 to state 10; thus, $P(10|01) = q^2$. The polynomials at the diagonal terms occur such that the sum of the terms in each column is equal to one. Consider the case when a desired digit change has a probability p = 0.99 and an undesired digit change has a probability q = 0.001. In this case, the preceding conditional probability matrix will have the following numeric values:

$P(00\|00) = 0.0179$	$P(00\|01) = 0.99$	$P(00\|10) = 0.001$	$P(00\|11) = 0.000001$
$P(01\|00) = 0.001$	$P(01\|01) = 0.008999$	$P(01\|10) = 0.9801$	$P(01\|11) = 0.001$
$P(10\|00) = 0.001$	$P(10\|01) = 0.000001$	$P(10\|10) = 0.0179$	$P(10\|11) = 0.99$
$P(11\|00) = 0.9801$	$P(11\|01) = 0.001$	$P(11\|10) = 0.001$	$P(11\|11) = 0.008999$

The conditional average randomnesses of the four columns of the conditional probability matrix are

$$\mathbb{H}(A|00) = 0.1522433743 \text{ binary bits}$$
$$\mathbb{H}(A|01) = 0.0854976701 \text{ binary bits}$$
$$\mathbb{H}(A|10) = 0.1522433743 \text{ binary bits}$$
$$\mathbb{H}(A|11) = 0.0854976701 \text{ binary bits}$$

It is assumed that the probabilities Q_{UV} of the initial states of the counter are all equal to 0.25. Thus, the noise $\mathbb{N}$ according to Equation 7.20b is

$$\mathbb{N} = \mathbb{H}(A|00)Q_{00} + \mathbb{H}(A|01)Q_{01} + \mathbb{H}(A|10)Q_{10} + \mathbb{H}(A|11)Q_{11}$$

or

$$\mathbb{N} = 0.1188705222 \text{ binary bits}$$

As it was assumed that the probabilities Q_{UV} of the initial states of the counter are all equal to 0.25, the initial information $\mathbb{H}_B$ will be equal to 2 according to Equation 9.1. The information $\mathbb{I}$ in the straight binary counter subject to the possible errors that can occur during counting is

$$\mathbb{I} = \mathbb{H}_B - \mathbb{N}$$

or

$$\mathbb{I} = 1.881129478 \text{ binary bits}$$

The maximum information $\mathbb{I}_{max}$ in this binary counter is equal to 2. Thus, the counter has an efficiency η of:

$$\eta \equiv \frac{\mathbb{I}}{\mathbb{I}_{max}}$$

$$\eta = 94.05647389\%$$

A better counter capable of counting up to 4 would be one that uses Gray code rather than the straight binary code just illustrated. In the Gray code, only one digit changes at a time. The probabilities that the counter was initially in various states are

Q_{00} is the probability that the counter was initially in state 00.
Q_{01} is the probability that the counter was initially in state 01.
Q_{11} is the probability that the counter was initially in state 11.
Q_{10} is the probability that the counter was initially in state 10.

Similar to the case discussed earlier, the conditional probabilities P(XY|UV) that the counter will make a transition to state XY provided it was in state UV are

$P(00\|00) = 1 - q - q^2 - p$	$P(00\|01) = p$	$P(00\|11) = q^2$	$P(00\|10) = q$
$P(01\|00) = q$	$P(01\|01) = 1 - q - q^2 - p$	$P(01\|11) = p$	$P(01\|10) = q^2$
$P(11\|00) = q^2$	$P(10\|11) = q$	$P(11\|11) = 1 - q - q^2 - p$	$P(11\|10) = p$
$P(10\|00) = p$	$P(10\|01) = q^2$	$P(10\|11) = q$	$P(10\|10) = 1 - q - q^2 - p$

As in the case of the straight binary counter, p is the probability that a desired digit change will occur, and q is the probability that an undesired digit change will occur. The digit change is either from a 1 to a 0 or from a 0 to a 1. Again, here a desired digit change has a probability p = 0.99, and an undesired digit change has a probability q = 0.001. In this case the foregoing conditional probability matrix will have the following numeric values:

$P(00\|00) = 0.008999$	$P(00\|01) = 0.99$	$P(00\|10) = 0.000001$	$P(00\|11) = 0.001$
$P(01\|00) = 0.001$	$P(01\|01) = 0.008999$	$P(01\|10) = 0.99$	$P(01\|11) = 0.000001$
$P(10\|00) = 0.000001$	$P(10\|01) = 0.001$	$P(10\|10) = 0.008999$	$P(10\|11) = 0.99$
$P(11\|00) = 0.99$	$P(11\|01) = 0.000001$	$P(11\|10) = 0.001$	$P(11\|11) = 0.008999$

In this case, all the conditional average randomnesses of the four columns of the conditional probability matrix are equal:

$$\mathbb{H}(A|UV) = 0.0854976701 \text{ binary bits}$$

Note that the noise of the gray code counter is less than the straight binary counter. As before, it is assumed that the probabilities Q_{UV} of the initial states of the counter are all equal to 0.25; the initial information $\mathbb{H}(A)$ is then equal to 2, according to Equation 9.5. The noise according to Equation 9.4 is

$$\mathbb{N} = 0.0854976701 \text{ binary bits}$$

The information $\mathbb{I}$ in the gray code binary counter subject to the possible errors that can occur during counting is

$$\mathbb{I} = 1.91450233 \text{ binary bits}$$

The maximum information $\mathbb{I}_{max}$ in this binary counter is equal to 2. Thus, the counter has an efficiency η of:

$$\eta = 95.72511649\%$$

Indeed, as expected, the Gray code counter is more efficient than the straight binary counter.

9.4 INFORMATION TRANSMISSION OF CONTINUOUS OR ANALOG SYSTEMS

Consider two sets of continuous random variables, $S_A\{-a < x \leq a\}$ and $S_B\{-b < y \leq b\}$. For example, the values of x could stand for the values of an analog output voltage from some electronic transmission system, whereas the values of y could stand for the values of an analog input voltage to the electronic transmission system. Here, p(x|y) is the conditional probability density that the random variable u with values between x and x + dx has been received provided an analog message that the random variable v with values between y and y + dy was transmitted. Here q(y) is the probability density that an analog message that the random variable v has values between y and y + dy has been transmitted. The noise associated with the conditional probabilities p(x|y) is

$$\mathbb{N} = \int_{-b}^{b} \mathbb{H}(A|y)q(y)dy \tag{9.11}$$

where the conditional average randomness $\mathbb{H}(A|y)$ is:

$$\mathbb{H}(A|y) = -B\int_{-a}^{a} p(x|y)\ln\left[\lambda p(x|y)\right]dx \tag{9.12}$$

and where $B \equiv 20/\ln 10$ for the case when the random variable is an analog voltage or electromagnetic field component and $B \equiv 10/\ln 10$ for the case when the random variable is electric or optical power. In this case the conditional average randomness is measured in decibels, or dB.

Thus far, three values of the quantity $1/\ln Q$ were encountered. It had a value of $-1/\ln 2$ for the case when the average randomness or information is in binary bits, it took the value of –k, Boltzmann's constant, for real physical systems, and $B \equiv 10/\ln 10$ or $B \equiv 20/\ln 10$ for the case when the information is measured in dB. The transmitted information $\mathbb{H}_B$ is

$$\mathbb{H}_B = -B\int_{-b}^{b} dy q(y)\ln\left[\lambda q(y)\right] \tag{9.13}$$

The received analog information $\mathbb{I}$ in dB is the difference between the transmitted information $\mathbb{H}_B$ and the noise $\mathbb{N}$.

$$\mathbb{I} = \mathbb{H}_B - \mathbb{N} \tag{9.14}$$

For example, consider an analog signal source. The signal has equal probability of having voltages U between $-V_o$ and V_o. Therefore, the probability density q(U) is

$$q(U) = \frac{1}{2V_o}\left[u(U+V_o) - u(U-V_o)\right] \tag{9.15}$$

Suppose the conditional probability density that a voltage between V and V + dV has been received provided a voltage with values between U and U + dU was transmitted is

$$p(V|U) = \frac{1}{\sqrt{2\pi\sigma}}\exp\left[-\frac{(V-U)^2}{2\sigma^2}\right] \tag{9.16}$$

where V can have values between minus and plus infinity and σ is the root mean square (RMS) noise voltage amplitude. The transmitted information $\mathbb{H}_B$ in dB according to Equations 9.13 and 9.15 is

$$\text{a)}\ \ \mathbb{H}_B = -20\frac{\ln\lambda}{\ln 10} + 20\frac{\ln(2V_o)}{\ln 10}\,\text{dB} \quad \text{or} \quad \text{b)}\ \ \mathbb{H}_B = 20\log(2V_o) - 20\log(\lambda)\,\text{dB} \tag{9.17}$$

where log(x) is the logarithm to the base 10 of x. The conditional average randomness $\mathbb{H}(A|U)$ from Equations 9.12 and 9.16 is

$$\mathbb{H}(A|U) = 10\log(2\pi) + 20\log(\sigma) - 20\log(\lambda) + \frac{10}{\sqrt{2\pi}\ln 10}\int_{-\infty}^{\infty}\frac{(V-U)^2}{\sigma^2}\exp\left[-\frac{(V-U)^2}{2\sigma^2}\right]d\frac{V}{\sigma} \tag{9.18}$$

To integrate Equation 9.18, one makes the following transformation of variables:

$$\text{a)}\ \ \eta \equiv \frac{V-U}{\sigma} \quad \text{and} \quad \text{b)}\ \ d\eta = d\frac{V}{\sigma} \tag{9.19}$$

Substituting Equations 9.19 into Equation 9.18:

$$\mathbb{H}(A|U) = 10\log(2\pi) + 20\log(\sigma) - 20\log(\lambda) + \frac{10}{\sqrt{2\pi}\ln 10}\int_{-\infty}^{\infty}\eta^2\exp\left[-\frac{\eta^2}{2}\right]d\eta \tag{9.20}$$

The integral appearing in Equation 9.20 is equal to $\sqrt{2\pi}$. Thus, Equation 9.20 reduces to

$$\mathbb{H}(A|U) = 20\left[\frac{\log(2\pi)}{2} + \log(\sigma) - \log(\lambda) + \frac{1}{2\ln 10}\right] \tag{9.21}$$

As all values in Equation 9.21 are constant, observe from Equation 9.11 that the noise N is also equal to Equation 9.21.

$$\mathbb{N} = 20\left[\frac{\log(2\pi e)}{2} + \log(\sigma) - \log(\lambda)\right] \text{dB} \tag{9.22}$$

where the first and last terms of Equation 9.21 were combined. To calculate the received information $\mathbb{I}$ in dB, Equations 9.17 and 9.22 are substituted into Equation 9.14.

$$\mathbb{I} = 20\log\frac{2V_o}{\sigma} - 10\log(2\pi e) \text{ in dB} \tag{9.23}$$

As it was assumed that λ has the same value in Equations 9.17 and 9.22, the term containing λ will cancel. The information $\mathbb{I}$ in dB is 20 times the logarithm to the base 10 of the ratio of the voltage range $2V_o$ of the signal divided by the RMS noise voltage amplitude σ minus a factor associated with the Gaussian conditional probabilities of the transmission medium. The factor associated with the Gaussian conditional probabilities of the transmission medium is equal to 12.3247435 dB.

Analog information can also be encrypted. Usually, an equation such as the equation of an ellipse which is a function of two variables, say x and y can be used. This equation is used to encrypt the variable x in terms of the variable y. For every value of the original variable x, there is a variable y.

9.5 THE MAXIMUM INFORMATION AND OPTIMUM TRANSMISSION RATES OF DISCRETE SYSTEMS

Consider, again, two sets $S_A\{A_1, A_2, ..., A_n, ..., A_N\}$ and $S_B\{B_1, B_2, ..., B_m, ..., B_N\}$ of random events as were used in Section 9.3 to describe a noisy transmission system. One can calculate the optimum values of the probabilities Q_m or the rates at which to transmit the various parts m of information over a noisy transmission system. These optimum values of the transmission rates Q_m of the events B_m allow us to transmit the maximum information over the noisy transmission system. A similar problem was treated by C. A. Fuchs and C. M. Caves.[8]

Thus, to find the optimum values of the rates Q_m at which to transmit the various parts m of information, the information $\mathbb{I}$ is maximized subject to the fact that the sum of all probabilities Q_m, is equal to one. By substituting Equations 9.8 and 9.9 into Equation 9.10, one obtains for the information $\mathbb{I}$:

$$\mathbb{I} = \frac{1}{\ln Q}\sum_{m=1}^{m=M}\left[Q_m \ln Q_m - (\ln Q)\,\mathbb{H}(A|B_m)Q_m\right] \tag{9.24}$$

The sum of all the probabilities Q_m must be equal to one.

$$\sum_{m=1}^{m=M} Q_m - 1 = 0 \tag{9.25}$$

Multiplying Equation 9.25 by a constant $\alpha/\ln Q$ and adding the results to Equation 9.24:

$$\mathbb{I} = \frac{1}{\ln Q}\sum_{m=1}^{m=M}\left[Q_m \ln Q_m - (\ln Q)\mathbb{H}(A|B_m)Q_m + \alpha Q_m\right] - \frac{\alpha}{\ln Q} \tag{9.26}$$

The information $\mathbb{I}$ will be a maximum where all the

$$\frac{\partial \mathbb{I}}{\partial Q_k} = 0 \tag{9.27}$$

Substituting Equation 9.26 into Equation 9.27:

$$\ln Q_k + 1 + \alpha - \ln Q\mathbb{H}(A|B_k) = 0 \tag{9.28}$$

One can solve for Q_m to obtain:

$$Q_m = e^{-1-\alpha} e^{\ln Q\mathbb{H}(A|B_m)} \tag{9.29}$$

To evaluate the Lagrange multiplier $\alpha/\ln Q$, Equation 9.29 is substituted into Equation 9.25.

$$1 = e^{-1-\alpha}\sum_{m=1}^{m=N} e^{\ln \mathbb{H}(A|B_m)} \tag{9.30}$$

By solving Equation 9.26 for e^{-1-a} and substituting the result into Equation 9.29, the optimum value $Q_{m(Optimum)}$ of the transmission rates of the events B_m is obtained:

$$Q_{m(Optimum)} = \frac{e^{\ln Q\mathbb{H}(A|B_m)}}{\sum_{k=1}^{k=N} e^{\ln Q\mathbb{H}(A|B_k)}} \tag{9.31}$$

By substituting Equation 9.31 into Equation 9.24, one obtains for the optimum information $\mathbb{I}_{Optimum}$ transmitted through the system:

$$\mathbb{I}_{Optimum} = \frac{-1}{\ln Q}\ln \sum_{k=1}^{k=N} e^{\ln Q\mathbb{H}(A|B_k)} \tag{9.32}$$

The equation for the optimum value $Q_{m(opt.)}$ of the transmission rates of the events B_m, Equation 9.27, for the case where all the conditional average randomness are equal reduces to:

$$Q_{m(Optimum)} = \frac{1}{N} \quad \text{for all } \mathbb{H}\left(A \middle| B_m\right) \text{ equal} \tag{9.33}$$

For this case the optimum information $\mathbb{I}_{Optimum}$ of Equation 9.32 is equal to:

$$\mathbb{I}_{Optimum} = \frac{\ln N}{\ln \frac{1}{Q}} - \mathbb{H}\left(A \middle| B\right) \tag{9.34}$$

For an ideal information transmission system where the conditional probabilities $P(A_n|B_m)$ that message n was received provided message m was sent are equal to one for n = m and to zero otherwise, the maximum information in binary bits is equal to its maximum value:

$$\mathbb{I}_{Optimum} = \frac{\ln N}{\ln \frac{1}{Q}} \tag{9.35}$$

On the other hand, for the case when the conditional probabilities $P(A_n|B_n)$ that message A_n was received provided message B_m was sent are all equal to 1/N, the conditional average randomness $\mathbb{H}\left(A \middle| B_m\right) = -\ln N/\ln Q$. In this case, the information $\mathbb{I}$ is equal to zero; that is, the information conveyed is completely lost in the noise.

Consider a very simple transmission system that can only transmit either a ONE or a ZERO at a time as shown in Figure 9.5. This transmission system has a bad connection in the line that transmits ZEROs. It is characterized by the following conditional probabilities (Figure 9.5):

P(ONE Rec'd.|ONE Trans'd.) = 0.99 P(ONE Rec'd.|ZERO Trans'd.) = 0.2

P(ONE Rec'd.|ONE Trans'd.) = 0.01 P(ONE Rec'd.|ZERO Trans'd.) = 0.8

FIGURE 9.5 Binary transmission system with an imperfect connection in the line that transmits ZEROs.

The optimum information $\mathbb{I}_{Opt}$ from Equation 9.32 that can be transmitted by this system on the average in binary bits is

$$\mathbb{I}_{Opt} = 1/\ln 2\ \ln[\exp(0.99\ln 0.99 + 0.01\ln 0.001) + \exp(0.2\ln 0.2 + 0.8\ln 0.8)]$$

$$\mathbb{I}_{Opt} = 0.6339652347 \text{ binary bits.}$$

It is interesting to calculate from Equation 9.31 the probabilities Q_{ONE} and Q_{ZERO} at which the information must be sent to achieve this modest maximum.

$$Q_{ONE} = \frac{\exp\left[0.99\ln 0.99 + 0.01\ln 0.01\right]}{\exp\left[0.99\ln 0.99 + 0.01\ln 0.01\right] + \exp\left[0.2\ln 0.2 + 0.8\ln 0.8\right]}$$

and

$$Q_{XERO} = \frac{\exp\left[0.2\ln 0.2 + 0.8\ln 0.8\right]}{\exp\left[0.99\ln 0.99 + 0.01\ln 0.01\right] + \exp\left[0.2\ln 0.2 + 0.8\ln 0.8\right]}$$

or

$$Q_{ONE} = 0.6093071738 \qquad Q_{ZERO} = 0.3906928264$$

That is, transmitting information that contains ONEs $\frac{0.6093071738}{0.390692864} = 1.55955557$ times as often as ZEROs will achieve the modest optimum average information transmission of $\mathbb{I}_{Opt} = 0.633965234$ binary bits.

9.6 THE MAXIMUM INFORMATION AND OPTIMUM TRANSMISSION RATES OF CONTINUOUS OR ANALOG SYSTEMS

Next, the optimum probability density $p_{Opt.}(v)$ that the random variable y has values between v and v + dv of a continuous system will be found subject to knowing the conditional average randomnesses $\mathbb{H}(A|y)$ of the transmission system and other parameters. One such example is a classical model of a molecule in a gas. The molecule can be characterized by a velocity vector **v** and a position vector **r**. To construct a model of the molecules in the gas, the interaction of the molecules in two sets of events is described. Set $S_A\{\mathbf{v}_1, \mathbf{r}_1, \mathbf{v}_2, \mathbf{r}_2, \ldots, \mathbf{v}_n, \mathbf{r}_n \ldots\}$ describes all but the m-th molecule and a set $S_B\{\mathbf{v}_m, \mathbf{r}_m\}$ describes the m-th molecule. The interaction between molecules in a gas can be described by a conditional probability density $p(\mathbf{v}_1, \mathbf{r}_1, \mathbf{v}_2, \mathbf{r}_2, \ldots, \mathbf{v}_n, \mathbf{r}_n \ldots|\mathbf{v}_m, \mathbf{r}_m)$ that an interaction was detected by all the molecules in the gas provided it was radiated by molecule m. The conditional average randomness or conditional entropy $\mathbb{H}(A|\mathbf{v}_m, \mathbf{r}_m)$ is:

$$\mathbb{H}\left(A|\mathbf{r}_m, \mathbf{v}_m\right) = -k\sum_{n\neq m}\int_{\mathbf{r}_n} d^3\mathbf{r}_n \int_{\mathbf{v}_n} d^3\mathbf{v}_n p\left(\mathbf{r}_1, \mathbf{v}_1, \mathbf{r}_2, \mathbf{v}_2, \ldots|\mathbf{r}_m, \mathbf{v}_m\right) \ln \left[\lambda p\left(\mathbf{r}_1, \mathbf{v}_1, \mathbf{r}_2, \mathbf{v}_2, \ldots|\mathbf{r}_m, \mathbf{v}_m\right)\right] \tag{9.36}$$

where k is Boltzmann's constant and λ is a constant ensuring that the information is positive. Let us for simplicity assume that the gas is sufficiently dilute and that the molecules can be assumed to be more or less freely moving through the gas. In this case, one can approximate the energies of the molecules by their kinetic energy. As stated before, the interaction of the molecules is described by the conditional probabilities. Thus, each molecule can be characterized by just the magnitude of its velocity.

The probability densities $p_m(\mathbf{v}_m)$ will be calculated here by maximizing the information subject to what is known about the system. As stated earlier, it is only a function of the magnitude **vm** of the velocity of the molecule. The information $\mathbb{I}$ associated with the communication between the molecule m and the rest of the gas is:

$$\mathbb{I} = -k4\pi\int_0^\infty v_m^2 dv_m p(v_m)\ln\left[\lambda p(v_m)\right] - 4\pi\int_0^\infty v_m^2 dv_m \mathbb{H}(A|v_m)p(v_m) \quad (9.37)$$

Note that all points on a spherical shell with radius $\mathbf{v}_m$ in $\mathbf{v}_m$ space with volume corresponds to the same energy and, thus, the same probability density. One will also need the average randomness $\mathbb{H}_m$ associated with the m-th molecule.

$$\mathbb{H}_m = -k4\pi\int_0^\infty v_m^2 dv_m p(v_m)\ln\left[\lambda p(v_m)\right] \quad (9.38)$$

The temperature T of the gas is known. It is defined as the change of the average energy U with the average randomness or entropy $\mathbb{H}_m$.

$$T = \frac{\partial U}{\partial \mathbb{H}} \quad (9.39)$$

where the subscript of the average randomness $\mathbb{H}_m$ of the m-th molecule was dispensed with. The probability density per unit velocity of a molecule has to be normalized.

$$4\pi\int_0^\infty v_m^2 dv_m p(v_m) - 1 = 0 \quad (9.40)$$

Again, one cannot use the equation for the temperature, Equation 9.39, in the information-maximizing process. The average energy U of molecule m will be used.

$$4\pi\frac{m}{2}\int_0^\infty v_m^4 dv_m p(v_m) - U = 0 \quad (9.41)$$

Multiplying Equations 9.40 and 9.41 by convenient constants $-k\alpha$ and $-k\beta$ and adding the result to Equation 9.37:

$$\mathbb{I} = -k4\pi \int_0^\infty v^2 dv \left\{ \begin{array}{l} p(v,\eta)\ln\left[\lambda p(v,\eta)\right] + \frac{\mathbb{H}(A|v)}{k} p(v,\eta) + \\ \alpha p(v,\eta) + \frac{\beta m}{2} v^2 p(v,\eta) \end{array} \right\} + \tag{9.42}$$

$$k\alpha + k\beta U$$

where the subscript m was dispensed with and a dummy variable η was introduced that will allow the variation in the probability densities p(v,η) to maximize the information $\mathbb{I}$. One would like to find the probability densities p(v) for which the information $\mathbb{I}$ is an optimum. This will occur when

$$\frac{\partial \mathbb{I}}{\partial \eta} = 0 \tag{9.43}$$

Substituting Equation 9.42 into Equation 9.43:

$$0 = -k4\pi \int_0^\infty v^2 dv \left[\ln\lambda + \ln p(v,\eta) + 1 + \frac{\mathbb{H}(A|v)}{k} + \alpha + \frac{\beta m}{2} v^2 \right] \frac{\delta p(v,\eta)}{\delta \eta} \tag{9.44}$$

The fundamental lemma of the calculus of variation requires that, because the derivative of p(v,η) with respect to η is arbitrary, the integral will be equal to zero if the square bracket in Equation 9.44 will equal zero.

$$\ln p(v) + \ln\lambda + 1 + \alpha + \frac{\beta m}{2} v^2 + \frac{\mathbb{H}(A|v)}{k} = 0 \tag{9.45}$$

Solving for p(v):

$$p(v) = \frac{1}{\lambda} \exp\left[-1-\alpha\right] \exp\left[-\frac{\beta m}{2} v^2 - \frac{\mathbb{H}(A|v)}{k} \right] \tag{9.46}$$

One can solve for the first Lagrange multiplier α by substituting Equation 9.46 into the constraint Equation 9.40.

$$1 = \frac{1}{\lambda} \exp\left[-1-\alpha\right] 4\pi \int_0^\infty v^2 dv \exp\left[-\frac{\beta m}{2} v^2 - \frac{\mathbb{H}(A|v)}{k} \right] \tag{9.47}$$

By solving Equation 9.47 for $1/\lambda \exp[-\alpha - 1]$ and substituting the resulting expression into Equation 9.46, one obtains for the probability density p(v):

$$p(v) = \frac{\exp\left[-\frac{\beta m}{2}v^2 - \frac{\mathbb{H}(A|v)}{k}\right]}{4\pi\int_0^\infty y^2 dy \exp\left[-\frac{\beta m}{2}y^2 - \frac{\mathbb{H}(A|y)}{k}\right]} \quad (9.48)$$

where both the constants λ and α have been eliminated. To start the process for eliminating the second Lagrange multiplier, β, one takes the logarithm of Equation 9.48:

$$\ln p(v) = -\ln\left\{4\pi\int_0^\infty y^2 dy \exp\left[-\frac{\beta m}{2}y^2 - \frac{\mathbb{H}(A|y)}{k}\right]\right\} - \frac{\beta m}{2}v^2 - \frac{\mathbb{H}(A|v)}{k} \quad (9.49)$$

Substituting Equation 9.49 into Equation 9.39 and making use of Equation 9.41:

$$\mathbb{H} = k\ln\left\{4\pi\int_0^\infty y^2 dy \exp\left[-\frac{\beta m}{2}y^2 - \frac{\mathbb{H}(A|y)}{k}\right]\right\} + \beta kU + 4\pi$$
$$\int_0^\infty v^2 dv \mathbb{H}(A|v)p(v) - k\ln\lambda \quad (9.50)$$

By solving Equation 9.50 for the average energy U of a molecule and substituting the resulting expression into Equation 9.39, one obtains for the second Lagrange multiplier β:

$$\text{a) } T = \frac{1}{\beta k} \quad \text{or} \quad \text{b) } \beta = \frac{1}{kT} \quad (9.51)$$

A final expression for the probability density p(v) can now be obtained by substituting Equation 9.47 into Equation 9.44 for β.

$$p(v) = \frac{\exp\left[-\frac{m}{2kT}v^2 - \frac{\mathbb{H}(A|v)}{k}\right]}{4\pi\int_0^\infty y^2 dy \exp\left[-\frac{m}{2kT}y^2 - \frac{\mathbb{H}(A|y)}{k}\right]} \quad (9.52)$$

To obtain a closed-form expression for the probability p(v), the form of the conditional average randomness $\mathbb{H}$(A,v) would have to be known.

One can calculate the average energy U of a molecule by substituting Equation 9.48 into Equation 9.41.

$$U = \frac{4\pi \frac{m}{2} \int_0^\infty v^4 dv \exp\left[-\frac{m}{2kT} v^2 - \frac{\mathbb{H}(A|v)}{k} \right]}{4\pi \int_0^\infty y^2 dy \exp\left[-\frac{m}{2kT} y^2 - \frac{\mathbb{H}(A|y)}{k} \right]} \tag{9.53}$$

To be able to evaluate the expression for the average energy U per molecule, one needs a form for the conditional average randomness $\mathbb{H}(A|v)$. Expanding the conditional average randomness to first order in the velocity v:

$$\mathbb{H}(A|v) \approx \mathbb{H}_0 + k\rho v + \ldots \tag{9.54}$$

For convenience, the expansion coefficient was chosen as kρ. Here, ρ has units of seconds per meter. Substituting Equation 9.54 into Equation 9.53 and canceling common factors in the numerator and denominator:

$$U = \frac{\frac{m}{2} \int_0^\infty v^4 dv \exp\left[-\frac{mv^2}{2kT} - \rho v \right]}{\int_0^\infty y^2 dy \exp\left[-\frac{my^2}{2kT} - \rho y \right]} \tag{9.55}$$

To perform the integration, one makes the following transformation of variables:

$$\begin{gathered} \text{a)}\quad x^2 \equiv \frac{mv^2}{2kT} \quad \text{b)}\quad v = \sqrt{\frac{2kT}{m}}\, x \quad \text{c)}\quad dv = \sqrt{\frac{2kT}{m}}\, dx \\ \text{d)}\quad v^2 = \frac{2kT}{m} x^2 \quad \text{e)}\quad v^4 = \frac{4k^2T^2}{m^2} x^4 \end{gathered} \tag{9.56}$$

Substituting the transformation of variables into Equation 9.55:

$$U = kT \frac{\int_0^\infty x^4 \exp\left[-x^2 - \sqrt{\frac{m}{2kT}}\, \rho x \right] dx}{\int_0^\infty \xi^2 \exp\left[-\xi^2 - \sqrt{\frac{m}{2kT}}\, \rho \xi \right] d\xi} \tag{9.57}$$

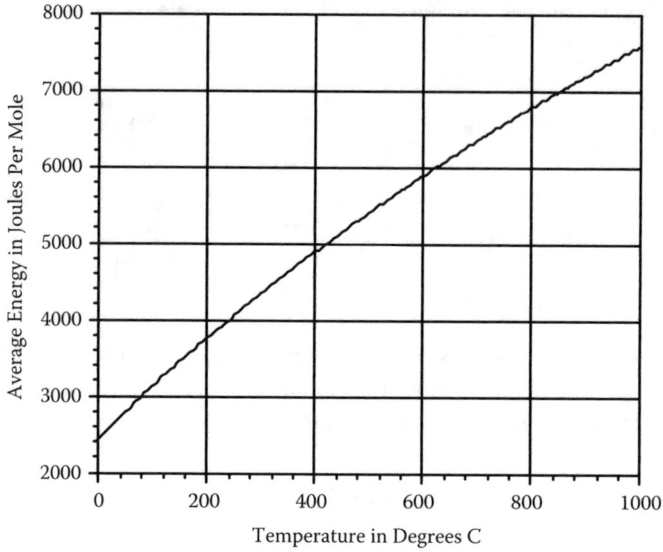

FIGURE 9.6 The average energy U per mole calculated from Equation 9.53.

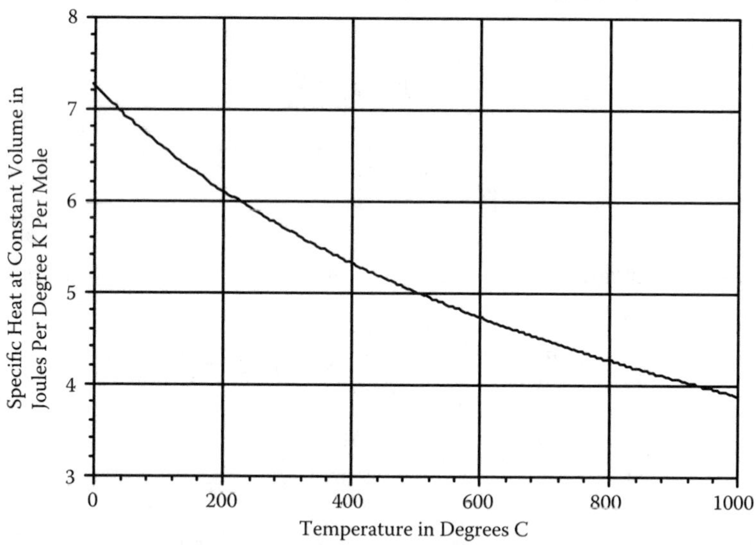

FIGURE 9.7 The specific heat at constant volume per degree C per mole calculated from Equation 9.53.

where ξ was used instead of x in the integral in the denominator. Equation 9.57 can be integrated using functions. A numeric function will be used here. The average energy and its temperature derivative, the specific heat at constant volume, are plotted in Figures 9.6 and 9.7. Here, $\rho\sqrt{2/m} = 1.55900717 \times 10^{10}$ was used.

9.7 THE BIT ERROR RATE

Recall the case of the jumping chair discussed in Chapter 1. Here is a practical case of such a small probability having an important effect. A small randomly varying noise signal is usually associated with every analog signal. In a digital signal, a ONE is represented by, say, a voltage of 0.8 V and a ZERO is represented by 0 V. Actually, any voltage above 0.5 V is counted as a ONE, and any voltage below 0.3 V is counted as a ZERO. It would appear that such an arrangement would eliminate the effect of small noise voltages in the few mV range. This is not so. Only the average value of noise voltage has an amplitude of a few mV. There is a small probability that a noise voltage spike will exceed 0.5 V when a ZERO signal is sent or that there will be a negative noise voltage spike with a magnitude below 0.3 V when a ONE is being transmitted. This is illustrated in Figure 9.8. Thus, a digital signal that can only have two values, ZERO and ONE, is still subject to noise, as will be demonstrated. As has been shown earlier, in a stochastic transmission system there exists a conditional probability P(ZERO|ONE) that a ONE was transmitted and a ZERO received and a conditional probability P(ONE|ZERO) that a ZERO was transmitted and a ONE received. Obviously, these cases represent errors. The rate at which errors are generated, the bit error rate (BER), is

$$\frac{1}{\text{BER}} = P(\text{ONE}|\text{ZERO})P_{\text{ZERO}} + P(\text{ZERO}|\text{ONE})P_{\text{ONE}} \tag{9.58}$$

where P_{ONE} and P_{ZERO} are the probabilities that a ONE and a ZERO are transmitted, respectively.

The BER is best illustrated by an example. Consider the optical signal illustrated in Figure 9.8. A ONE digital light signal nominally corresponds to a power level, or brightness, of $U_1 = 10$ mW, and a ZERO digital light signal nominally corresponds to a power level of $U_0 = 4$ mW. However, because the signals are noisy, one interprets light with power greater than $U_{UP} = 8$ mW as a ONE and light with a power level less

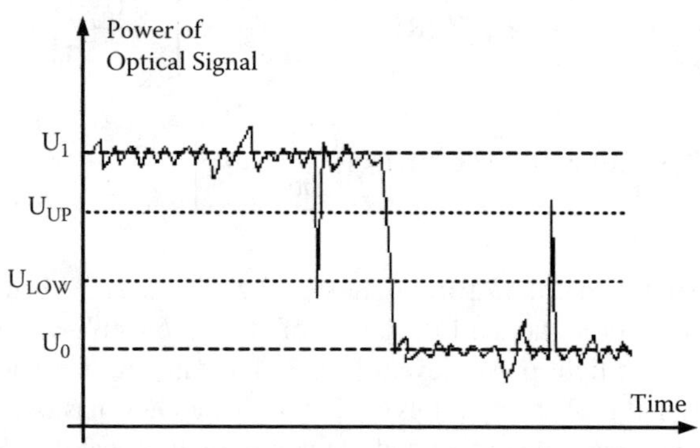

FIGURE 9.8 A ONE digital light signal nominally corresponds to a power level, or brightness, U_1 and a ZERO digital light signal nominally corresponds to a power level U_0. However, because the signal is noisy, light with power greater than U_{UP} is interpreted as a ONE signal and light with a power level less than U_{LOW} as a ZERO signal.

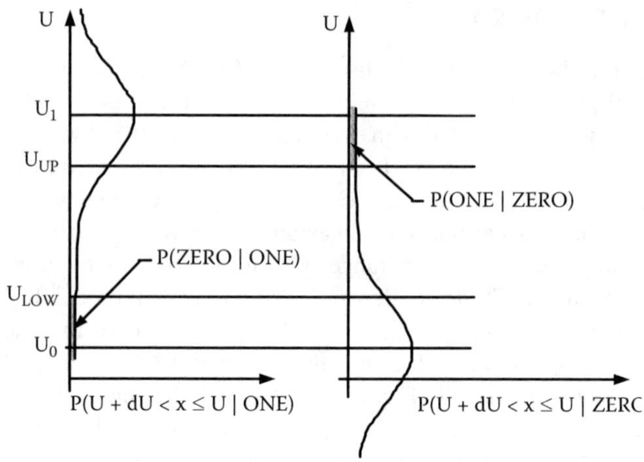

FIGURE 9.9 Gaussian noise of a digital system. The shaded areas are the conditional probability P(ZERO|ONE) that a ONE has been transmitted and a ZERO received and the conditional probability P(ONE|ZERO) that a ZERO has been transmitted and a ONE received.

than $U_{LOW} = 6$ mW as a ZERO. There is a finite but small probability that a ZERO signal will produce a spike large enough to be interpreted as a ONE. Similarly, there is a finite but small probability that a ONE signal will produce a negative spike large enough to be interpreted as a ZERO.

Assume that the noise light power has a Gaussian probability density as shown in Figure 9.9; that is, the conditional probability $P(U + dU < x \leq U|ZERO)$ that the light signal power x will have values between U and U + dU provided a ZERO has been transmitted and the conditional probability $P(U + dU < x \leq U|ONE)$ that the light signal power x will have values between U and U + dU provided a ONE has been transmitted are

$$\text{a)}\quad P\left(U + dU > x \geq U|ZERO\right) = \sqrt{\frac{1}{2\pi\sigma^2}}\exp\left[-\frac{\left(U - U_0\right)^2}{2\sigma_0^2}\right]dU$$

$$\text{b)}\quad P\left(U + dU > x \geq U|ONE\right) = \sqrt{\frac{1}{2\pi\sigma^2}}\exp\left[-\frac{\left(U - U_1\right)^2}{2\sigma_1^2}\right]dU \tag{9.59}$$

where it was assumed that the ZERO light signal has an average brightness or light power level of U_0 and a standard deviation of $\sigma_0 = 0.64$ mW and the ONE light signal has an average light power level of U_1 and a standard deviation of $\sigma_1 = 0.48$ mW. The conditional probability P(ONE|ZERO) that a ONE has been received provided a ZERO has been transmitted and the conditional probability P(ZERO|ONE) that a ZERO has been received provided a ONE has been transmitted are equal to the probabilities that the light power will have values above U_{UP} and below U_{LOW}, respectively, as shown in Figure 9.8:

$$\text{a)}\quad P(\text{ONE}|\text{ZERO}) = \sqrt{\frac{1}{2\pi\sigma^2}}\int_{U_{UP}}^{\infty}\exp\left[-\frac{(U-U_0)^2}{2\sigma_0^2}\right]dU$$

$$\text{b)}\quad P(\text{ZERO}|\text{ONE}) = \sqrt{\frac{1}{2\pi\sigma^2}}\int_{-\infty}^{U_{LOW}}\exp\left[-\frac{(U-U_1)^2}{2\sigma_1^2}\right]dU \tag{9.60}$$

The BER in this case is equal to the reciprocal of the sum of the conditional probability that a ONE was received provided a ZERO was transmitted and the conditional probability that a ZERO was received provided a ONE was transmitted times the probabilities P_{ONE} and P_{ZERO} that a ONE and a ZERO have actually been transmitted, as shown in Figure 9.9.

$$\frac{1}{BER} = \frac{1}{2}\left\{\begin{array}{l}\sqrt{\frac{1}{2\pi\sigma^2}}\int_{U_{UP}}^{\infty}\exp\left[-\frac{(U-U_0)^2}{2\sigma_0^2}\right]dU + \\ \sqrt{\frac{1}{2\pi\sigma^2}}\int_{-\infty}^{U_{LOW}}\exp\left[-\frac{(U-U_1)^2}{2\sigma_1^2}\right]dU\end{array}\right\} \tag{9.61}$$

where both the probabilities P_{ONE} that a ONE was transmitted P_{ZERO} that a ZERO was transmitted are equal to ½.

The integrals of Equation 9.57 need to be evaluated. Note that both $U_{UP} - U_0/\sigma_0$ and $U_1 - U_{LOW}/\sigma_1$ are much larger than one. Defining a variable z as:

$$\text{a)}\ z \equiv U - U_{UP}\quad \text{b)}\ dz = dU\quad \text{c)}\ \text{at } U = U_{UP} + z = 0 \text{ and d) at } U = \infty\ z = \infty \tag{9.62}$$

Substituting Equation 9.62 into the first integral of Equation 9.61:

$$\sqrt{\frac{1}{2\pi\sigma^2}}\int_0^{\infty}\exp\left[-\frac{(U-U_0)^2}{2\sigma^2}\right]dU = \sqrt{\frac{1}{2\pi\sigma^2}}\exp\left[-\frac{(U_{UP}-U_0)^2}{2\sigma^2}\right]\times$$
$$\int_0^{\infty}\exp\left[-\frac{z(U_{UP}-U_0)}{\sigma^2}\right]\exp\left[-\frac{z^2}{2\sigma^2}\right]dx \tag{9.63}$$

The exponentials will only have significant values near $z = 0$. Therefore, the last exponential in the integral of Equation 9.63 can be neglected. By integrating, one obtains for Gaussian noise:

$$\sqrt{\frac{1}{2\pi\sigma_0^2}}\int_{U_{UP}}^{\infty}\exp\left[-\frac{(U-U_0)^2}{2\sigma_0^2}\right]dU \approx \sqrt{\frac{1}{2\pi}}\frac{\sigma_0}{U_{Up}-U_0}\exp\left[-\frac{(U_{UP}-U_0)^2}{2\sigma_0^2}\right] \tag{9.64}$$

By using a similar approximation for the second integral of Equation 9.61, one obtains for the BER:

$$\frac{1}{\text{BER}} \approx \frac{1}{2}\left\{\sqrt{\frac{1}{2\pi}}\frac{\sigma_o}{(U_{UP}-U_o)}\exp\left[-\frac{(U_{UP}-U_o)^2}{2\sigma_o^2}\right]+\right.$$
$$\left.\sqrt{\frac{1}{2\pi}}\frac{\sigma_1}{(U_1-U_{LOW})}\exp\left[-\frac{(U_1-U_{LOW})^2}{2\sigma_1^2}\right]\right\} \tag{9.65}$$

By substituting numerically into Equation 9.60, one obtains for the BER = 4.756461 $\times$ 10^9; that is, there will be on the average one error in 4.756461 $\times$ 10^9 pulses. Suppose the system operates at a rate of 1 $\times$ 10^9 pulses per second. Thus, there will on the average be one error every 4.756461 s.

REFERENCES

1. "Black Hole Computers" by Seth Lloyd and Y. Jack Ng, Scientific American, Vol. 291, No. 5, pp. 53–61, November 2004.
2. "The Mathematical Theory of Communication" by C. E. Shannon, Bell System Technical Journal, Vol. 27, pp. 379–423, and pp. 623–656 (1948).
3. "The Mathematical Theory of Communication" by C. E. Shannon and W. Weaver University of Illinois Press, Urbana, Illinois (1964).
4. www.ugrad.cs.jhu.edu/~russell/classes/enigma/www.codesandciphers.org.uk/enigma/
5. "Voynich Manuscript Mini-FAQ" by Dennis J. Stallings, http://www.geocities.com
6. http://elonka.com./kryptos/
7. "How to Count" by Brian Hays, American Scientist Vol. 89, No. 2, March–April 2001, pp. 110–114.
8. "Ensemble-Dependent Bounds for Accessible Information in Quantum Mechanics" by Christopher A. Fuchs and Carlton M. Caves, Physical Review Letters Vol. 73, No. 23, pp. 3047–3050, 5 December 1994.

PROBLEMS

Problem 9.1

Given:

Consider the following paragraph containing approximately 611 characters:

> For more than 40 years after Hall and Heroult in 1886 independently discovered a commercial feasible method of producing aluminum, the metal was scarcely more than mentioned in most textbooks on chemistry or metallurgy. The first comprehensive treatise on aluminum from mining and production through fabrication and use was not published until 1930. This was "The Aluminum Industry", by Edwards, Frary and Jeffries. Since then, several hundred other books have been published reflecting the aluminum industry; although still young it has benefited from an unexpected number of research projects of high quality.

The paragraph contains the following symbols: The 26 letters; the numbers 1, 2, 3, 4, 5, 6, 7, 8, 9, 0; the space; the period, comma, colon, semicolon, and the open and close quotation marks. Here, no distinction is made between upper- and lowercase letters. Thus, there are 43 different symbols. The probability P_n that the n-th symbol will occur in the paragraph is equal to the number of times N_n the n-th symbol occurs in the paragraph divided by the total number of symbols N_{total}.

$$P_n = \frac{N_n}{N_{total}}$$

Problem:

a) Calculate the 43 probabilities P_n of the various symbols occurring.
b) Calculate the information content in the paragraph in binary bits.
c) Calculate the efficiency η of the information conveyed in the paragraph; that is, calculate the ratio of the information conveyed by the paragraph divided by the maximum information it is possible to transmit by using the 43 symbols.

Problem 9.2

Given:

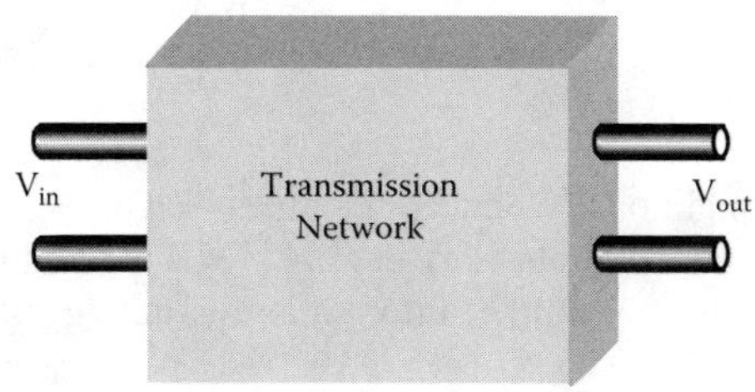

An analog transmission system is characterized by a conditional probability $p(V_{out}|V_{in})$ of the following form:

$$p\left(V_{out}V_{in}\right) = \frac{5}{V_{RMS}\Gamma\left(\frac{1}{10}\right)}\exp\left[-\left(\frac{V_{out} - V_{in}}{V_{RMS}}\right)^{10}\right]$$

where $-\infty \le V_{out} \le \infty$ and $-15\ V \le V_{out} \le 15\ V$.

We define two sets of events $S_A\{-\infty \le V_{out} \le \infty\}$ and $S_B\{-15\ V \le V_{out} \le 15\ V\}$.

Note:

Consider the integral:

$$\int_0^{\infty} \exp\left[-Au^N\right]du$$

We make the following transformation of variables:

$$w = Au^N$$

By solving for u we obtain:

$$u = \frac{1}{A^{\frac{1}{N}}} w^{\frac{1}{N}} \quad \text{and} \quad du = \frac{1}{NA^{\frac{1}{N}}} w^{\frac{1}{N}-1} dw$$

and by substituting the transformation of variables into the integral, we obtain:

$$\int_0^\infty \exp\left[-Au^N\right] du = \frac{1}{NA^{\frac{1}{N}}} \int_0^\infty w^{\frac{1}{N}-1} \exp\left[-w\right] dw$$

The integral on the right is equal to the gamma function of 1/N.

$$\int_0^\infty \exp\left[-Au^N\right] du = \frac{\Gamma\left(\frac{1}{N}\right)}{NA^{\frac{1}{N}}}$$

Problem:

a) Calculate the conditional average randomness $\mathbb{H}(\text{Out}|V_{in})$ in dB.
b) Calculate the optimum probability density $p(V_{in})$ per unit input voltage that will maximize the transmitted information assuming the system is most random.
c) Calculate the optimum information $\mathbb{I}_{opt}$ in dB.

Problem 9.3

Given:

A hot object radiates electromagnetic waves. This is known as *black body radiation.* This radiation passes through a scattering medium. The conditional average randomness $\mathbb{H}(\text{Received}|\text{m photons radiated})$ that characterizes the scattering medium is proportional to the logarithm of the number m of photons radiated:

$$\mathbb{H}\left(\text{Received}|\text{m photons transmitted}\right) = \rho k|nm$$

where $\rho \approx 0.01$ is a small constant and $k = 1.38066244 \times 10^{-23}$ J per °K is Boltzmann's constant. The energy U_m of m photons is

$$U_m = \hbar\omega m$$

The absolute temperature T = 298 °K (25°C) of the system is known:

$$T = \frac{\partial < u_m >}{\partial \mathbb{S}}$$

where $\langle U_m \rangle$ is the average energy and $\mathbb{S}$ is the transmitted entropy:

$$\mathbb{S} = -k \sum_{m=0}^{m=\infty} P_m \ln P_m$$

Problem:

a) Calculate the probabilities P_m of transmitting m photons assuming the system is most random (20 points).
b) Calculate the average energy $\langle U_m \rangle$. Multiply the average energy $\langle U_m \rangle$ by $(\hbar\omega)^2$, a density of states factor, and plot the result as a function of the frequency ω for ρ = 0 and ρ = 0.01 for frequencies ω from 0 to 3.1211 × 10^{14} radians per second. Determine the shift in frequency of the peak of the curves from ρ = 0 to ρ = 0.01 (30 points).

Problem 9.4

Given:

The conditional probability density p(V|ZERO) per unit voltage that a spontaneous emitted voltage of magnitude V will occur provided the nominal output signal of the device is 0 V is

$$p(V|ZERO) = \frac{e}{kT} \frac{\sin^2\left[\frac{\pi e}{kT}(V - V_0)\right]}{\left[\frac{\pi e}{kT}(V - V_0)\right]^2}$$

Here, $V_o = 0$ and

$$\int_{-\infty}^{\infty} dx \frac{\sin^2(\pi x)}{(\pi x)^2} = 1$$

The conditional probability density p(V|ONE) per unit voltage that a spontaneous emitted voltage of a magnitude V will occur provided the nominal output voltage V_1 of the device is 2 V is:

$$p(V|ONE) = \frac{e}{kT} \frac{\sin^2\left[\frac{\pi e}{kT}(V - V_1)\right]}{\left[\frac{\pi e}{kT}(V - V_1)\right]^2}$$

A voltage less than 0.75 V is counted as a ZERO, and a voltage greater than 1.25 V is counted as a ONE.

Problem:

a) Calculate the bit error rate.
b) Calculate the pulse data rate so that the time between error pulses is at least 1 min.

Problem 9.5

Given:

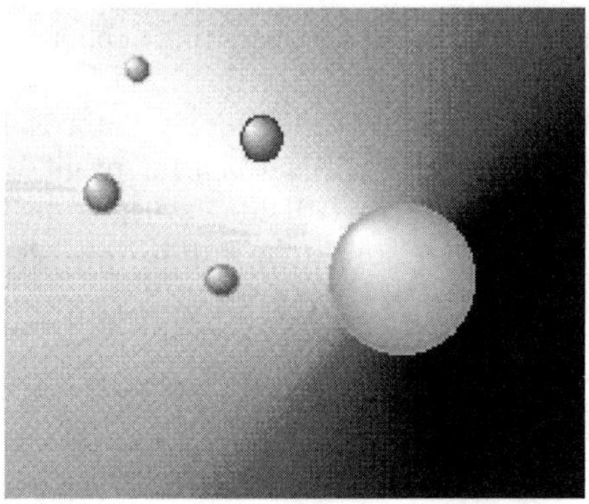

Consider an object in space emitting electromagnetic radiation ranging in frequency from radio to gamma ray frequencies. The radiation is scattered by various objects in space. Electromagnetic radiation comes in energy quanta $\hbar\omega$ called photons. Some of the radiation can be absorbed by other objects in space, but it also can conceivably be amplified by passing near other radiation sources, a natural laser of some kind out in space. This radiation finally reaches the Earth and is detected.

To construct a mathematical model, we define two sets of events: $S_A\{A_0, A_1, A_2, A_3, \ldots, A_n, \ldots\}$, where A_n is the event of receiving n photons on Earth, and $S_B\{B_0, B_1, B_2, B_3, \ldots, B_n \ldots\}$, where B_n is the event of m photons being transmitted. There is a conditional probability P(n photons received|m photons transmitted) that n photons were received on Earth provided by m photons radiated by some object in space. There is a conditional average randomness $\mathbb{H}(B|n)$ associated with these conditional probabilities.

$$\mathbb{H}\left(B\middle|n\right)=-k\sum_{m=0}^{m=\infty}P\left(B_m\middle|A_n\right)\ln P\left(B_m\middle|A_n\right)$$

where $k = 1.380661998 \times 10^{-23}$ J per degree K is Boltzmann's constant. We assume in this case that this conditional average randomness $\mathbb{H}(B|n)$ is

$$\mathbb{H}(B|n) = \rho \ln n$$

where $0 < \rho \leq 0.4$. The temperature T of the radiation is known.

$$T = \frac{\partial < \langle \hbar\omega n \rangle}{\partial H}$$

where $\langle \hbar\omega n \rangle$ is the average energy and the received information $\mathbb{H}(A)$ is

$$\mathbb{H}(A) = -k \sum_{n=0}^{n=\infty} P(A_n) \ln P(A_n)$$

At constant frequency ω, one can transfer information by transmitting signals with a different number of photons; that is, by transmitting signals of different brightness.

Problem:

a) Calculate the probabilities P_n that n photons were received by maximizing the information $\mathbb{I}$ associated by transmitting different numbers of photons (30 points).
b) Calculate the average energy $\langle \hbar\omega n \rangle$ of the photons received on Earth. Multiply the result by a density of states function D

$$D = G_0 \frac{\hbar^2 \omega^2}{k^3 T^3}$$

The resulting expression, the black body radiation curve, is a function of $x = \hbar\omega / kT$. Calculate the value of x at the maximum of the black body radiation curve for different values of ρ. Plot the ratio x/x_o as a function of ρ, where x_0 is the value of x at the maximum of the black body radiation curve at $\rho = 0$. All these calculations would have to be done numerically (20 points).

PROBLEM 9.6

Given:

Analog information transmission system.

Consider an analog information transmission system. To construct a model of a particular transmission system, we postulate the following two continuous sets of events: set $S_A\{-V_o < U < V_o\}$ is the set of all values of the input voltages U and SB$\{-\infty < V < \infty\}$ is the set of all values of the output voltages V.

The conditional probability density p(V|U) per unit output voltage V that a voltage V will appear at the output provided a voltage U was applied to the input is a super-Gaussian:

$$p(V|U) = \frac{5}{s\Gamma\left(\frac{1}{10}\right)} \exp\left[-\frac{(V-U)^{10}}{s^{10}}\right]$$

where $s\sqrt{\Gamma(3/10)/\Gamma(1/10)} = 0.94294423711s$ is the RMS noise voltage.

Problem:

a) Calculate the conditional average randomness $\mathbb{H}(A|U)$ in voltage dB (20 points).
b) Calculate the probability density q(U) per unit input voltage, assuming the information $\mathbb{I}$ is an optimum (20 points).
c) Calculate the optimum information $\mathbb{I}_{opt}$ in voltage dB (10 points).

10 Random Processes

10.1 RANDOM PROCESSES

In the previous chapters, time-independent probabilistic systems were discussed. Here, time-dependent stochastic systems will be analyzed. A time-dependent function of a random variable is referred to as a *random process*. For example, the height of a burning candle, as shown in Figure 10.1, can be thought of as a process that progresses with time. Because of random air motions and imperfect consistency of the candle wax, the rate at which the candle burns cannot be precisely predicted. Thus, the candle height is a random process. On the other hand, because one can exactly predict the values of the function $f(t) = \sin\omega t$ illustrated in Figure 10.2 at any time, this function is a *deterministic process*.

There are processes where the value of the function cannot be predicted exactly at any time. Here, with the best forecasting skills one can only predict that at any given instant of time, the function or process $f_n(t)$ will have any of a number of values; that is, the function $f_n(t)$ is a function not only of the time t but also of a random variable n, as illustrated in Figure 10.3. The random variable n could be the number of one of the values that the function can take at a given time. Say, the function $f_n(t)$ can have values number 1, 2, 3, or 4 at each instant of time, as shown in Figure 10.3. Such a time-dependent stochastic function is one type of random process. In particular, the random process $f_n(t)$ illustrated in Figure 10.3 is a function of a discrete random variable n and of the time t. This is known as a *discrete random process*.

The random process $f(x;t)$ could have a range of continuous values at every instant of time, as illustrated in Figure 10.4. Such a function is known as a *continuous random process*. The aforementioned functions are only a few examples of random processes. Indeed, other examples of random processes will be described in the following text.

10.2 RANDOM WALK AND THE FAMOUS CASE OF SCENT MOLECULES EMERGING FROM A PERFUME BOTTLE

The mathematical model of a mechanical clock can, without difficulty, be made to run either forward or backward. Indeed, clocks in barber shops that are meant to be viewed in a mirror appear to run backward (Figure 10.5). Real clocks, consisting of material that are made of atomic- and subatomic-scale particles that execute random motions, of course, progress only forward in time. However, the probability that the scent molecules escaping from a perfume bottle will reassemble back into the bottle is exceedingly small. The scent molecules are much more randomly distributed after they leave the bottle than they were in the bottle (see Figure 10.5). Therefore, their collective motion exhibits a *unique arrow of time*.

FIGURE 10.1 The decreasing height with time of a burning candle is a random process.

A crude model of molecules escaping from a perfume bottle is as follows. Assume a number of molecules are aligned in a line next to one another. Each particle subsequently executes a random walk[1,2] as shown in Figure 10.6. In the example illustrated in Figure 10.6, it is assumed that 100 particles are, initially, aligned in a row representing the top surface of the liquid perfume in the bottle. One assigns a two-digit

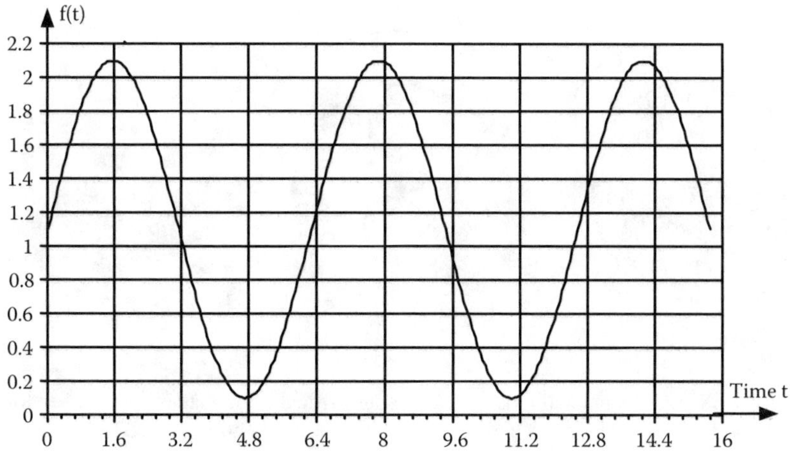

FIGURE 10.2 A deterministic process consisting of a fully specified function f(t) of the time t.

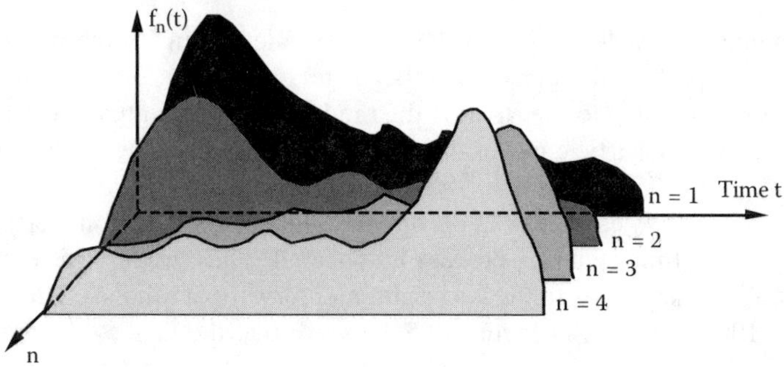

FIGURE 10.3 A random process consisting of a function $f_n(t)$ of a discrete random variable n and of the time t. At each instant of time the function $f_n(t)$ can have four different values.

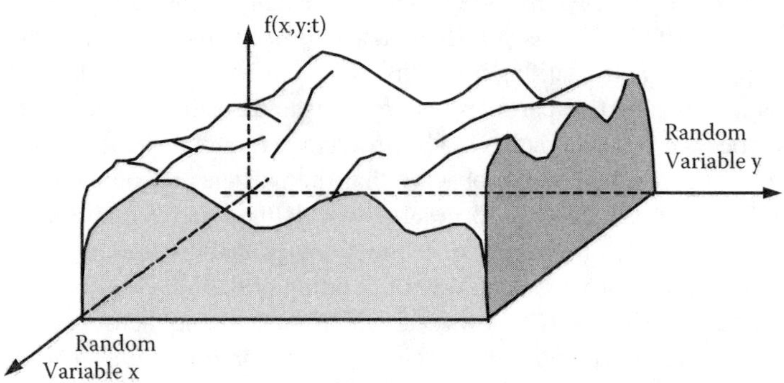

FIGURE 10.4 A random process consisting of a function f(x;t) of a continuous random variable x and of the time t.

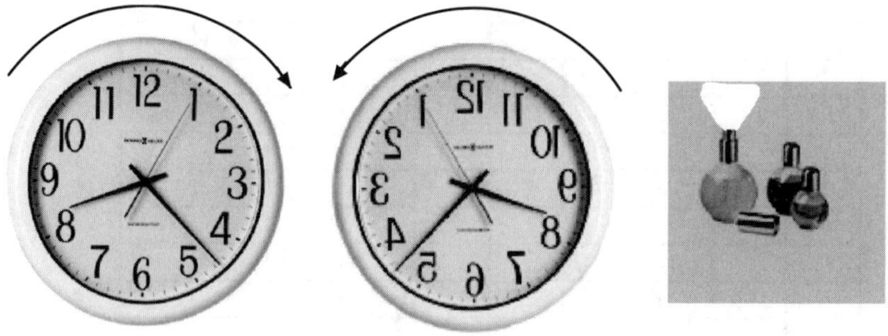

FIGURE 10.5 A mechanical clock can, without difficulty, be made to run either forward or backward. Indeed, clocks in barber shops that are meant to be viewed in a mirror run backward. However, the probability that the scent molecules escaping from a perfume bottle will reassemble back into the bottle is exceedingly small. The scent molecules are much more random after they leave the bottle than when they were in the bottle. Thus, the random motion of the perfume molecules has a unique arrow of time.

random number to each particle. If the number is less than 25, the particle moves forward one step. If the random number is greater or equal to 25 but less than 50, the particle moves one step to the right. If the random number is greater or equal to 50 but less than 75, the particle moves one step to the left, and if the random number is greater or equal to 75 but less than 100, the particle remains in place. For simplicity it is assumed that two particles can occupy the same place. This completes one time step. In the next time step this process is repeated. Thus, each particle executes a random walk, taking steps to the left, right, and forward at random. This process is shown for 100 time steps in Figure 10.5. Observe that the molecules spread out as they emerge from the bottle.

The random walk of the perfume particles used here is an example of a discrete random process. The process illustrated here is a very simple example of a random walk. More complicated random walk models with various conditions imposed on the motion of the particles are often used. For example, instead of "stepping" the position of the particles as was done here, one can "step" the velocity of the particles while still plotting their resulting positions. Also, collisions of particles can be taken into account so that no two particles occupy the same position at the same time.

This process can be characterized by an average randomness associated with the location of the particles. Indeed, observe that the particles become more randomly distributed as time progresses. To be able to calculate an average randomness as given in Equation 7.1, one needs to define some probabilities $P_m(t_n)$ at each time step t_n of this random process. One way of defining probabilities associated with the location of the particles is as follows: At each time step one measures the horizontal distance between neighboring particles on each horizontal line. The probability $P_m(t_n)$ that a particular distance between neighboring particles will occur is equal to the number of times N_m that this distance occurred divided by the total number N_{total} of horizontal distances between particles.

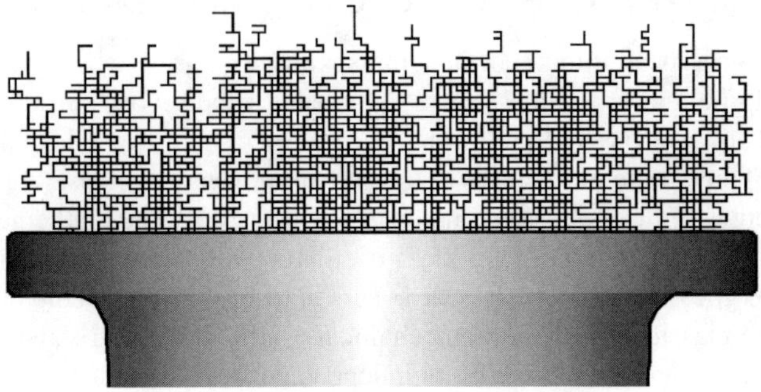

FIGURE 10.6 Random walk of scent molecules emerging from a perfume bottle.

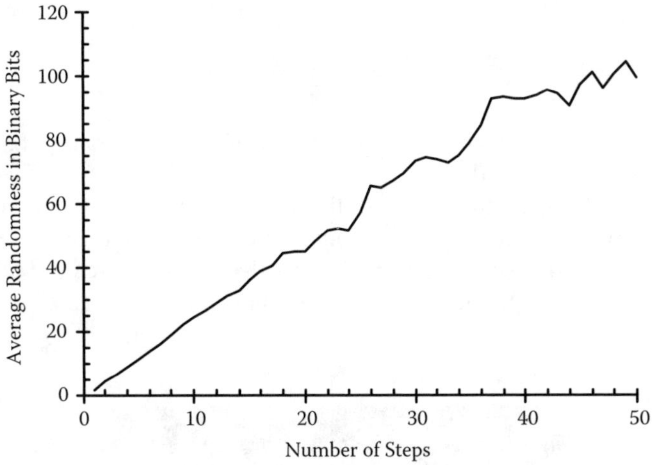

FIGURE 10.7 The average randomness of the spatial positions of molecules emanating from a perfume bottle. Note that the average randomness increases monotonically.

$$P_m\left(t_n\right) \equiv \frac{N_m}{N_{total}} \tag{10.1}$$

This is similar to the definition of the probabilities of a symbol occurring in the text paragraphs analyzed in Chapter 9. The probabilities $P_m(t_n)$ and Equation 7.1 are used to calculate the average randomness $\mathbb{H}(t_n)$ in binary bits at time step t_n. A plot of the average randomness of the positions of molecules emanating from a perfume bottle of Figure 10.6 is shown in Figure 10.7. Note that the average randomness associated with the spatial positions of the particles increases monotonically.

For example, in a crystal lattice where all horizontal distances between lattice points are equal, the probability that the distance between neighboring points is equal to the lattice spacing is one, and the probability of any other spacing occurring is equal to zero. Thus, the spatial average randomness for the crystal lattice would be equal to zero.

10.3 THE SIMPLE STOCHASTIC OSCILLATOR AND CLOCKS

Oscillators such as pendulums and, later, masses and springs have been used to measure time. Indeed, the quartz watches in use at the end of the second millennium employ an oscillating quartz crystal to tell time. The quartz crystal expands and contracts as it oscillates. Thus, some of the mass of the crystal moves in response to its inertia and the elastic forces of the expanding and contracting crystal. The mathematical models used to describe the motion of an ideal oscillator work equally well regardless of whether time goes forward or backward. This is true whether the model is quantum mechanical or classical. A quantum mechanical oscillator, of course, has some probabilistic aspects. However, its motion is independent of the direction of time. On the other hand, the motion of the mass of the simple stochastic oscillator (SSO) described here is different for time going in the forward and backward directions.

For example, the position x of the mass of an ideal oscillator such as the sphere and spring is

$$x = A\sin\omega t \tag{10.2}$$

If at some point it is decided to reverse the flow of time (t becomes −t), the sphere would retrace its path exactly.

Now consider, again, the motion of the mass at position x of an oscillator, this time, subject to a spring force and a damping force as shown in Figure 10.8. The sum of the three forces, the inertial force $m(d^2x/dt^2)$, the damping force $D(dx/dt)$, and the force due to the spring Kx acting on the mass shown in Figure 10.8 must be equal to zero.

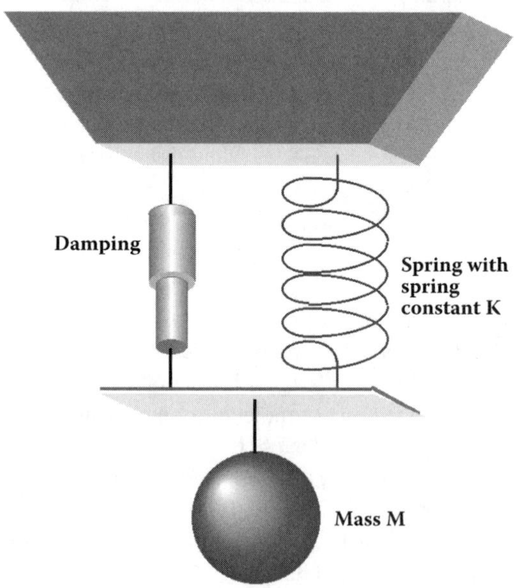

FIGURE 10.8 Oscillator consisting of a mass, a spring, and a damping mechanism.

$$m\frac{d^{2=0}x}{dt^2}+D\frac{dx}{dt}+Kx=0 \qquad (10.3)$$

where m is the mass, D is a damping constant, K is the spring constant, and t is the time. This differential equation has the following general solution:

$$x = Ae^{-\alpha_1 t} + Be^{-\alpha_2 t} \qquad (10.4)$$

where A and B are constants that can be determined from the initial conditions, the position at t = 0 and the velocity at t = 0. Each term of Equation 10.4 represents a mode of the oscillation. By substituting each term of Equation 10.4 into the differential equation, Equation 10.3, one obtains for α_1 and α_2:

$$m\alpha_{1,2}^2 - D\alpha_{1,2} + K = 0 \qquad (10.5)$$

The solutions for α_1 and α_2 are

$$\text{a)}\quad \alpha_1 = \frac{D}{2m} + j\sqrt{\frac{K}{m}-\frac{D^2}{4m^2}} \quad \text{and} \quad \text{b)}\quad \alpha_2 = \frac{D}{2m} - j\sqrt{\frac{K}{m}-\frac{D^2}{4m^2}} \qquad (10.6)$$

where it was assumed that K/m is larger than $D^2/4m^2$. Substituting Equation 10.6 back into the equation for the displacement, Equation 10.4:

$$x(t) = \exp\left[-\frac{D}{2m}t\right]\left[A\exp\left(j\sqrt{\frac{K}{m}-\frac{D^2}{4m^2}}t\right)+B\exp\left(-j\sqrt{\frac{K}{m}-\frac{D^2}{4m^2}}t\right)\right] \qquad (10.7)$$

It is true that adding the nonconservative damping force gives a strange result for negative time. The amplitude of the oscillation grows with increasing negative time. This does not seem to be reasonable. However, this result cannot be proved to be wrong, because nobody has ever observed negative time.

Another application of a random walk problem is the calculation of a one-dimensional harmonic oscillator subject to a random bombardment by air molecules, as shown in Figure 10.9. The motion of the mass of the oscillator will be calculated using a discrete time step model. In each time step, the motion of the mass due to the spring force only is first calculated and then the effect of the collision of the oscillator mass with the gas molecules is determined. A random walk model is used to model the effect of the collisions.

The effect of the spring on the mass M is modeled by a discretized time Lagrangian sum. The Lagrangian in classical mechanics is equal to the difference between the kinetic and potential energies. The discrete time approximation of the kinetic energy of the oscillator mass of Figure 10.8 is

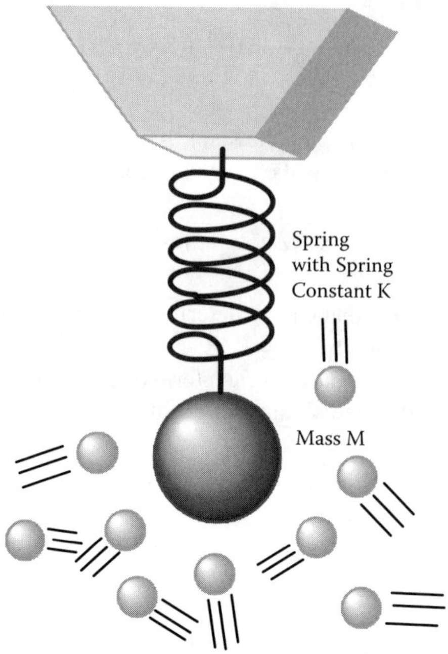

FIGURE 10.9 A one-dimensional oscillator consisting of a spring with spring constant K and a ball with mass m. The ball is being bombarded at random by other balls.

$$\text{Kinetic Energy } \approx \frac{1}{2} M \left(\frac{x_n - x_{n-1}}{\Delta t} \right)^2 \tag{10.8}$$

where the difference in the positions at time steps n and n – 1 divided by the time difference Δt is a discrete approximation of the oscillator mass velocity. The potential energy stored in the spring is

$$\text{Potential Energy } = \frac{1}{2} M \omega^2 x_n^2 \tag{10.9}$$

where the radial frequency squared ω^2 is equal to the spring constant K divided by the oscillator mass M. The Lagrangian sum L of the discrete time model is equal to the sum of the classical mechanics Lagrangians over all time steps.

$$L = \sum_{n=-\infty}^{n=\infty} \left[\text{Kinetic energy at time step n} - \text{ Potential energy at time step n} \right]$$

or (10.10)

$$L = \sum_{n=-\infty}^{n=\infty} \left[\frac{1}{2} M \frac{(x_n - x_{n-1})^2}{\Delta t^2} - \frac{1}{2} M \omega^2 x_n^2 \right]$$

The equation of motion for the discrete Lagrangian sum is

$$\frac{\partial L}{\partial x_n} = 0 \tag{10.11}$$

Substituting Equation 10.10 into Equation 10.11:

$$M\frac{x_{n+1} - 2x_n + x_{n-1}}{\Delta t^2} + M\omega^2 x_n = 0 \tag{10.12}$$

The first term on the left-hand side of Equation 10.12 represents the inertial force on the oscillator mass. It is equal to the mass M times the discrete second difference, an approximation of the second derivative of the oscillator mass position. This equation can be simplified by multiplying the equation by Δt^2 and dividing by the mass M.

$$x_{n+1} - 2x_n + x_{n-1} + \omega^2 \Delta t^2 x_n = 0 \tag{10.13}$$

This is a difference equation that can be solved for x_{n+1} by iteration. The initial values of the oscillator mass position at time steps 0 and 1, x_0 and x_1 have to be specified. One can define the discrete versions of the momentum P_{S1} of the oscillator mass before the collision with an air molecule as

$$P_{S1} \equiv M\frac{x_{n+1} - x_n}{\Delta t} \tag{10.14}$$

where the position x_{n+1} at time step n + 1 was calculated in Equation 10.13.

Next, the effect on the motion of the oscillator mass of the collision with the air molecules is determined. To do this the conservation of momentum and energy before and after the collision is used. The conservation of momentum is

$$P_{S1} + P_{A1} = P_{S2} + P_{A2} \tag{10.15}$$

where P_{S1} is the momentum of the oscillator mass before the collision with the air molecules as given by Equation 10.14, P_{S2} is the momentum of the oscillator mass after the collision with the air molecules, P_{A1} is the momentum of an air molecule before the collision with the oscillator mass, and P_{A2} is the momentum of the air molecule after the collision with the oscillator mass. The initial momentum P_{A1} of the air molecule is determined by a random walk process. Its value is determined by a random number generator. The conservation of the energy before and after the collision with the air molecules is

$$\frac{P_{S1}^2}{2M} + \frac{P_{A1}^2}{2m} = \frac{P_{S2}^2}{2M} + \frac{P_{A2}^2}{2m} \tag{10.16}$$

Solving Equation 10.15 for the momentum of an air molecule after the collision P_{A2} and substituting the result into Equation 10.16:

$$\frac{P_{S1}^2}{2M}+\frac{P_{A1}^2}{2m}=\frac{P_{S2}^2}{2M}+\frac{P_{S1}^2}{2m}+\frac{P_{A1}^2}{2m}+\frac{P_{S2}^2}{2m}=\frac{P_{S1}P_{A1}}{m}-\frac{P_{S1}P_{S2}}{m}-\frac{P_{A1}P_{S2}}{m} \tag{10.17}$$

Collecting terms and solving for P_{S2}^2:

$$P_{S2}^2-2P_{S2}\left(P_{S1}+P_{A1}\right)\frac{M}{M+m}+P_{S1}^2\frac{M-m}{M+m}+2P_{S1}P_{A1}\frac{M}{M+m}=0 \tag{10.18}$$

This is a quadratic equation that can be solved for the momentum P_{S2} of the oscillator mass after the collision.

$$P_{S2}=\frac{M}{M+m}\left[P_{S1}+P_{A1}\pm\left(\frac{m}{M}P_{S1}-P_{A1}\right)\right] \tag{10.19}$$

This equation has two solutions:

a) $P_{S2}=P_{S1}$ where $P_{A2}=P_{A1}$

and (10.20)

b) $P_{S2}=\frac{M-m}{M+m}P_{S1}+\frac{2M}{M+m}P_{A1}$ where $P_{A2}=\frac{2m}{M+m}P_{S1}-\frac{M-m}{M+m}P_{A1}$

The momentum P_{S2} can be used to calculate the velocity V_{n+1} of the oscillator mass at the n + 1 time step after the collision with an air molecule:

$$V_{n+1}=P_{S2}M \tag{10.21}$$

One can use the velocity to calculate the position x_{n+1} at time step n + 1 of the oscillator mass after the collision.

$$x_{n+1}=V_{n+1}\Delta t+x_n \tag{10.22}$$

This value of the position x_{n+1} of the oscillator mass after the collision with the air molecules at time step n + 1 is used in the next iteration employing Equation 10.13 to calculate the position x_{n+2}.

The positions of the oscillator mass were calculated for 249 time steps. At time step 249, the direction of time was reversed. The position of the oscillator mass was calculated for another 151 time steps in the reverse time direction. This path of the oscillator mass was recalculated 500 times. Each path is slightly different because the collision with air molecules is a random process. Recall that a random number

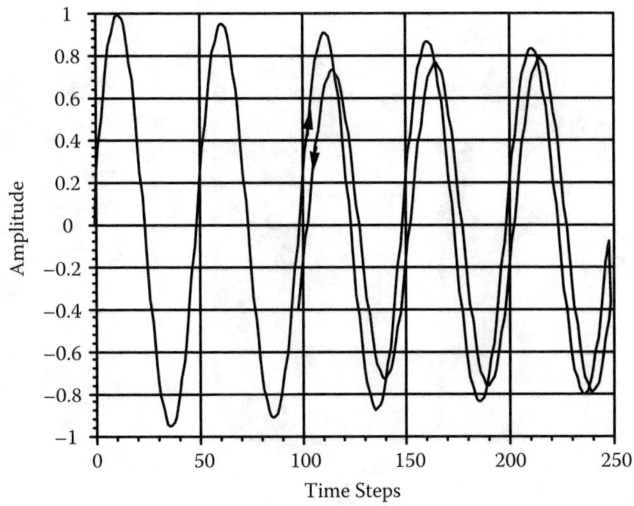

FIGURE 10.10 The average path of the oscillator mass is plotted. At time step 249, the direction of time suddenly reverses. Note that the oscillation amplitude decreases for both forward- and backward-moving time. Thus, the probabilistic model used here for the motion of the oscillator mass exhibits a unique arrow of time. The path is the average of 500 runs.

generator was used to determine the momentum of the air molecules at each collision. The average path of the oscillator mass is plotted in Figure 10.10. The path shown in Figure 10.10 is the average of 500 runs. Note that the oscillation amplitude decreases for both forward- and backward-moving time. Thus, the probabilistic model used here for the motion of the oscillator mass exhibits a *unique arrow of time*.

One way to think of this process is to postulate that there are many parallel spaces, as shown in Figure 10.11. In each space the oscillator mass follows a particular sequence of positions. At the end of each time step, the ball can hop between spaces.

The time dependence of the average randomness associated with the probabilities of the discrete positions of the oscillator mass is shown in Figure 10.12. The plot is the average of five sets of average randomness runs. This corresponds to a total of 500 runs of the oscillator mass positions. Note that the average randomness increases more or less monotonically with time. Thus, the system is becoming more random with time. *It has been experimentally observed that the average randomness, or entropy, of the world seems to be monotonically increasing with time*. However, locally, the average randomness can be decreased as long as the overall average randomness increases. Thus, the average randomness exhibits an *arrow of time*.[1] It increases monotonically with time. This can also be illustrated by analyzing the time evolution of scent molecules escaping from a perfume bottle, as shown in Figure 10.6. Observe that the scent molecules will escape from the perfume bottle when the bottle is opened. The scent molecules are considerably more random after they leave the bottle than when they were in the bottle. The probability that the scent molecules escaping from a perfume bottle will reassemble back into

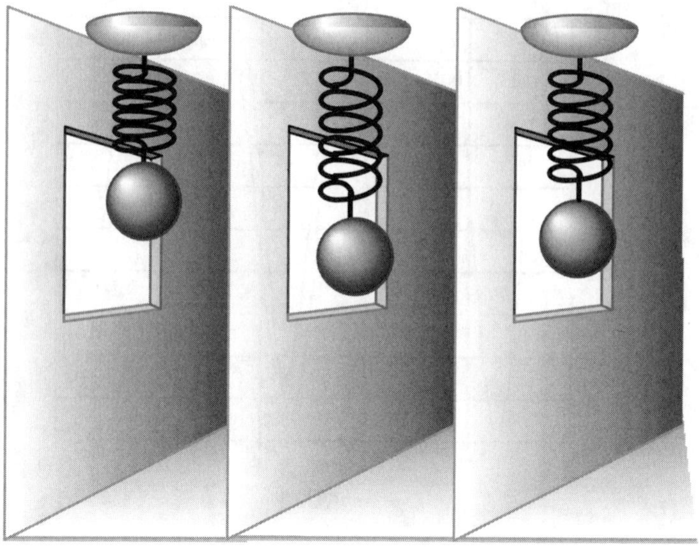

FIGURE 10.11 One way to think of the motion of the ball of the SSO is to postulate that there are many parallel spaces. Three such spaces are illustrated here. In each space the ball has a particular position.

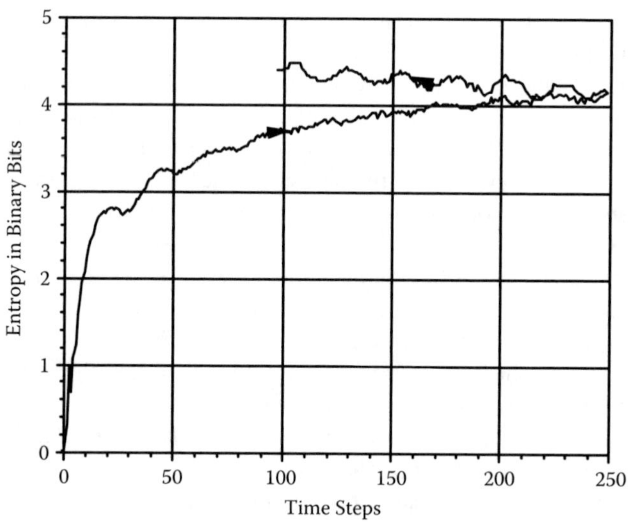

FIGURE 10.12 The time evolution of the average randomness associated with the positions of the oscillator mass in the 100 different runs. At time step 249, the direction of time suddenly reverses. The plot is the average of five sets of averaged randomness runs. This corresponds to a total of 500 runs of the oscillator mass position. Note that the average randomness increases more or less monotonically with both positive and negative time.

the bottle is exceedingly small. Thus, the randomness of the scent molecules will increase.

Many physical systems, including all living animals, age with time. A study[3] using middle-aged mice has shown that the changes brought on by aging are detectable in individual cells that drive the body's biological clock.

People's daily rhythms deteriorate with age, which causes sleep problems. The same is true of mice. Research has linked this difference to the suprachiasmatic nucleus (SCN), the brain's daily timekeeper. To see whether this disruption arises at the level of individual neurons or results from their interaction, neuroscientist Gene Block and colleagues at the University of Virginia recorded the electrical activity of mouse SCN neurons cultured in such low concentrations that they were unlikely to communicate with one another. In the September 2000 issue of *Neuroscience*, they reported that 8 neurons from middle-aged mice beat irregularly for days in the dish, whereas the 12 young adult cells remained robustly rhythmic. The research suggests that overall decline in SCN behavior is driven in large part by the deterioration of individual neurons. Paradoxically, researchers observed that aging appears to disrupt the rhythm of single cells more severely than it does the animal's overall behavior as measured by their sleep patterns. This raises an interesting question: "How do you get reliability from unreliable components?"

The simplest such system is a simple aging oscillator (SAO) located on a barren rotating planet. An observer resides on this planet. Consider the ball and spring oscillator shown in Figure 10.13. The spring of this oscillator would weaken with time. Therefore, the period of oscillation of this SAO will increase with time. Thus, there will be fewer oscillations in each revolution of the planet. *If the observer were to use the SAO as a clock, he or she would observe that according to the clock the days as well as the seasons of the planet will grow shorter as both the SAO clock and the observer age.*

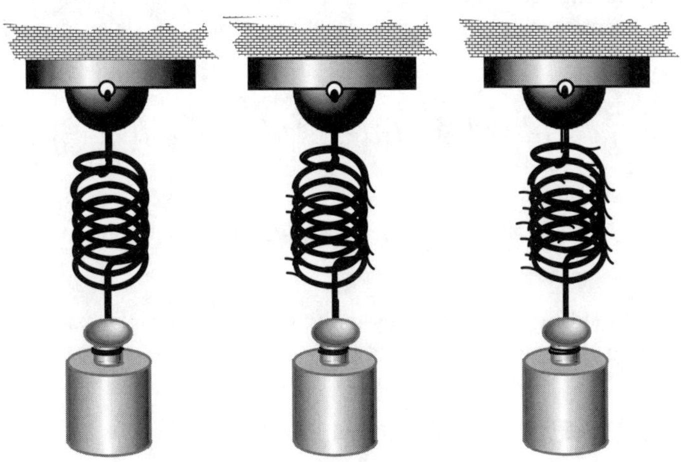

FIGURE 10.13 An example of an aging oscillator. Note the deteriorating springs.

10.4 CORRELATION FUNCTIONS OF RANDOM PROCESSES

The correlation $R_{fg}(t_2,t_1)$ of two random processes $f(x;t_1)$ and $g(x;t_2)$ is the average of the product of the two random processes, each of which occurs at different times t_1 and t_2.

$$R_{fg}(t_1,t_2)=\langle f^*(x;t_1)g(x;t_2)\rangle \tag{10.23}$$

where * denotes the complex conjugate. It is used if f(x,t) is complex. Two random processes with zero correlation are *orthogonal.* Two random processes with correlation $R_{fg}(t_1,t_2)$ equal to the product $\langle f(x,t_1)\rangle\langle g(x,t_2)\rangle$ of the individual average values $\langle f(x,t_1)\rangle$ and $\langle g(x,t_2)$of the random processes are *uncorrelated.*

A perhaps more-often-used correlation function is the *autocorrelation function* $R_{ff}(t,t+\tau)$, the correlation of a random process f(x;t) with its delayed version $f(x;t+\tau)$

$$R_{ff}(t,t+\tau)\equiv\langle f^*(x;t)f(x;t+\tau)\rangle \tag{10.24}$$

The *covariance* $C_{fg}(t,t+\tau)$ is a "central" correlation of two random processes f(x;t) and g(x;t) as follows:

$$\text{a)}\quad C_{fg}(t,t+\tau)\equiv\left\langle\left[f^*(x;t)-\langle f^*(x;t)\rangle\right]\left[g(x;t+\tau)-\langle g(x;t+\tau)\rangle\right]\right\rangle$$

or (10.25)

$$\text{b)}\quad C_{fg}(t,t+\tau)\equiv R_{fg}(t,t+\tau)-\langle f^*(x;t)\rangle\langle g(x;t+\tau)\rangle$$

The second equation is obtained by multiplying out the right-hand side of the first equation and making use of Equation 10.23. Similarly, the *autocovariance* $C_{ff}(t,t+\tau)$ is a "central" correlation of a random process f(x;t) with its delayed version $f(x;t+\tau)$ is

$$C_{ff}(t,t+\tau)\equiv\left\langle\left[f^*(x;t)-\langle f^*(x;t)\rangle\right]\left[f(x;t+\tau)-\langle f(x;t+\tau)\rangle\right]\right\rangle \tag{10.26}$$

or

$$C_{ff}(t,t+\tau)\equiv R_{ff}(t,t+\tau)-\langle f^*(x;t)\rangle\langle f(x;t+\tau)\rangle \tag{10.27}$$

where one obtained Equation 10.27 by multiplying out the right-hand side of equation and making 10.26 and made use of Equation 10.23.

10.5 STATIONARITY OF RANDOM PROCESSES

Next, the stationarity of random processes is discussed. A random process is said to be *stationary to first order* if the density function p(x;t) as well as the random process f(x;t) is invariant under time translation; that is:

$$\text{a)}\quad p(x;t) = p(x;t+\tau) \quad \text{and} \quad \text{b)}\quad f(x;t) = f(x;t+\tau) \tag{10.28}$$

where x is a random variable and t is the time. A consequence of this is that the average value of a random process that is stationary to first order is constant.

A random process is said to be *stationary to second order,* or *wide-sense stationary* (WSS), if the average value $\langle f(x;t)\rangle$ of the random process is constant and its autocorrelation function $R_{ff}(\tau)$ depends only on the time difference τ.

$$\langle f(x;t)\rangle = \text{constant} \quad \text{and} \quad R(t,t+\tau) = R(\tau) \tag{10.29}$$

that is, a random process is said to be stationary to second order, or WSS, if:

1. The autocorrelation must depend only on the time difference.
2. The average value of the random process is a constant.

The autocorrelation of a wide-sense stationary process has the following properties:

1. $R_{ff}(\tau) \le R_{ff}(0)$
2. $R_{ff}(-\tau) = R_{ff}(\tau)$
3. $R_{ff}(0) = \langle |f(x;t)|^2 \rangle$
4. If $\langle f(x;t)\rangle$ is equal to zero and f(x;t) has no periodic part then $(\lim/|\tau| \to \infty) R_{ff}(\tau) = 0$
5. If f(x;t) has a periodic component, then $R_{ff}(\tau)$ will have a periodic component with the same period.

The *correlation* or *cross-correlation* of WSS processes has the following properties:

1. $R[f(x,t)g(x;t) - \tau)] = R[f(x,t+\tau)g(x;t)]$
2. $|R_{fg}(\tau)| \le \sqrt{R_{ff}(\tau)R_{gg}(\tau)}$
3. $|R_{fg}(\tau)| \le 1/2\ [(R_{ff}(\tau) + R_{gg}(\tau)]$

The covariance or cross-covariance $C_{fg}(\tau)$ of two random processes that are wide-sense stationary have the following form:

$$C_{fg}(\tau) \equiv R_{fg}(\tau) - \langle f^*(x;t)\rangle \langle g(x;t+\tau)\rangle \tag{10.30}$$

where, of course, $\langle f(x;t)\rangle$ and $\langle g(x;t) + \tau)\rangle$ are constants in this case. Similarly, the autocovariance $C_{ff}(\tau)$ of a random process that is WSS has the following form:

$$C_{ff}(\tau) \equiv R_{ff}(\tau) - \left[\langle f(x;t)\rangle^2 + \text{constant}\right] \tag{10.31}$$

where $\langle f(x;t) \rangle$ is a constant. In the limit as the time difference τ goes to zero, the autocovariance of a WSS process becomes equal to the variance of the random process.

$$\lim_{\tau \to 0} C_{ff}(\tau) = \sigma^2\left[f(x;t)\right] \tag{10.32}$$

EXAMPLE 10.1

A simple example of a random process has the following form:

$$f(f,\theta) = A\cos(\omega t - \theta) + B \tag{EX10.1.1}$$

where A, B, and ω are constants and θ is a uniformly distributed random variable.

$$p(\theta) = \frac{1}{2\pi} \tag{EX10.1.2}$$

where $\pi > \theta \geq -\pi$. First, the average value $\langle f(\theta;t) \rangle$ of the random process $f(t,\theta)$ is calculated.

$$\left\langle f(t,\theta) \right\rangle = \frac{A}{2\pi}\int_{-\pi}^{\pi} \cos(\omega t - \theta)\,d\theta + \frac{B}{2\pi}\int_{-\pi}^{\pi} d\theta \tag{EX10.1.3}$$

By integrating Equation EX10.1.3, one obtains for the average value $\langle f(\theta;t) \rangle$ of the random process $f(\theta;t)$:

$$\left\langle f(\theta;t) \right\rangle = B \tag{EX10.1.4}$$

Note that the average value $\langle f(t,\theta) \rangle$ of the random process $f(t,\theta)$ is a constant equal to B. Next, the autocorrelation function $R(t,t+\tau)$ is calculated.

$$\begin{aligned} R(t,t+\tau) = {} & \frac{A^2}{2\pi}\int_{-\pi}^{\pi} \cos(\omega t + \theta)\cos\left[\omega(t+\tau) + 0\right]d\theta + \frac{AB}{2\pi}\int_{-\pi}^{\pi} \cos(\omega t + \theta)\,d\theta + \\ & \frac{AB}{2\pi}\int_{-\pi}^{\pi} \cos\left[\omega(t+\tau) + \theta\right]d\theta + \frac{B^2}{2\pi}\int_{-\pi}^{\pi} d\theta \end{aligned} \tag{EX10.1.5}$$

The integrals in the second and third terms of Equation EX10.1.5 are equal to zero, and the product of the cosines in the first integral results in a term that does not depend on θ.

$$R(t,t+\tau) = \frac{A^2}{4\pi}\cos\omega\tau\int_{-\pi}^{\pi} d\theta + \frac{A^2}{4\pi}\int_{-\pi}^{\pi} \cos\left[\omega(2t+\tau) + 2\theta\right]d\theta + B^2 \tag{EX10.1.6}$$

The second integral of Equation EX10.1.6 is also equal to zero.

$$R(\tau) = \frac{A^2}{2}\cos\omega\tau + B^2 \quad \text{(EX10.1.7)}$$

Observe from Equation EX10.1.7 that the autocorrelation function $R(\tau)$ is only a function of the time difference τ. Because the average value $\langle f(t,\theta) \rangle$ of the random process $f(t,\theta)$ is a constant and the autocorrelation function $R(\tau)$ is only a function of the time difference τ, this random process is WSS.

EXAMPLE 10.2

Another example is a joint Gaussian random process. A simple version of a joint Gaussian random process f(x,y,t) is

$$f(t,x,y) = x\cos\omega t + y\sin\omega t \quad \text{(EX10.2.1)}$$

where x and y are Gaussian random variables with the following joint Gaussian probability density:

$$p(x \cap y) = \frac{1}{2\pi\sigma^2\sqrt{1-\rho^2}}\exp\left[-\frac{(x-a)^2 - 2\rho(x-a)(y-b) + (y-b)^2}{2\sigma^2(1-\rho^2)}\right] \quad \text{(EX10.2.2)}$$

where $-\infty < x \le \infty$ and $-\infty < y \le \infty$

For simplicity the variances σ^2 of the two Gaussian random variables x and y are chosen to be identical. One will require the various averages of the two joint Gaussian random variables x and y. These averages have been calculated in Chapter 8, Equations 8.119, 8.123, and 8.127; they are

$$\begin{aligned} &\text{a) } \langle x \rangle = a \quad \text{b) } \langle y \rangle = b \quad \text{c) } \langle xy \rangle = ab + \rho\sigma^2 \\ &\text{d) } \langle x^2 \rangle = \sigma^2 + a^2 \quad \text{e) } \langle y^2 \rangle = \sigma^2 + b^2 \end{aligned} \quad \text{(EX10.2.3)}$$

As was done in Example 10.1, one first calculates the average value $\langle f(t,x,y) \rangle$ of the random process f(t,x,y).

$$\langle f(t,x,y) \rangle = \langle x \rangle \cos\omega t + \langle y \rangle \sin\omega t \quad \text{(EX10.2.4)}$$

By using Equations EX10.2.3a and EX10.23b in Equation EX10.2.4, the average value $\langle f(t,x,y) \rangle$ of the random process f(t,x,y) is obtained:

$$\langle f(t,x,y)\rangle = a\cos\omega t + b\sin\omega t \tag{EX10.2.5}$$

Next, the autocorrelation function R(t,t + t) is calculated.

$$R(t,t+\tau) = \langle [x\cos\omega t + y\sin\omega t][x\cos\omega(t+\tau) + \sin\omega(t+\tau)]\rangle \tag{EX10.2.6}$$

Multiplying out Equation EX10.2.6:

$$\begin{aligned} R(t,t+\tau)\langle x^2\rangle \cos\omega t\cos\omega(t+\tau) + \langle xy\rangle[\sin\omega t x\cos\omega(t+\tau) + \\ \cos\omega t\sin\omega t(t+\tau)] + \langle y^2\rangle \sin\omega t\sin\omega(t+\tau) \end{aligned} \tag{EX10.2.7}$$

Substituting Equations EX10.2.3 into Equation EX10.2.7:

$$\begin{aligned} R(t,t+\tau) = \left[\sigma^2 + \frac{a^2+b^2}{2}\right]\cos\omega\tau + \frac{a^2-b^2}{2}\cos[2\omega t + \omega\tau] + \\ [ab + \rho\sigma^2]\sin[2\omega t + \omega\tau] \end{aligned} \tag{EX10.2.8}$$

This process is not stationary in any sense.

10.6 THE TIME AVERAGE AND ERGODICITY OF RANDOM PROCESSES

In relativistic calculations no distinction is made between the spatial and time coordinates. In that case the time average and the average over the spatial random variables is treated in the same way. In the relativistic case, one can simply consider time to be the fourth coordinate and treat it similarly to the other "random" coordinates. The probability density p(t,**r**) in this case would be a function of both the random spatial and temporal coordinates. However, here the time average {f(x;t)} of a random process is defined as follows:

$$\{f(x;t)\} \equiv \lim_{T\to\infty} \frac{1}{2T}\int_{-T}^{T} f(x;t)\,dt \tag{10.33}$$

Note that the time average {f(x;t)} is still a function of the random variable x. The time average defined previously will be used in the remaining document. The time autocorrelation $R_{ff}(x,\tau)$ of the random process f(x,t) can be written as follows:

$$R_{ff}(x;\tau) \equiv \{f(x;t)f[x;t+\tau]\} \tag{10.34}$$

The time correlation $R_{ff}(x;\tau)$ is still a function of the random variable x.

Next, the average value of Equation 10.34 is taken, and it is assumed that the random process f(x;t) is a WSS process.

$$\text{a)}\ \langle R_{ff}(x;\tau)\rangle = R_{ff}(\tau) \quad \text{and} \quad \text{b)}\ \langle\{f(x,t)\rangle = \text{constant} \tag{10.35}$$

where it was assumed that one can interchange the averaging over time and the random variable x.

An *ergodic* random process is a system for which the average over a random variable is equal to its time average.

$$\{f(x;t)\} = \langle f(x;t)\rangle \tag{10.36}$$

It is often assumed that some real physical systems are ergodic, although this assumption is very seldom correct.

10.7 PARTIALLY COHERENT LIGHT RAYS AS RANDOM PROCESSES.

By coherence one means the ability to soothsay the future instantaneous amplitude of the light electromagnetic field at some distance from the light source from the knowledge of its current value at the light source. Since interference effects can be observed in thin oil films illuminated by incoherent light sources the light from these sources must have some coherence properties. The concept of a random process was used to derive a model of monochromatic but partially coherent light. This type of light can be generated by a Light Emitting semiconductor Diode or a gas discharge lamp. It is true that the light electromagnetic field is a vector. However, experience has shown that a scalar model of the light electromagnetic field that is simpler to calculate than the vector model gives sufficient accurate results for most applications. A Green's function derivation of diffraction theory has traditionally been used. A Green's function that is a random process was utilized here. That is, a scalar Green's function $G(t,r_{A1},\theta)$ that besides being a function of position vector r_{A1} is also a function of the time t and a function of a random variable θ was used.

$$G(t,r_{A1},\theta) = \frac{1}{j\lambda r_{A1}} \exp\left[j\left(\frac{2\pi}{\lambda}+\frac{\theta}{\eta}\right)(ct-r_{A1})\right]\cos\phi(\mathbf{n},\mathbf{r}_{A1}) \tag{10.37}$$

where λ is the wavelength of the light, c is the speed of light, θ is a random phase, η is the coherence length, r_{A1} is the distance between a point on the object plane and an observation point in the image plane, $\phi(\mathbf{n},r_{01})$ is the angle between a normal $\mathbf{n}$ to the surface of the object plane and r_{A1}, see Figure 10.14. The distance r_{A1} between a point on the object plane and an observation point in the image plane is:

$$r_{A1} = \sqrt{z^2 + (\mathbf{r}_A - \mathbf{r}_1)^2} \tag{10.38}$$

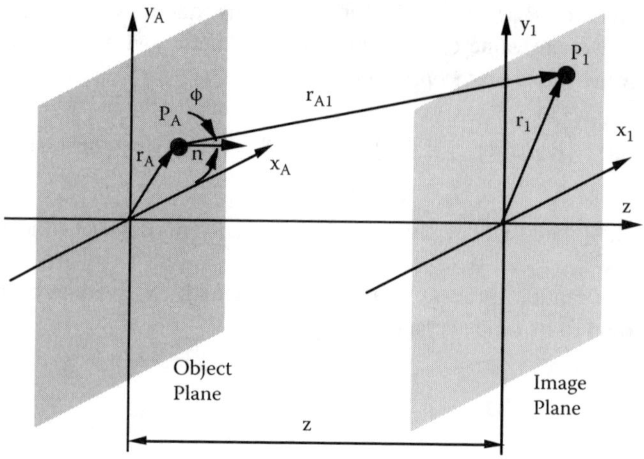

FIGURE 10.14 A light ray from the point P_A on the object plane impinging at the point P_1 on the image plane.

where $\mathbf{r}_A$ is a vector in the object plane, $\mathbf{r}_1$ is a vector in the image plane, and z is the distance between the object and image planes along the optical axis, see Figure 10.14.

One starts with a diffraction equation for the scalar light electromagnetic field $u(t,\mathbf{r}_1,\theta)$ at an observation point P_1 in the image plane using the Green's function $G(t,r_{A1},\theta)$ of equation 10.37, see Figure 10.14. The diffraction equation is derived from Equations 4-1 and 4-2 on page 58 of "Introduction to Fourier Optics" by Joseph Goodman, McGraw-Hill Publishing Co.

$$u(t,\mathbf{r}_1,\theta) = \int_{O.P.} ds_A u(\mathbf{r}_A) \frac{\exp\left[j\left(\frac{2\pi}{\lambda}+\frac{\theta}{\eta}\right)\left(ct - z\sqrt{1+\frac{(\mathbf{r}_1-\mathbf{r}_A)^2}{z^2}}\right)\right]}{j\lambda z\sqrt{1+\frac{(\mathbf{r}_1-\mathbf{r}_A)^2}{z^2}}} \cos\phi(\mathbf{n}, r_{A1}) \tag{10.39}$$

where equation 10.49 was used for r_{A1} with z factored out of the square root, and the integration is over the object plane abbreviated O.P.

One assumes that the distance z between planes is much larger than distances r_A and r_1 in the object and image planes considered. This is known as the paraxial approximation. With this approximation Equation 10.39 can be simplified as follows:

$$u(t,\mathbf{r}_1,\theta) \approx \frac{\exp\left[j\frac{2\pi}{\lambda}(ct-z)\right]}{j\lambda z}\int_{O.P.} ds_A u(\mathbf{r}_A)\exp\left[-j\frac{\pi(\mathbf{r}_1-\mathbf{r}_A)^2}{\lambda z}\right]\times \tag{10.40}$$

$$\exp\left\{j\frac{\theta}{\eta}\left[ct - z - \frac{(\mathbf{r}_1-\mathbf{r}_A)^2}{2z}\right]\right\}$$

where one expanded the square roots to first order in the small term $(\mathbf{r}_1 - \mathbf{r}_a)^2/z^2$. This is the Fresnel approximation of the scalar light electric field. The small term $(\mathbf{r}_1 - \mathbf{r}_A)^2/2z$ is kept in the last exponential of Equation 10.40 at this point.

Note that the light electric field $u(t,\mathbf{r}_1,\theta)$ on the image plane is a random process. Assume that all values of the random phase θ between −π and π are equally likely. Thus, the probability density per unit phase p(θ) is

$$p(\theta) = \frac{1}{2\pi} \quad \text{for} \quad -\pi < \theta \le \pi \tag{10.41}$$

One next calculates the average value $\langle u(t,\mathbf{r}_1,\theta)\rangle$ of the scalar light electric field on the image plane $u(t,\mathbf{r}_1,\theta)$.

$$u(t,\mathbf{r}_1,\theta) \approx \frac{\exp\left[j\frac{2\pi}{\lambda}(ct-z)\right]}{j\lambda z}\int_{O.P.} ds_A u(\mathbf{r}_A)\exp\left[-j\frac{\pi(\mathbf{r}_1-\mathbf{r}_A)^2}{\lambda z}\right]\times \tag{10.42}$$
$$\frac{1}{2\pi}\int_{-\pi}^{\pi} d\theta \exp\left[j\frac{\theta}{\eta}(ct-z)\right]$$

where the small term $(\mathbf{r}_1 - \mathbf{r}_A)^2/2z$ was neglected in the integral over the random phase. By integrating one obtains:

$$\langle u(t,\mathbf{r}_1,\theta)\rangle \approx \frac{\exp\left[j\frac{2\pi}{\lambda}(ct-z)\right]}{j\lambda z}\int_{O.P.} ds_A u(\mathbf{r}_A)\exp\left[-j\frac{\pi(\mathbf{r}_1-\mathbf{r}_A)^2}{\lambda z}\right]$$
$$\frac{\exp\left[j\frac{\pi(ct-z)}{\eta}\right]-\exp\left[-j\frac{\pi(ct-z)}{\eta}\right]}{2j\frac{\pi(ct-z)}{\eta}} \tag{10.43}$$

or

$$\langle u(t,\mathbf{r}_1,\theta)\rangle \approx \frac{\exp\left[j\frac{2\pi}{\lambda}(ct-z)\right]}{j\lambda z}\int_{O.P.} ds_A u(\mathbf{r}_A)\exp\left[-j\frac{\pi(\mathbf{r}_1-\mathbf{r}_A)^2}{\lambda z}\right]\frac{\sin\frac{\pi(ct-z)}{\eta}}{\frac{\pi(ct-z)}{\eta}} \tag{10.44}$$

Note that as time t increases, the average value $\langle u(t,\mathbf{r}_1,\theta)\rangle$ of the scalar light electric field on the image plane $u(t,\mathbf{r}_1,\theta)$ becomes very small.

$$\lim_{t \to \infty} \left\langle u\left(t, \mathbf{r}_l, \theta\right) \right\rangle = 0 \tag{10.45}$$

The crosscorrelation function $R_{uu}(t,t + \tau)$ is:

$$R_{uu}\left(t, t+\tau\right) = \left\langle u^*\left(t, \mathbf{r}_l - \mathbf{r}_A, \theta\right) u\left(t+\tau, \mathbf{r}_l - \mathbf{r}_B, \theta\right) \right\rangle \tag{10.46}$$

where $u^*(t,\mathbf{r}_l - \mathbf{r}_A,\theta)$ is the complex conjugate of $u(t,\mathbf{r}_l - \mathbf{r}_A,\theta)$. Substituting Equation 10.44 into Equation 10.46:

$$R_{uu}\left(t, t+\tau\right) = \frac{\exp\left[j\frac{2\pi}{\lambda}c\tau\right]}{\lambda^2 z^2 2\pi} \int_{O.P} ds_A \int_{O.P} ds_B u^*\left(\mathbf{r}_A\right) u\left(\mathbf{r}_B\right)$$
$$\exp\left[-j\frac{\pi\left(\mathbf{r}_l - \mathbf{r}_A\right)^2 + \pi\left(\mathbf{r}_l - \mathbf{r}_B\right)^2}{\lambda z}\right] \times \int_{-\pi}^{\pi} d\theta \exp\left\{j\frac{\theta}{\eta}\left[c\tau - \frac{\left(\mathbf{r}_l - \mathbf{r}_A\right)^2 - \left(\mathbf{r}_l - \mathbf{r}_B\right)^2}{2z}\right]\right\} \tag{10.47}$$

Let:

$$\text{a)}\quad \mathbf{r}_{lA} = \mathbf{r}_l - \mathbf{r}_A \quad \text{and} \quad \text{b)}\quad -\mathbf{r}_{lB} = \mathbf{r}_l - \mathbf{r}_B \tag{10.48}$$

By integrating Equation 10.47 over the random phase and making use of Equations 10.48, one obtains by a process similar to the one leading to Equation 10.44:

$$R_{uu}\left(t, t+\tau\right) = -\frac{\exp\left[j\frac{2\pi}{\lambda}c\tau\right]}{\lambda^2 z^2} \int_{O.P} ds_A \int_{O.P} ds_B u^*\left(\mathbf{r}_A\right) u\left(\mathbf{r}_B\right)$$
$$\exp\left[-j\frac{\pi\left(\mathbf{r}_l - \mathbf{r}_A\right)^2 + \pi\left(r_l - r_B\right)^2}{\lambda z}\right] \times \frac{\sin\left[\frac{\pi}{\eta}c\tau - \frac{\pi\left(r_{lA}^2 - r_{lB}^2\right)}{2\eta z}\right]}{\frac{\pi}{\eta}c\tau - \frac{\pi\left(r_{lA}^2 - r_{lB}^2\right)}{2\eta z}} \tag{10.49}$$

Note that for a sufficiently long time t the average value of the scalar electric field is equal to a constant, zero, and the crosscorrelation function $R_{uu}(\tau)$ is only a function of the time difference τ. Therefore, the scalar electric field is a wide-sense stationary random process.

Even though the stochastic power P_{uu} has not yet been discussed it will be used here. The stochastic power will be discussed in Chapter 11. The stochastic power here, is proportional to the electrical power. The stochastic power of a wide-sense stationary random process is equal to the limit as the time difference τ goes to zero of the crosscorrelation function $R_{uu}(\tau)$.

$$P_{uu} = \frac{-1}{\lambda^2 z^2}\int_{O.P} ds_A \int_{O.P} ds_B u^*(\mathbf{r}_A)u(\mathbf{r}_B)\exp\left[-j\frac{\pi(\mathbf{r}_l - \mathbf{r}_A)^2 + \pi(\mathrm{r}_l - \mathrm{r}_B)^2}{\lambda z}\right]\times \frac{\sin\left[\frac{\pi\left(r_{1A}^2 - r_{1B}^2\right)}{2\eta z}\right]}{\frac{\pi\left(r_{1A}^2 - r_{1B}^2\right)}{2\eta z}} \tag{10.50}$$

Note that the sink function will have zeros for:

$$\frac{\pi\left(r_{1A}^2 - r_{1B}^2\right)}{2\eta z} = \pi \tag{10.51}$$

By cross mutiplying equation 10.51 one obtains that the difference in area between a disc with an area πr_{1A}^2 and a disc with an area πr_{1B}^2 is equal to $2\pi\eta z$

$$\pi r_{1A}^2 - \pi r_{1B}^2 = 2\pi\eta z \tag{10.52}$$

That is, the light radiated from annular rings of area $2\pi\eta z$ centered at a vectorial distance $\mathbf{r}_l$ from the optical axis in the object plane will be coherent at a point in the image plane that has the same vectorial location $\mathbf{r}_l$ relative to the optical axis as the object point as shown in Figure 10.15. The radius $\mathbf{r}_B$ of these rings can have any value including zero. Since the area of the rings is constant rings with larger radii will be narrower. For example a ring with inner radius $\mathbf{r}_{1B}$ equal to zero will have an outer radius $\mathbf{r}_{1A}$ of 0.72111 mm for a coherence length η of 1040 nm and a distance z of 25 cm from object plane to image plane. This is much larger than the coherence length.

10.8 STOCHASTIC ASPECTS OF TRANSITIONS BETWEEN STATES

None of the mathematical models of nature such as classical mechanics, quantum mechanics, electromagnetic theory, etc., distinguish between transitions from a lower energy to a higher energy and a transition from a higher energy to a lower energy. However, the stochastic mathematical model of nature does distinguish between transitions from a lower energy to a higher energy and a transition from a higher energy to a lower energy. A derivation similar to the one used to derive Nernst's theorem

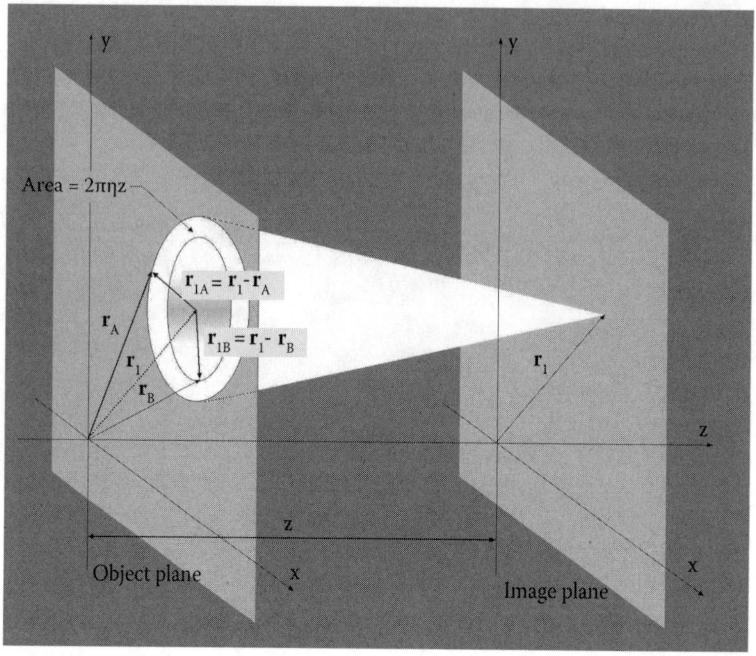

FIGURE 10.15 The light radiated from annular rings of area $2\pi\eta z$ centered at a vectorial distance $\mathbf{r}_1$ from the optical axis in the object plane will be coherent at a point in the image plane that has the same vectorial location $\mathbf{r}_1$ relative to the optical axis as the object point. Here η is the coherence length.

at the end of Chapter 8 to analyze the stochastic aspects of transitions between two energy levels will be used.

Consider a system that at various times has either an energy E_1 or a larger energy E_2, as shown in Figure 10.16. Two sets of events are postulated. Set $S_A\{A_1,A_2\}$ consists of event A_1, which is the system having an energy E_1 at time t_a, and event A_2, which is the system having an energy E_2 at time t_a. Set $S_B\{B_1,B_2\}$ consists of event B_1, which is the system having an energy E_1 at time t_b and event B_2, which is the system having an energy E_2 at time t_b.

The description of the motion of this system that at various times has either an energy E_1 or E_2 involves four joint probabilities:

$P(A_1 \cap B_1) = P(A_1|B_1)Q_1$ is the joint probability that the particle has an energy E_1 at time t_a and an energy E_1 at time t_b.

$P(A_2 \cap B_1) = P(A_2|B_1)Q_1$ is the joint probability that the particle has an energy E_2 at time t_a and an energy E_1 at time t_b.

$P(A_1 \cap B_2) = P(A_1|B_2)Q_2$ is the joint probability that the particle has an energy E_1 at time t_a and an energy E_2 at time t_b.

$P(A_2 \cap B_2) = P(A_2|B_2)Q_2$ is the joint probability that the particle has an energy E_2 at time t_a and an energy E_2 at time t_b.

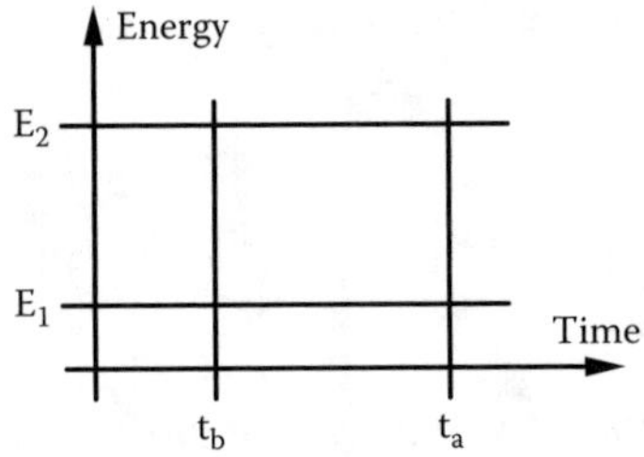

FIGURE 10.16 The evolution of time of a system with two energy states, E_1 and E_2.

The preceding conditional probabilities are

$P(A_1|B_1)$ is the conditional probability that the particle has an energy E_1 at time t_a provided it had an energy E_1 at time t_b.

$P(A_2|B_1)$ is the conditional probability that the particle has an energy E_2 at time t_a provided it had an energy E_1 at time t_b.

$P(A_1|B_2)$ is the conditional probability that the particle has an energy E_1 at time t_a provided it had an energy E_2 at time t_b.

$P(A_2|B_2)$ is the conditional probability that the particle has an energy E_2 at time t_a provided it had an energy E_2 at time t_b.

Q_1 is the probability the particle has an energy E_1 at time t_b and Q_2 is the probability the particle has an energy E_2 at time t_b.

P_1 is the probability the particle has an energy E_1 at time t_a and P_2 is the probability the particle has an energy E_2 at time t_a. From Chapter 3, Section 3.1, these quantities are related as follows:

$$P_1 = P(A_1|B_1)Q_1 + P(A_1|B_2)Q_2 \tag{10.53}$$

$$P_2 = P(A_2|B_1)Q_1 + P(A_2|B_2)Q_2 \tag{10.54}$$

The information $\mathbb{I}$ conveyed by the motion of this system is

$$\mathbb{I} = \mathbb{S}(B) - \left[\mathbb{S}(A|B_1)Q_1 + \mathbb{S}(A|B_2)Q_2\right] \tag{10.55}$$

where the conditional entropies $\mathbb{S}(A|B_1)$ and $\mathbb{S}(A|B_1)$ as described by Equation 7.21 of Chapter 7 are given by:

$$\begin{aligned} &\text{a)}\quad \mathbb{S}(A|B_1) \equiv -k\left[P(A_1|B_1)\ln P(A_1|B_1) + P(A_2|B_1)\ln P(A_2|B_1)\right] \quad \text{and} \\ &\text{b)}\quad \mathbb{S}(A|B_1) \equiv -k\left[P(A_1|B_2)\ln P(A_1|B_2) + P(A_2|B_2)\ln P(A_2|B_2)\right] \end{aligned} \tag{10.56}$$

where k is Boltzmann's constant. The entropy $\mathbb{S}(B)$ associated with the set of events B is

$$\mathbb{S}(B) = k\left[Q_1 \ln Q_1 + Q_2 \ln Q_2\right] \tag{10.57}$$

As was shown in Chapter 8, the various probabilities describing the system depend on what is known about the system. The following fact about the stochastic system is known: The sum of the conditional probabilities with the same condition must be equal to one.

$$\begin{aligned} &\text{a)}\quad k\alpha_2 Q_1[P(A_1|B_1) + P(A_1|B_1) - 1] = 0 \\ &\text{b)}\quad k\alpha_1 Q_2[P(A_1|B_2) + P(A_2|B_2) - 1] = 0 \end{aligned} \tag{10.58}$$

where the normalization conditions of the conditional probabilities were multiplied by Lagrange multiplier constants $k\alpha_1 Q_1$ and $k\alpha_2 Q_2$ that will be used at a later step in the derivation. The temperature T of this system is defined in a conventional way, as was done in Equation 7.40 in Chapter 7.

$$T \equiv \frac{\partial U}{\partial \mathbb{S}(B)} \tag{10.59}$$

where U is the average energy of the particle at time t_a. The average energy U at time t_a is

$$U = E_1 P_1 + E_2 P_2 \tag{10.60}$$

Substituting Equations 10.43 and 10.44 for P_1 and P_2 into Equation 10.50:

$$-k\beta[E_1P(A_1|B_1)Q_1 + E_1P(A_1|B_2)Q_2 + E_2P(A_2|B_1)Q_1 + E_2P(A_2|B_2)Q_2 - U] = 0 \tag{10.61}$$

where $-k\beta$ is another Lagrange multiplier. Because Equations 10.58 and 10.61 are equal to zero, one can add them to the joint information $\mathbb{I}$ of the system without altering its value. Using Equations 10.56 for the conditional entropies to express the joint entropy in terms of the conditional probabilities:

$$\begin{aligned} \mathbb{I} = \mathbb{S}(B) + k\{&P(A_1|B_1)\ln P(A_1|B_1)Q_1 + P(A_2|B_1)\ln P(A_2|B_1)Q_1 + \\ &P(A_1|B_2)\ln P(A_1|B_2)\, Q_2 + P(A_2|B_2)\ln P(A_2|B_2)Q_2 + \alpha_1 P(A_1|B_1)Q_1 + \\ &\alpha_1 P(A_2|B_1)Q_1 + \alpha_2 P(A_1|B_2)Q_2 + \alpha_2 P(A_2|B_2) + \beta E_1 P(A_1|B_1)Q_1 + \\ &\beta E_1 P(A_1|B_2)Q_2 + \beta E_2 P(A_2|B_1)Q_1 + \beta E_2 P(A_2|B_2)Q_2\} - kQ_1\,\alpha_1 \\ &- k\, Q_2\, \alpha_2 - k\beta U \end{aligned} \tag{10.62}$$

The various processes acting on a system always tend to increase its entropy or information. The conditional probabilities will be used to maximize the information to determine the properties of the transition processes. The joint information $\mathbb{I}$ will be a maximum when all its partial derivatives with respect to the conditional probabilities will be equal to zero.

$$\frac{\partial \mathbb{I}}{\partial P\left(A_n \middle| B_m\right)} = 0 \tag{10.63}$$

Substituting Equation 10.55 into Equation 10.63:

$$\ln P(A_1|B_1) + 1 + \alpha_1 + \beta E_1 = 0 \tag{10.64}$$

$$\ln P(A_2|B_1) + 1 + \alpha_1 + \beta E_2 = 0 \tag{10.65}$$

$$\ln P(A_1|B_2) + 1 + \alpha_2 + \beta E_1 = 0 \tag{10.66}$$

$$\ln P(A_2|B_2) + 1 + \alpha_2 + \beta E_2 = 0 \tag{10.67}$$

Equations 10.64, 10.65, 10.66, and 10.67 can be solved for the conditional probabilities:

$$P\left(A_1 \middle| B_1\right) = \exp\left[-1-\alpha_1\right]\exp\left[-\beta E_1\right] \tag{10.68}$$

$$P\left(A_2 \middle| B_1\right) = \exp\left[-1-\alpha_1\right]\exp\left[-\beta E_2\right] \tag{10.69}$$

$$P\left(A_1 \middle| B_2\right) = \exp\left[-1-\alpha_2\right]\exp\left[-\beta E_1\right] \tag{10.70}$$

$$P\left(A_2 \middle| B_2\right) = \exp\left[-1-\alpha_2\right]\exp\left[-\beta E_2\right] \tag{10.71}$$

To evaluate the Lagrange multiplier constants α_1 and α_2, Equations 10.68 and 10.69 are substituted into Equation 10.58a, and Equations 10.70 and 10.71 are substituted into Equation 10.58b. Solving for $\exp[-1 - \alpha_1]$ and $\exp[-1 - \alpha_1]$ and substituting the resulting expressions back into Equations 10.68, 10.69, 10.70, and 10.71:

$$P\left(A_1 \middle| B_1\right) = \frac{\exp\left[-\beta E_1\right]}{\exp\left[-\beta E_1\right] + \exp\left[-\beta E_2\right]} \tag{10.72}$$

$$P\left(A_2 \middle| B_1\right) = \frac{\exp\left[-\beta E_2\right]}{\exp\left[-\beta E_1\right] + \exp\left[-\beta E_2\right]} \tag{10.73}$$

$$P\left(A_1\middle|B_2\right)=\frac{\exp\left[-\beta E_1\right]}{\exp\left[-\beta E_1\right]+\exp\left[-\beta E_2\right]} \tag{10.74}$$

$$P\left(A_2\middle|B_2\right)=\frac{\exp\left[-\beta E_2\right]}{\exp\left[-\beta E_1\right]+\exp\left[-\beta E_2\right]} \tag{10.75}$$

The Lagrange multipliers α_1 and α_2 have thus been eliminated. Next, the last Lagrange multiplier β has to be evaluated. For this, it is convenient to first calculate the logarithm of the conditional probabilities:

$$\ln P\left(A_1\middle|B_1\right)=\sim\beta E_1-\ln\left\{\exp\left[-\beta E_1\right]+\exp\left[-\beta E_2\right]\right\} \tag{10.76}$$

$$\ln P\left(A_2\middle|B_1\right)=\sim\beta E_2-\ln\left\{\exp\left[-\beta E_1\right]+\exp\left[-\beta E_2\right]\right\} \tag{10.77}$$

$$\ln P\left(A_1\middle|B_2\right)=\sim\beta E_1-\ln\left\{\exp\left[-\beta E_1\right]+\exp\left[\beta E_2\right]\right\} \tag{10.78}$$

$$\ln P\left(A_2\middle|B_2\right)=\sim\beta E_2-\ln\left\{\exp\left[-\beta E_1\right]+\exp\left[-\beta E_2\right]\right\} \tag{10.79}$$

Substituting Equations 10.76, 10.77, 10.78, and 10.79 into Equations 10.56, substituting the resulting expressions into Equation 10.55 and making use of Equation 10.61:

$$\mathbb{I}=\mathbb{S}\left(B\right)+k\ln\left\{\exp\left[-\beta E_1\right]+\exp\left[-\beta E_2\right]\right\}-\beta kU \tag{10.80}$$

Solving Equation 10.80 for the average energy U:

$$U=\frac{\mathbb{S}(B)}{\beta k}-\frac{\mathbb{I}}{\beta k}+\frac{1}{\beta}\ln\left\{\exp[-\beta E_1]+\exp[-\beta E_2]\right\} \tag{10.81}$$

The last Lagrange multiplier β can be evaluated by substituting Equation 10.81 into Equation 10.70.

$$\text{a)}\,T=\frac{1}{k\beta}\quad \text{b)}\,\beta=\frac{1}{kT} \tag{10.82}$$

Only Equations 10.73 and 10.74 represent transitions. Substituting Equation 10.82b into Equations 10.73 and 10.74:

$$P(A_2|B_1) = \frac{\exp\left[-\frac{E_2 - E_1}{kT}\right]}{1 + \exp\left[-\frac{E_2 - E_1}{kT}\right]} \tag{10.83}$$

$$P(A_1|B_2) = \frac{1}{1 + \exp\left[-\frac{E_2 - E_1}{kT}\right]} \tag{10.84}$$

where the numerators and denominators were multiplied by $\exp[E_1/kT]$. Note that $P(A_2|B_1)$ is the conditional probability that the system has now, at time t_a, an energy E_2 provided it previously at time t_b had an energy E_1. Thus, $P(A_2|B_1)$ represents a transition $p_{1\to 2}$ from energy state E_1 to energy state E_2. Similarly, note that $P(A_1|B_2)$ is the conditional probability that the system has now, at time t_a, an energy E_1 provided it previously at time t_b had an energy E_2. Thus, $P(A_2|B_1)$ represents a transition $p_{2\to 1}$ from energy state E_2 to energy state E_1. The energy E_2 is larger than energy E_1. Thus, in the limit when the temperature goes to zero, the transition $p_{1\to 2}$ from energy state E_1 to energy state E_2 goes to zero and the transition $p_{2\to 1}$ from energy state E_2 to energy state E_2 becomes 1.

Consider a typical example of transitions resulting in the generation of photons with, say, a wavelength of 1 μm. This corresponds to an energy difference $E_2 - E_1$ of 1.9865×10^{-19} J. Boltzmann's constant $k = 1.380662 \times 10^{-23}$ J per °K. Let us assume that the temperature T is, say, 25°C or T = 298°C. Thus, in this case, one obtains for the transition probability $p_{2\to 1}$ from state E_2 to E_1:

$$p_{2\to 1} \approx 1$$

Similarly, one obtains for the transition probability $p_{1\to 2}$ from state E_1 to E_2:

$$p_{1\to 2} \approx 1.076229 \times 10^{-21}$$

Thus, the mathematical model of probability with the assumption that the entropy always increases, predicts that transitions from higher energies to lower energies are much more likely than transitions from lower energies to higher energies. This is in agreement with our everyday experience. This derivation is an extension of Nernst's theorem.

10.9 CANTOR SETS AS RANDOM PROCESSES

Consider a signal containing bursts of noise. For example, the reception of distant radio or TV signals exhibits this phenomenon. These types of signals usually consist of completely noise-free time intervals followed by bursts of noise. For example, watching the signal from a distant TV station, one observes that the picture is clear for a few seconds followed by a noisy period. During the clear period the signal is

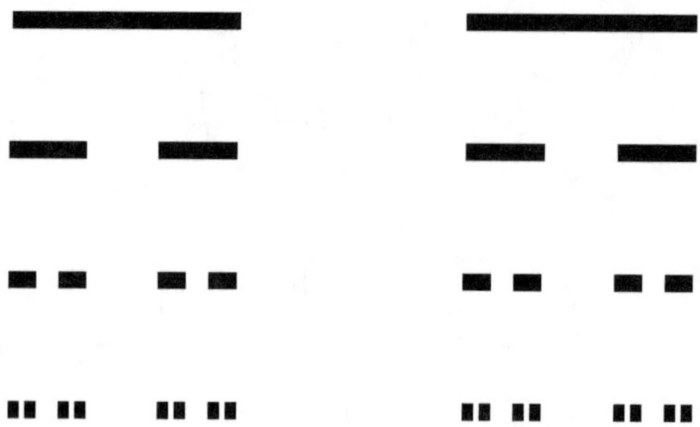

FIGURE 10.17 Cantor set. The same pattern is repeated on any scale.

noise free. However, the noise periods contain shorter periods of noise bursts and clear periods. If one observes these shorter noise bursts on an even shorter time scale, one observes that these noise bursts also consist of noise-free periods followed by short noise bursts, etc. B. Mandelbrot observed that the ratio of the time duration of the noise-free periods to the noisy periods remains the same on any time scale. Indeed, the noise seems to behave like the Cantor set illustrated in Figure 10.17. Thus, noise that behaves like a Cantor set is also a random process.

$$d = \sim \frac{\ln\left(\text{Factor by which the number of sections increases in each step}\right)}{\ln\left(\text{Factor by which each section decreases in length}\right)} \quad (10.85)$$

Objects similar to the Cantor set can be characterized by noninteger dimensions, so-called *fractal dimensions d*, given by Equation 10.85. For example, in the Cantor set of Figure 10.16 the number of line segments increases by a factor of 2 in each step, while the length of each section shrinks by a factor of ⅓. Thus, the Cantor set has a fractal dimension d of:

$$d = \sim \frac{\ln(2)}{\ln\left(\frac{1}{3}\right)} \quad \text{or} \quad d = 0.6309297536 \quad (10.86)$$

Observe that the Cantor set of Figure 10.16 has a fractal dimension that is less than one. The noise can be characterized by its fractal dimension. Here, the fractal dimension d is a macroscopic parameter of the stochastic noise system.

REFERENCES

1. *Time's Arrows* by Richard Morris, Simon & Schuster, New York (1985).
2. *Time's Arrows* by Richard Morris, Scientific American, (1975).
3. *Aging and the Single Cell,* Science, Vol. 293 No. 5535, page 1587, 31 August 2001.

PROBLEMS

Problem 10.1

Given:

The ion implantation in a semiconductor can be simulated by the following random walk model. Consider 10 particles that start out all in one row equally spaced 5 spatial units apart. All particles start with a velocity of 5 velocity units in the forward direction. At each time step, a velocity of one velocity step in either the backward direction, to the right or to the left is added. There is also a probability that the velocity will not change at that time step. Thus, there are four possibilities at each time step: add a velocity unit to the right, to the left, subtract a velocity unit, or keep the velocity the same. This can be accomplished by generating a two-digit (0 to 99) random number N_R for each particle at each time step:

If $0 \le N_R < 25$, add one velocity unit to the right.
If $25 \le N_R < 50$, add one velocity unit to the left.
If $50 \le N_R < 75$, subtract one velocity unit from the forward velocity.
If $75 \le N_R < 100$, there is no change in the velocity.

The particle stops when its velocity is equal to zero. Once the particle has reached zero velocity, do not add or subtract velocities. Calculate the position of each particle at the end of each time step. For simplicity, assume that more than one particle can be at the same position at the same time.

Problem:

Calculate and plot the path of each particle in the random walk until all particles achieve zero velocity. Each particle stops at the point where its velocity is zero.

Problem 10.2

Given:

A random process has the following form:

$$V(x,t) = \frac{x^2}{2}\exp\left[j\omega_o t\right] + ax\exp\left[-j\omega_o t\right]$$

where t is the time, a is a constant, ω_o is a constant angular frequency, and x is a random variable. The various moments of the random variable x are as follows:

$$\langle x \rangle = \frac{a}{2} \quad \sigma_{xx}^2 = \frac{3}{4}a^2 \quad \langle x^3 \rangle = \frac{a^3}{8} \quad \langle x^4 \rangle = 4a^4$$

Note:

$$\cos u = \frac{\exp[ju] + \exp[-ju]}{2} \qquad \sin u = \frac{\exp[ju] - \exp[-ju]}{2j}$$

$$\delta(\omega - \omega_o) = \frac{1}{2\pi}\int_{-\infty}^{\infty} \exp[\pm j(\omega - \omega_o)u]du$$

Problem:

a) Calculate the average value $\langle V(x,t) \rangle$ of the random process V(x,t) in terms of a.
b) Calculate the autocorrelation function $R_{VV}(t,t+\tau)$ of the random process in terms of a, where

$$R_{VV}(t,t+\tau) = \langle V(x,t)V^*(x,t+\tau) \rangle$$

and where * denotes the complex conjugate. The complex conjugate of a function is obtained by letting all j go to –j and all –j go to j in the function.
c) Determine the degree of stationarity of the random process V(x,t).

Problem 10.3

Given:

A random process V(x,y;t) has the following form:

$$V(x,y;t) = x \sin \omega_o t + y \cos \omega_o t + V_{D.C.}$$

where $V_{D.C.}$ is a D.C. voltage, ω_o is a constant angular frequency, t is the time, and x and y are random variables. The various moments of x and y are as follows:

$$\langle x \rangle = 0, \quad \sigma_{xx}^2 = 0.25, \quad \langle y \rangle = 0 \quad \sigma_{yy}^2 = 0.09, \quad \text{and} \quad \langle xy \rangle = 0.02$$

Problem:

a) Calculate the average value $\langle V(x,y;t) \rangle$ of the random process.
b) Calculate the autocorrelation function $R_{VV}(t,t+\tau)$ of the random process.
c) Determine the degree of stationarity of this random process.

Problem 10.4

Given:

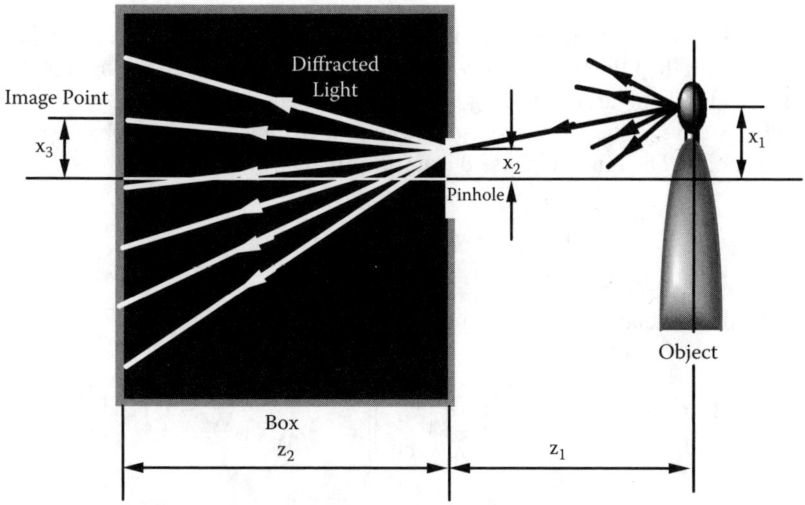

Pinhole camera.

It is thought that some of the paintings by the painters in the Renaissance were made by projecting an image on a wall using a pinhole camera arrangement.

They then traced the painting from this image. However, the operating principle of a pinhole camera was not understood until 150 to 200 years later. The pinhole camera uses white light with a coherence length of a few wavelengths. The following explains how a pinhole camera works.

Light is really an electromagnetic wave, and one should use the electric field vector and magnetic flux density pseudovector to calculate the propagation of the light from a point on the object to a point on the image. This is a very cumbersome calculation. It surely was difficult to do about 150 years ago when there were no computers. However, one can use a much simpler scalar theory for the case when all dimensions are much larger than a wavelength of light. This is the scalar diffraction theory, which assumes that at any point where light is incident, it is scattered with equal intensity in all directions. Using scalar diffraction theory, one obtains the following equation for the light electric field $u_2(x_2,z_2)$ from an object point with coordinates x_1 and z_1 impinging at an image point with coordinates x_2 and z_2.

$$u_2(x_2,t) = \int_{\text{Object}} u_1(x_1)\exp\left[j\left(\frac{2\pi}{\lambda}+\frac{\theta}{c\eta}\right)\left(ct_1-\sqrt{(x_2-x_1)^2+z_1^2}\right)\right]dx_1$$

The light electric field $u_3(x_3,t)$ from a point in the pinhole with coordinates x_2 and z_1 impinging at an image point with coordinates x_3 and z_2.

$$u_3(x_3,t)=\int_{\text{Pinhole}} u_2(x_2,t)\exp\left[j\left(\frac{2\pi}{\lambda}+\frac{\theta}{c\eta}\right)\left(ct-\sqrt{(x_3-x_2)^2+z_2^2}\right)\right]dx_2$$

where $c = 2.99792458 \times 10^8$ m per second is the speed of light in free space, $\lambda = 512$ nm is the wavelength of the light (yellow, $c\eta \approx 1500$ nm is the coherence length of the light and θ is a random variable having values between $-\pi$ and π. Here $z_1 = 10$ m, $z_2 = 2$ m, and $x_2 = 0.5$ m. The square roots represent the distance from the object point to the image point. The random phase θ is uniformly distributed.

$$p(\theta)=\frac{1}{2\pi} \quad \text{where} \quad -\pi<\theta\leq\pi$$

The horizontal differences z_1 and z_2 can be factored out.

$$u_2(x_2,t)=\int_{\text{Object}} u_1(x_1,t)\exp\left[j\left(\frac{2\pi}{\lambda}+\frac{\theta}{c\eta}\right)\left(ct_1-z_1\sqrt{1+\frac{(x_2-x_1)^2}{z_1^2}}\right)\right]dx_1$$

When the vertical distances x_1 and x_2 are much smaller than the horizontal distances z_1 and z_2, the square roots are approximated as follows:

$$u_{2a}(x_2,z_2,t)\approx u_1\exp\left[j\left(\frac{2\pi}{\lambda}+\frac{\theta}{c\eta}\right)\left(ct-z_1-\frac{x_a^2}{2z_1}-z_2-\frac{x_2^2}{2z_2}\right)\right]$$

$$u_{2b}(x_2,z_2,t)\approx u_1\exp\left[j\left(\frac{2\pi}{\lambda}+\frac{\theta}{c\eta}\right)\left(ct-z_1-\frac{x_b^2}{2z_1}-z_2-\frac{x_2^2}{2z_2}\right)\right]$$

where u_{2a} is the scalar electric of the light from the image point x_a, z_1 and u_{2b} is the scalar electric field origination from the image point x_b, z_1.

Problem:

a) Calculate the average values $\langle u_{2a}(x_a,z_2;t)\rangle$ and $\langle u_{2b}(x_b,z_2;t)\rangle$ of the random processes $u_{2a}(x_a,z_2;t)$ and $u_{2b}(x_b,z_2;t)$ in the limit as the time goes to infinity.
b) Calculate the cross-correlation function $R_{ab}(t,t+\tau)$ of the random processes in the limit as the time goes to infinity. Note that:

$$R_{ab}(t,t+\tau)=\left\langle u_{2a}(x_2,z_2,t)\,u_{2b}^{*}(x_2,z_2,t+\tau)\right\rangle$$

where * denotes the complex conjugate.

c) Determine the degree of stationarity of the random processes in the limit as the time goes to infinity.
d) Calculate the cross-stochastic power P_{ab} of the random process in the limit as time goes to infinity. Determine the distance between the two object points x_a and x_b for which the light at the object point x_2 is coherent; that is, determine how the pinhole camera works.

Problem 10.5

Given:

Two signals U(z,θ;t) and V(θ;t) used in a radar are random processes. The signal U(z,θ;t) that is reflected back from the target is

$$U(z,\theta;t) = A\cos\left[\omega_o\left(t-\frac{2z}{c}\right)+\frac{\theta}{\eta}\left(t-\frac{2z}{c}\right)\right]$$

and the reference signal V(θ;t) is

$$V(\theta;t) = A\cos\left(\omega_o t+\frac{\theta}{\eta}t\right)$$

where the angular frequency ω_o is equal to 6.2831853×10^{10} radians per second, c = 2.99792458×10^8 m per second is the speed of light in free space, z is the distance to the target, $\eta = 0.333564$ mS is a coherence time, and θ is a random phase. The probability density p(θ) per unit phase angle is

$$p\left(\theta = \frac{1}{2\pi}\right) \quad \text{for} \quad -\pi < \theta \le \pi$$

Problem:

a) Calculate the average values $\langle U(t,z,\theta)\rangle$ and $\langle V(t,\theta)\rangle$ of the two signals U(t,z,θ) and V(t,θ) in the limit as the time t goes to infinity (20 points).
b) Calculate the cross-correlation function $R_{UV}(t,t+\tau,z)$ in the limit as the time t goes to infinity where

$$R_{UV}(t,t+\tau,z) = \langle U(t,z,\theta)V(t+\tau,\theta)\rangle$$

and where τ is a time delay.
c) Determine the degree of stationarity of these signals in the limit as the time t goes to infinity.

11 Spectral Densities

11.1 STOCHASTIC POWER

The concept of *stochastic power* can be developed in analogy with electric power. Stochastic power is a macroscopic parameter. For example, the mean electrical power P_{Elect} is proportional to the time average of the square of the current I(t) or the voltage V(t). It is assumed that the values of the current or voltage are known exactly at every instant of time.

$$\text{a)}\quad P_{Elect} = \frac{\lim}{T\to\infty}\frac{1}{2T}\int_{-T}^{T} I^2(t)R\,dt \quad \text{or} \quad \text{b)}\quad P_{Elect} = \frac{\lim}{T\to\infty}\frac{1}{2T}\int_{-T}^{T}\frac{V^2(t)}{R}dt \tag{11.1}$$

where R is the electrical resistance (see Figure 11.1). Because both the current I(t) and the voltage V(t) are assumed to be completely known at every instant of time, these two functions are deterministic processes.

Consider the case of a sinusoidal voltage.

$$V(t) = V_o \sin(\omega t) \tag{11.2}$$

The mean electrical power P_{Elect} can be obtained by substituting Equation 11.2 into Equation 11.1b.

$$\begin{aligned} &\text{a)}\quad P_{Elect} = \frac{\lim}{T\to\infty}\frac{V_o^2}{2TR}\int_{-T}^{T}\sin^2(\omega t)\,dt \quad \text{or} \\ &\text{b)}\quad P_{Elect} = \frac{\lim}{T\to\infty}\frac{V_o^2}{4TR}\int_{-T}^{T}\left[1-\cos(2\omega t)\right]dt \end{aligned} \tag{11.3}$$

Integrating Equation 11.3b and taking the limit as T goes to infinity;

$$\text{a)}\quad P_{Elect} = \frac{V_o^2}{2R} \quad \text{or} \quad \text{b)}\quad P_{Elect} = \frac{V_{RMS}^2}{R} \quad \text{where} \quad V_{RMS} = \frac{V_o}{(2)^{\frac{1}{2}}} \tag{11.4}$$

where V_{RMS} is the *root mean square* (RMS) voltage (see Figure 11.2).

In analogy with the mean electrical power, one defines the *average stochastic power* P_{ff} as the time average of the average value of the square of a random process f(x;t).

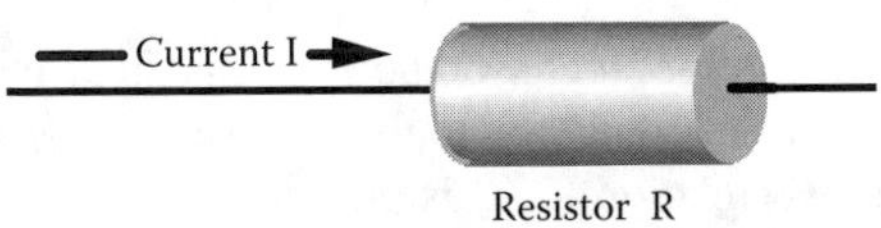

FIGURE 11.1 Resistor with current passing through it.

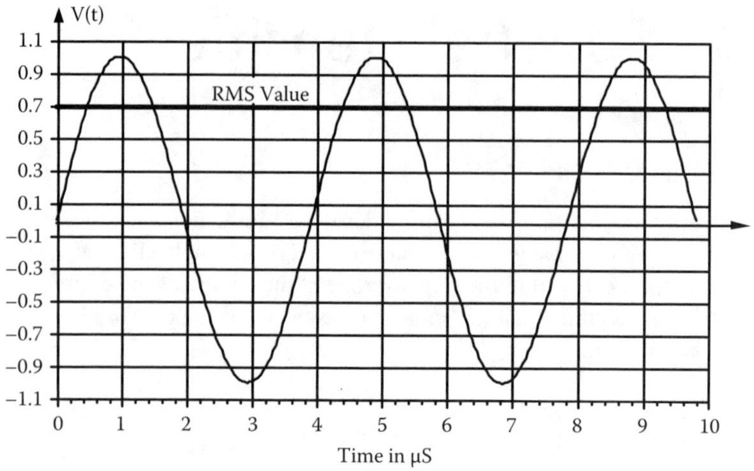

FIGURE 11.2 A sinusoidal voltage and its RMS value.

$$P_{ff} \equiv \lim_{T \to \infty} \frac{1}{2T} \int_{-T}^{T} \left\langle f^2(x;t) \right\rangle dt \tag{11.5}$$

Here, f(x;t) is a continuous random process. However, a discrete random process $f_n(t)$ could also be used here. Equation 11.5 can be rewritten as follows:

$$P_{ff} = \lim_{\tau \to 0} \rho_{ff}(\tau) \tag{11.6}$$

where

$$\rho_{ff}(\tau) \equiv \lim_{T \to \infty} \frac{1}{2T} \int_{-T}^{T} \left\langle f(x;t) f(x;t+\tau) \right\rangle dt \tag{11.7}$$

The definition of the average stochastic power P_{ff} given by Equation 11.5 is, of course, much simpler than the one described by Equations 10.6 and 11.7. However, Equation 11.7 allows one to make use of the autocorrelation $R_{ff}(t, t + \tau)$.

$$\rho_{ff}(\tau) \equiv \lim_{T \to \infty} \frac{1}{2T} \int_{-T}^{T} R_{ff}(t, t+\tau) dt \tag{11.8}$$

In the general case, the average stochastic power P_{ff} is given by Equation 10.6 where $\rho_{ff}^{(\tau)}$ is given by Equation 11.7 or 11.8. However, for a wide-sense stationary (WSS) process where the autocorrelation $R_{ff}(\tau)$ is only a function of the time difference τ, thus:

$$\rho_{ff}(\tau) = R_{ff}(\tau) \tag{11.9}$$

Therefore, for a WSS process, the average stochastic power P_{ff} is equal to the autocorrelation $R_{ff}(\tau)$ evaluated as the time difference τ equal to zero.

$$P_{ff} = R_{ff}(\tau = 0) \tag{11.10}$$

One can similarly define a *stochastic cross-power* P_{fg}.

$$P_{fg} \equiv \frac{\lim}{T \to \infty} \frac{1}{2T} \int_{-T}^{T} \langle f(x;t) g(y;t) \rangle dt \tag{11.11}$$

Here, f(x;t) and g(y;t) are continuous random processes. However, discrete random processes $f_n(t)$ and $g_m(t)$ could also be used here. Equation 11.11 can be rewritten just as Equations 11.6 and 11.7 were rewritten:

$$P_{fg} = \frac{\lim}{\tau \to 0} \rho_{fg}(\tau) \tag{11.12}$$

where

$$\rho_{fg}(\tau) \equiv \frac{\lim}{T \to \infty} \frac{1}{2T} \int_{-T}^{T} \langle f(x;t) g(y;t+\tau) \rangle dt \tag{11.13}$$

The function inside the integral of Equation 11.13 is the cross-correlation $R_{fg}(t,t+\tau)$.

$$\rho_{fg}(\tau) \equiv \frac{\lim}{T \to \infty} \frac{1}{2T} \int_{-T}^{T} R_{gf}(t, t+\tau) dt \tag{11.14}$$

Similarly, for a wide-sense stationary process where the cross-correlation $R_{fg}(\tau)$ is only a function of the time difference τ:

$$\rho_{fg}(\tau) = R_{fg}(\tau) \tag{11.15}$$

Therefore, for a wide-sense stationary process, the average stochastic cross-power P_{fg} is equal to the cross-correlation $R_{fg}(\tau)$ evaluated at the time difference τ equal to zero:

$$P_{fg} = R_{fg}(\tau = 0) \tag{11.16}$$

11.2 THE POWER SPECTRUM AND CROSS-POWER SPECTRUM

For the case when $\rho_{ff}(\tau)$ is periodic in τ, one can expand the function $\rho_{ff}(\tau)$ in a Fourier series.

$$\rho_{ff}^{(\tau)} = \sum_{n=0}^{n=\infty} c_n^2 \cos\frac{2n\pi}{T}\tau + \sum_{n=1}^{n=\infty} b_n^2 \sin\frac{2n\pi}{T}\tau \tag{11.17}$$

where the power spectrum components $c_n^2(\omega_n)$ and $b_n^2(\omega_n)$ are the amplitudes of the various frequency components of the function $\rho_{ff}(\tau)$. The power spectrum components $c_n^2(\omega_n)$ and $b_n^2(\omega_n)$ are also macroscopic parameters. They can be computed by the same method that is used for calculating the components of a Fourier series:

$$\text{a)}\quad c_n^2 = \frac{1}{T}\int_0^T \rho_{ff}(\tau)\cos\frac{2n\pi}{T}\tau\, d\tau \quad\text{and}\quad \text{b)}\quad b_n^2 = \frac{1}{T}\int_0^T \rho_{ff}(\tau)\sin\frac{2n\pi}{T}\tau\, d\tau \tag{11.18}$$

For a periodic wide-sense stationary process where $R_{ff}(\tau) = R_{ff}(-\tau)$:

$$R_{ff}(\tau) = \sum_{n=0}^{n=\infty} c_n^2 \cos\frac{2n\pi}{T}\tau \tag{11.19}$$

In general, the *power density spectrum* $s_{ff}(\omega)$ consists of the complex amplitudes of the various frequency components of $\rho_{ff}(\tau)$.

$$\rho_{ff}(\tau) = \frac{1}{2\pi}\int_{-\infty}^{\infty} s_{ff}(\omega)e^{j\omega\tau}d\omega \tag{11.20}$$

The power density spectrum is a macroscopic parameter. The power density spectrum $s_{ff}(\omega)$ can be calculated by using the Fourier transform of $\rho_{ff}(\tau)$.

$$s_{ff}(\omega) \equiv \int_{-\infty}^{\infty} \rho_{ff}(\tau)e^{-j\omega\tau}d\tau \tag{11.21}$$

Indeed, Equation 11.20 is the inverse Fourier transform of the power density spectrum $s_{ff}(\omega)$. Similarly, for a wide-sense stationary process, the power density spectrum $s_{ff}(\omega)$ consists of the complex amplitudes of the various frequency components of the autocorrelation $R_{ff}(\tau)$.

$$R_{ff}(\tau) = \frac{1}{2\pi}\int_{-\infty}^{\infty} s_{ff}(\omega)e^{j\omega\tau}d\omega \tag{11.22}$$

The power density spectrum $s_{ff}(\omega)$ of a wide-sense stationary process can be calculated from the Fourier transform of the autocorrelation $R_{ff}(t)$.

$$s_{ff}(\omega) \equiv \int_{-\infty}^{\infty} R_{ff}(\tau) e^{-j\omega\tau} d\tau \tag{11.23}$$

The power density spectrum of a wide-sense stationary process has the following properties:

1. $s_{ff}(\omega) = s_{ff}(-\omega)$ for real $R_{ff}(\tau)$.
2. $s_{ff}(\omega) \geq 0$.
3. $s_{ff}(\omega)$ is real.

EXAMPLE 11.1

Consider the following random process:

$$f(c_0, c_1, c_2, \ldots, b_1, b_2, \ldots; t) = \sum_{n=0}^{n=\infty} c_n \cos \omega_n t + \sum_{n=1}^{n=\infty} b_n \sin \omega_n t$$

where the c_n's and b_n's are zero-mean uncorrelated random variables, and ω_o is equal to zero.

$$\langle c_n \rangle = 0 \text{ and } \langle b_n \rangle = 0$$

$$\langle c_n c_m \rangle = \begin{cases} \langle c_n \rangle \langle c_m \rangle & \text{for } m \neq n \\ \sigma_{nn}^2 & \text{for } m = n \end{cases}$$

$$\langle b_n b_m \rangle = \begin{cases} \langle b_n \rangle \langle b_m \rangle & \text{for } m \neq n \\ \sigma_{nn}^2 & \text{for } m = n \end{cases}$$

and

$$\langle c_n b_m \rangle = \langle c_n \rangle \langle b_m \rangle \text{ for both } m = n \text{ and } m \neq n$$

First, the average value of $f(c_0, c_1, c_3, \ldots, b_1, b_2, \ldots; t)$ is calculated:

$$\langle f(c_0, c_1, c_2, \ldots, b_1, b_2, \ldots; t) \rangle = \sum_{n=0}^{n=\infty} \langle c_n \rangle \cos \omega_n t + \sum_{n=1}^{n=\infty} \langle b_n \rangle \sin \omega_n t$$

or

$$\langle f(c_0, c_1, c_3, \ldots, b_1, b_2, \ldots; t) \rangle = 0$$

Note that the average value of the random process is a constant. It is equal to zero. Next, the autocorrelation $R_{ff}(t, t + \tau)$ is calculated:

$$R_{ff}(t, t+\tau) = \sum_{n=0}^{n=\infty} \sum_{m=0}^{n=\infty} \left[\langle c_n c_m \rangle \cos\omega_n t \cos\omega_m (t+\tau) + \right.$$

$$\langle b_n c_m \rangle \sin\omega_n t \cos\omega_m (t+\tau) + \langle c_n b_m \rangle \cos\omega_n t \sin\omega_m (t+\tau) +$$

$$\left. \langle b_n b_m \rangle \sin\omega_n t \sin\omega_m (t+\tau) \right]$$

Note the following trigonometric identities:

$$\cos A \cos B = \frac{1}{2}\left[\cos(A-B) + \cos(A+B)\right]$$

and

$$\sin A \sin B = \frac{1}{2}\left[\cos(A-B) - \cos(A+B)\right]$$

Making use of the properties of the random variables and the trigonometric identities:

$$R_{ff}(\tau) = \sum_{n=0}^{n=\infty} \sigma_n^2 \cos\omega_n \tau$$

The stochastic power P_{ff} is equal to the autocorrelation evaluated at the time difference t equal to zero.

$$P_{ff} = \sum_{n=0}^{n=\infty} \sigma_n^2$$

Note that the autocorrelation is only a function of the time difference τ. Because the average value of the random process is a constant and autocorrelation is only a function of the time difference, $f(c_0, c_1, c_3, \ldots, b_1, b_2, \ldots, t)$ is a wide-sense stationary random process.

The power density spectrum can be found by taking the Fourier transform of the autocorrelation $R_{ff}(\tau)$.

$$S_{ff}(\omega) = \frac{1}{2}\sum_{n=0}^{n=\infty} \sigma_n^2 \int_{-\infty}^{\infty} \left[e^{-j(\omega-\omega_n)\tau} + e^{-j(\omega+\omega_n)\tau} \right] d\tau$$

Integrating, the power density spectrum is

$$s_{ff}(\omega) = \pi \sum_{n=0}^{n=\infty} \sigma_n^2 \left[\delta(\omega - \omega_n) + \delta(\omega + \omega_n)\right]$$

Similar to the power density spectrum, the cross-power density spectrum $s_{fg}(\omega)$ is the complex amplitude of the various frequency components of $\rho_{fg}(\tau)$.

The *cross-power density spectrum* is a macroscopic parameter. Similar to the power density spectrum, the cross-power density spectrum $s_{fg}(\omega)$, can be calculated by using the Fourier transform of $\rho_{fg}(\tau)$.

$$s_{fg}(\omega) \equiv \int_{-\infty}^{\infty} \rho_{fg}(\tau) e^{-j\omega\tau} d\tau \tag{11.25}$$

Similarly, for a wide-sense stationary process, the cross-power density spectrum $s_{fg}(\omega)$ consists of the complex amplitudes of the various frequency components of the cross-correlation $R_{fg}(\tau)$.

$$R_{fg}(\tau) = \frac{1}{2\pi} \int_{-\infty}^{\infty} s_{fg}(\omega) e^{j\omega\tau} d\omega \tag{11.26}$$

One can calculate the cross-power density spectrum $s_{fg}(\omega)$ of a wide-sense stationary process from the Fourier transform of the cross-correlation $R_{fg}(\tau)$.

$$s_{fg}(\omega) \equiv \int_{-\infty}^{\infty} R_{fg}(\tau) e^{-j\omega\tau} d\tau \tag{11.27}$$

The cross-power density spectrum has the following properties:

1. $s_{fg}(\omega) = s_{gf}(-\omega)$.
2. $s_{fg}(\omega) = s_{gf}{}^*(\omega)$.
3. $s_{fg}(\omega) = 0$ and $s_{gf}(\omega) = 0$ if f(x;t) and g(y;t) are orthogonal.

11.3 THE EFFECTS OF FILTERS ON THE AUTOCORRELATION FUNCTION AND THE POWER SPECTRAL DENSITY

Many practical communication systems use filters. The effect of filtering a signal $f_{in}(t, \theta)$ with a filter having an impulse response h(t) is equal to the convolution of the signal with the impulse response of the filter, as shown in Figure 11.3.

$$f_{out}(\theta; t) = \int_{-\infty}^{\infty} f_{in}(\theta; u) h(t - u) du \tag{11.28}$$

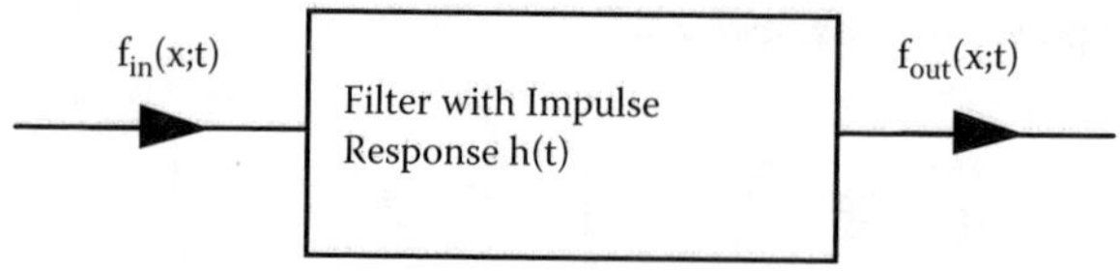

FIGURE 11.3 Signal $f_{in}(x;t)$ passing through a filter with impulse response h(t).

One can calculate the output autocorrelation function $R_{out,ff}$ of the signal $f_{out}(\theta;t)$ by using Equation 11.28 in Equation 11.25, where $g(\theta;t + \tau)$ is equal to $f_{in}(\theta;t + \tau)$:

$$R_{out,ff}(t,t+\tau)=\left\langle \int_{-\infty}^{\infty} du \int_{-\infty}^{\infty} dv f_{in}(\theta;u)h(t-u)f_{in}(\theta;v)h(t+\tau-v)\right\rangle \quad (11.29)$$

where the random variable θ is averaged. One makes the following transformations of variables:

$$a)\ v=u+\eta \quad \text{and} \quad b)\ d\eta=dv \quad (11.30)$$

Substituting Equation 11.30 into Equation 11.29:

$$R_{out,ff}(t,t+\tau)=\left\langle \int_{-\infty}^{\infty} du \int_{-\infty}^{\infty} d\eta f_{in}(\theta;u)h(t-u)f_{in}(\theta;u+\eta)h(t+\tau-u-\eta)\right\rangle \quad (11.31)$$

The averaging is performed over θ:

$$R_{out,ff}(t,t+\tau)=\int_{-\infty}^{\infty} du \int_{-\infty}^{\infty} d\eta R_{in,ff}(u,u+\eta)h(t-u)h(t+\tau-u-\eta) \quad (11.32)$$

where $R_{in,ff}(u,u + \eta)$ is the autocorrelation function of the input signal. One can make the following transformation of variables:

$$a)\ q=t-u \quad b)\ u=t-q \quad c)\ dq=-du \quad (11.33)$$

Substituting Equation 11.33 into Equation 11.32 for u and du:

$$R_{out,ff}(t,t+\tau)=\int_{-\infty}^{\infty} dq \int_{-\infty}^{\infty} d\eta R_{in,ff}(t-q,t-q+\eta)h(q)h(\tau+q-\eta) \quad (11.34)$$

For the case where the signal $f_{in}(\theta;t)$ is wide-sense stationary, Equation 11.34 would be independent of t and t – q.

$$R_{out,ff}(\tau)=\int_{-\infty}^{\infty} dq \int_{-\infty}^{\infty} d\eta R_{in,ff}(\eta)h(q)h(\tau+q-\eta) \quad (11.35)$$

Next, one calculates the output power density spectrum $s_{out,ff}(\omega)$. However, before calculating the power density spectrum, the time average of the output autocorrelation function must be taken if it is time dependent, as is the case in Equation 11.34. One obtains, similar to Equation 11.8:

$$\rho_{out,ff}(\tau) = \int_{-\infty}^{\infty} dq \int_{-\infty}^{\infty} d\eta \rho_{in,ff}(\eta) h(q) h(\tau + q - \eta) \tag{11.36}$$

The output power density spectrum $s_{out,ff}(\omega)$ can be obtained by taking the Fourier transform of Equation 11.35 for the wide-sense stationary case or Equation 11.36 for the general case.

$$s_{out,ff}(\omega) = \int_{-\infty}^{\infty} d\tau \int_{-\infty}^{\infty} dq \int_{-\infty}^{\infty} d\eta e^{-j\omega\tau} \rho_{in,ff}(\eta) h(q) h(\tau + q - \eta) \tag{11.37}$$

One performs a final transformation of variables:

$$\text{a)} \quad p = \tau + q - \eta \quad \text{b)} \quad \tau = p - q + \eta \quad \text{and} \quad \text{c)} \quad d\tau = dp \tag{11.38}$$

Substituting Equation 11.38 into Equation 11.37 for t and dt:

$$s_{out,ff}(\omega) = \int_{-\infty}^{\infty} d\eta e^{-j\omega\eta} \rho_{in,ff}(\eta) \int_{-\infty}^{\infty} dp e^{-j\omega p} h(p) \int_{-\infty}^{\infty} dr e^{j\omega q} h(q) \tag{11.39}$$

For the wide-sense stationary case, the input autocorrelation function $R_{in,ff}(h)$ would appear in Equation 11.39 instead of $\rho_{in,ff}(\eta)$. Equation 11.39 is equal to:

$$s_{out,ff}(\omega) = s_{in,ff}(\omega) \left|H(\omega)\right|^2 \tag{11.40}$$

where $s_{in,ff}(\omega)$ is the power density spectrum of the input signal, and $H(\omega)$ is the filter transfer function.

11.4 THE BANDWIDTH OF THE POWER SPECTRUM

The *center frequency* ω_c of the power density spectrum $s_{ff}(\omega)$ can be calculated as an average value of sorts of the frequency ω using the power density spectrum as the probability density and considering positive frequencies only. Frequencies at which the power density spectrum has large values will contribute more to the center frequency than frequencies at which the power density spectrum has small values.

$$\omega_c \equiv \frac{\int_0^{\infty} \omega s_{ff}(\omega) d\omega}{\int_0^{\infty} s_{ff}(\omega) d\omega} \tag{11.41}$$

One can calculate the bandwidth B_w of the power density spectrum $s_{ff}(\omega)$ as a standard deviation of sorts of the frequency, ω using, again, the power density spectrum as the probability density and considering positive frequencies only.

$$\frac{B_w^2}{4} \equiv \frac{\int_0^\infty \omega^2 s_{ff}(\omega)\, d\omega}{\int_0^\infty s_{ff}(\omega)\, d\omega} - \left[\frac{\int_0^\infty \omega s_{ff}(\omega)\, d\omega}{\int_0^\infty s_{ff}(\omega)\, d\omega}\right]^2 \tag{11.42}$$

EXAMPLE 11.2

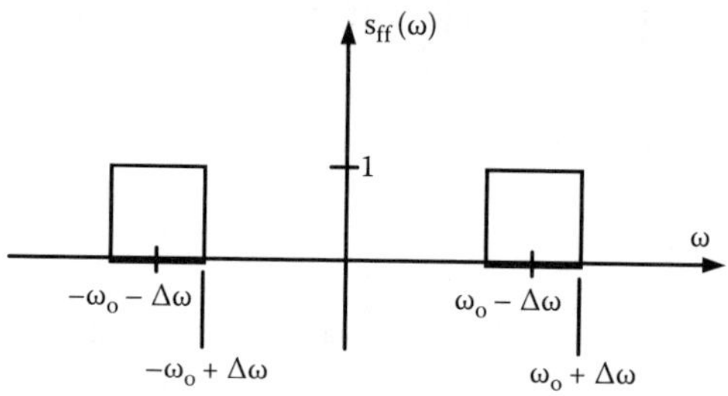

Consider the power density spectrum shown in the preceding diagram. The center frequency ω_c according to Equation 11.28 is

$$\omega_c = \frac{\int_{\omega_o-\Delta\omega}^{\omega_o+\Delta\omega} \omega\, d\omega}{\int_{\omega_o-\Delta\omega}^{\omega_o-\Delta\omega} d\omega}$$

Integrating:

$$\omega_c = \frac{1}{2}\frac{\omega_o^2 + 2\omega_o\Delta\omega + \Delta\omega^2 - \omega_o^2 + 2\omega_o\Delta\omega - \Delta\omega^2}{\omega_o + \Delta\omega - \omega_o + \Delta\omega}$$

or

$$\omega_c = \omega_o$$

To calculate the bandwidth B_w, we first calculate the average value of ω^2.

$$\left\langle \omega^2 \right\rangle = \frac{\int_{\omega_o-\Delta\omega}^{\omega_o+\Delta\omega} \omega^2 d\omega}{\int_{\omega_o-\Delta\omega}^{\omega_o-\Delta\omega} d\omega}$$

Integrating:

$$\langle \omega^2 \rangle = \frac{1}{3}\frac{\omega_o^3 + 3\omega_o^2\Delta\omega + 3\omega_o\Delta\omega^2 + \Delta\omega^3 - \omega_o^3 + 3\omega_o^2\Delta\omega - 3\omega_o\Delta\omega^2 + \Delta\omega^3}{\omega_o + \Delta\omega - \omega_o + \Delta\omega}$$

Collecting terms:

$$\langle \omega^2 \rangle = \omega_o^2 + \frac{\Delta\omega^2}{3}$$

The square of the bandwidth B_w^2 is

$$\frac{B_w^2}{4} = \langle \omega^2 \rangle - \omega_c^2$$

By substituting for $\langle \omega^2 \rangle$ and ω_c^2, we obtain for the bandwidth:

$$B_W = \frac{2\Delta\omega}{\sqrt{3}}$$

PROBLEMS

PROBLEM 11.1

Given:

Two signals $U(t,z,\theta)$ and $V(t,\theta)$ used in radar are random processes. The signal $U(t,z,\theta)$ that is reflected back from the target is

$$U(t,z,\theta) = A\cos\left[\omega_o\left(t - \frac{2z}{c}\right) + \frac{\theta}{\eta}\left(t - \frac{2z}{c}\right)\right]$$

and the reference signal $V(t,\theta)$ is

$$V(t,\theta) = B\cos\left[\omega_o t + \frac{\theta}{\eta} t\right]$$

where the angular frequency ω_o is equal to 6.2831853×10^{10} rad/s, $c = 2.99792458 \times 10^8$ m/s is the speed of light in free space, z is the distance to the target, $\eta = 0.333564$ mS is a coherence time, and θ is a random phase. The probability density $p(\theta)$ per unit phase angle is

$$p(\theta) = \frac{1}{2\pi} \quad \text{for} \quad -\pi < \theta \leq \pi$$

Problem:

a) Calculate the average values $\langle U(t,z,\theta) \rangle$ and $\langle V(t,\theta) \rangle$ of the two signals $U(t,z,\theta)$ and $V(t,\theta)$ in the limit as the time t goes to infinity. Calculate the cross-correlation function $R_{UV}(t,t + \tau,z)$ in the limit as the time t goes to infinity, where

$$R_{UV}\left(t,t+\tau,z\right)=\left\langle U\left(t,z,\theta\right)V\left(t+\tau,\theta\right)\right\rangle$$

and where τ is a time delay.

b) Determine the degree of stationarity of these signals in the limit as the time t goes to infinity.

c) Calculate the stochastic power $P_{UV}(z)$ in the limit as the time t goes to infinity. The range, the distance from the radar to the target, can be determined from the time required for the signal to travel to the target and return to the radar. The velocity of the target can be determined from the change in frequency of the return signal. To compare the frequencies of the signals, the signals have to be coherent. Determine the maximum range of this radar so that the signals $U(t,z,\theta)$ and $V(t,\theta)$ stay coherent. The frequency of the reference signal $V(t,\theta)$ has to be made equal to the return signal $U(t,z,\theta)$ to compare the two signals.

d) Calculate the power density spectrum $s_{UV}(\omega)$ for the case when the time t goes to infinity.

Problem 11.2

Given:

A random process has the following form:

$$V\left(x,t\right)=\frac{x^2}{2}\exp\left[j\omega_o t\right]+ax\exp\left[-j\omega_o t\right]$$

where t is the time, a is a constant, ω_o is a constant angular frequency, and x is a random variable. The various moments of the random variable x are as follows:

$$\left\langle x\right\rangle=\frac{a}{2}\quad \sigma_{xx}^2=\frac{3}{4}a^2\quad \left\langle x^3\right\rangle=\frac{a^3}{8}\quad \left\langle x^4\right\rangle=4a^4$$

Note:

$$\cos u=\frac{\exp\left[ju\right]+\exp\left[-ju\right]}{2}\qquad \sin u=\frac{\exp\left[ju\right]-\exp\left[-ju\right]}{2j}$$

$$\delta\left(\omega-\omega_o\right)=\frac{1}{2\pi}\int_{-\infty}^{\infty}\exp\left[\pm j\left(\omega-\omega_o\right)u\right]du$$

Problem:

a) Calculate the average value $\langle V(x,t) \rangle$ of the random process V(x,t) in terms of a.

b) Calculate the autocorrelation function $R_{VV}(t,t+\tau)$ of the random process in terms of a, where

$$R_{VV}(t,t+\tau) = \langle V(x,t)V^*(x,t+\tau) \rangle$$

and where * denotes the complex conjugate. The complex conjugate of a function is obtained by letting all j's go to −j and all −j go to j's in the function.

c) Determine the degree of stationarity of the random process V(x,t).

d) Calculate the statistic power P_{VV} of the random process V(x,t) in terms of a.

e) Calculate the power spectral density $s_{VV}(\omega)$ of the random process V(x,t) in terms of a.

Problem 11.3

Given:

A random process V(x,y;t) has the following form:

$V(x,y;t) = x \sin\omega_o t + y \cos\omega_o t + V_{D.C.}$

where $V_{D.C.}$ is a D.C. voltage, ω_o is a constant angular frequency, t is the time, and x and y are random variables. The various moments of x and y are as follows:

$$\langle x \rangle = 0, \quad \sigma_{xx}^2 = 0.25, \quad \langle y \rangle = 0, \quad \sigma_{yy}^2 = 0.09, \quad \text{and} \quad \langle xy \rangle = 0.02$$

Problem:

a) Calculate the average value $\langle V(x,y;t) \rangle$ of the random process.

b) Calculate the autocorrelation function $R_{VV}(t,t+\tau)$ of the random process.

c) Determine the degree of stationarity of this random process.

d) Calculate the stochastic power P_{VV} of the random process.

e) Calculate the power density spectrum $s_{VV}(\omega)$ of the random process.

12 Data Analysis

12.1 LEAST SQUARE DIFFERENCES

Real physical data can only be measured to within some limit of accuracy. Therefore, real physical data is a random function. There are often a large number of data points. Therefore, the data can be considered to be microscopic stochastic variables. As with other stochastic systems, it is often desirable to express the set of data by only a few parameters. This can be done by fitting the data $y_{n,data}$ to some curve. This curve could be one derived from a theoretical model that is thought to fit the data, or one could simply guess at some mathematically describable curve that would fit the data.

$$y_{n,data} - f_n\left(a_{on}, a_{1n}, a_{2n}, \ldots\right) = \Delta y_n \tag{12.1}$$

where the parameters a_o, a_{1n}, a_{2n}, etc., can be adjusted to best fit the data. For example, the curve described by a polynomial can be used to approximate some set of data shown in Figure 12.1.

$$y_{n,data} - \left[a_o + a_1 x_n + a_2 x_n^2 + a_3 x_n^3 + a_4 x_n^4 + \ldots\right] = \Delta y_n \tag{12.2}$$

The data consist of a set of numbers y_n and x_n where n = 1, 2, 3, …, enumerates the data points. There is a value of y_n for each value of x_n of the data. There might be a large number of data points y_n and x_n in the set; however, it is only necessary to know limited numbers a_o, a_1, a_2, a_3, etc., to describe the curve. The numbers a_o, a_1, a_2, a_3, etc., are the macroscopic parameters describing various aspects of the data. This example is illustrated in Figure 12.1. Here, $a_o = 1/20$, $a_1 = 1/3$, $a_2 = -1$, $a_3 = 3/4$, etc.

These mathematical models have *adjustable parameters* that can be varied to best fit a curve to the data. In general, the values a_o, a_1, a_2, a_3, etc., can be adjusted until the curve best fits the data. In this section, a method of fitting curves to various data is discussed.

To best fit the curve to the data, the adjustable parameters can be varied until the sum S of the squares of the differences Δy_n between the data values $y_{n,data}$ and the curve values $y_{n,curve}$ is a minimum. These differences are illustrated in Figure 12.1. This is known as the method of least square differences. It is sometimes also known as the χ squared method because the sum is often designated by the square of the Greek letter χ.

$$\chi^2 = \frac{1}{2}\sum_{n=1}^{n=N}\left[y_{n,data} - f_n\left(a_{on}, a_{1n}, a_{2n}, \ldots\right)\right]^2 \tag{12.3}$$

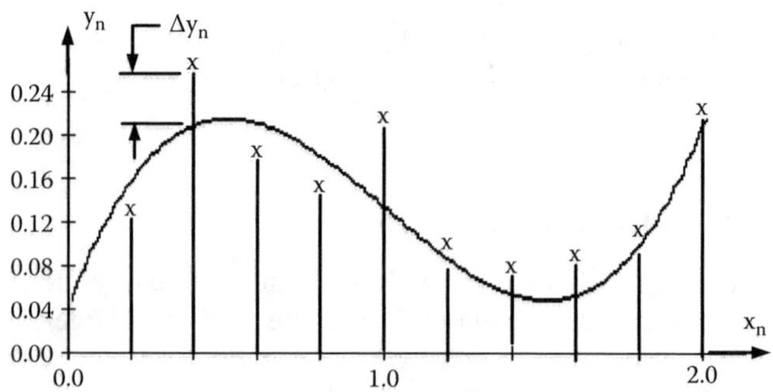

FIGURE 12.1 Plot of some data fitted by a curve in the form of Equation 12.1. Here, Δy_n is the difference between the data and the value the mathematical curve has at that data point.

For the case when the data are approximated by a polynomial:

$$\chi^2 = \frac{1}{N}\sum_{n=1}^{n=N}\left(y_{n,\text{data}} - \sum_{m=o}^{m=M} a_m x_n^m\right)^2 \tag{12.4}$$

The values a_m are adjustable parameters. The curve described by the adjustable parameters will fit the data best when χ^2 is a minimum. This will be the case when the derivatives of χ^2 with respect to each parameter a_m is equal to zero. The value of each adjustable parameter a_m can be found by setting the derivative of χ^2 with respect to each adjustable parameter a_m equal to zero.

$$\frac{\partial \chi^2}{\partial a_k} = 0 \tag{12.5}$$

Substituting Equation 12.4 into Equation 12.5:

$$0 = -2\left[\frac{1}{N}\sum_{n=1}^{n=N} y_{n,\text{data}}\, x_n^k - \sum_{m=0}^{m=M} a_m \frac{1}{2}\sum_{n=1}^{n=N} x_n^{m+k}\right] \tag{12.6}$$

This can be written as:

$$\begin{aligned}
\left\langle y_{n,\text{data}}\right\rangle &= a_o + a_1\left\langle x_n\right\rangle + a_2\left\langle x_n^2\right\rangle + a_3\left\langle x_n^3\right\rangle + \ldots \\
\left\langle y_{n,\text{data}} x_n\right\rangle &= a_o\left\langle x_n\right\rangle + a_1\left\langle x_n^2\right\rangle + a_2\left\langle x_n^3\right\rangle + a_3\left\langle x_n^4\right\rangle + \ldots \\
\left\langle y_{n,\text{data}} x_n^2\right\rangle &= a_o\left\langle x_n^2\right\rangle + a_1\left\langle x_n^3\right\rangle + a_2\left\langle x_n^4\right\rangle + a_3\left\langle x_n^5\right\rangle + \ldots
\end{aligned} \tag{12.7}$$

etc.

where

$$\text{a)}\quad \left\langle y_{n,data} x_n^m \right\rangle \equiv \frac{1}{N}\sum_{n=1}^{n=N} y_{n,data} x_n^m \quad \text{and} \quad \text{b)}\quad \left\langle x_n^m \right\rangle \equiv \frac{1}{N}\sum_{n=1}^{n=N} x_n^m \tag{12.8}$$

Indeed, Equations 12.8 has forms that are analogous to an average value with uniformly distributed random variables. This can be written in matrix form as follows:

$$\begin{pmatrix} \langle y_{0\,data} \rangle \\ \langle y_{n\,data} x_n \rangle \\ \langle y_{n\,data} x_n^2 \rangle \\ \langle y_{n\,data} x_n^3 \rangle \\ \cdots \end{pmatrix} = \begin{pmatrix} 1 & \langle x_n \rangle & \langle x_n^2 \rangle & \langle x_n^3 \rangle & \cdots \\ \langle x_n \rangle & \langle x_n^2 \rangle & \langle x_n^3 \rangle & \langle x_n^4 \rangle & \cdots \\ \langle x_n^2 \rangle & \langle x_n^3 \rangle & \langle x_n^4 \rangle & \langle x_n^5 \rangle & \cdots \\ \langle x_n^3 \rangle & \langle x_n^4 \rangle & \langle x_n^5 \rangle & \langle x_n^6 \rangle & \cdots \\ \cdots & \cdots & \cdots & \cdots & \cdots \end{pmatrix} \begin{pmatrix} a_0 \\ a_1 \\ a_2 \\ a_3 \\ \cdots \end{pmatrix} \tag{12.9}$$

The a_m's can be calculated by multiplying both sides of Equation 12.9 from the left by the inverse of the matrix. Of course, the inverse of the matrix in Equation 12.9 has to be determined first.

Recall that a similar procedure to maximize the average randomness with respect to individual probabilities P_n, was used. However, unlike the case of the average randomness, which has a convex function of the individual probabilities P_n, the sum $\chi^2(a_m)$ of the squares of the differences Δy_n might have minima as well as maxima and inflection points in the space of the adjustable parameters a_m where the partial derivatives are equal to zero. Therefore, it is necessary to check if the various second partial derivatives are positive.

12.2 THE SPECIAL CASE OF LINEAR REGRESSION

The simplest case is to approximate data by a linear equation. This is known as a *linear regression*, in which the data are approximated by a straight line.

$$\chi^2 = \frac{1}{N}\sum_{n=1}^{n=N} \left[y_{n,data} - (a_o + a_1 x_n) \right]^2 \tag{12.10}$$

Minimizing χ^2 as was done in the process leading to Equation 12.9:

$$\begin{pmatrix} \langle y_{0\,data} \rangle \\ \langle y_{n\,data} x_n \rangle \end{pmatrix} = \begin{pmatrix} 1 & \langle x_n \rangle \\ \langle x_n \rangle & \langle x_n^2 \rangle \end{pmatrix} \begin{pmatrix} a_0 \\ a_1 \end{pmatrix} \tag{12.11}$$

To solve for the adjustable parameters a_o and a_1, one must multiply Equation 11.11 by the inverse of the matrix in Equation 11.11.

$$\begin{pmatrix} a_0 \\ a_1 \end{pmatrix} = \frac{1}{\langle x_n^2 \rangle - \langle x_n \rangle^2} \begin{pmatrix} \langle x_n^2 \rangle & -\langle x_n \rangle \\ -\langle x_n \rangle & 1 \end{pmatrix} \begin{pmatrix} \langle y_{0\,data} \rangle \\ \langle y_{n\,data} x_n \rangle \end{pmatrix} \tag{12.12}$$

Multiplying out Equation 11.12, the adjustable parameters a_0 and a_1 of the linear approximation are

$$a_o = \frac{\langle x_n^2 \rangle \langle y_{0,data} \rangle - \langle x_n \rangle \langle y_{n,data} x_n \rangle}{\langle x_n^2 \rangle - \langle x_n \rangle^2} \tag{12.13}$$

and

$$\begin{pmatrix} a_0 \\ a_1 \end{pmatrix} = \frac{1}{\langle x_n^2 \rangle - \langle x_n \rangle^2} \begin{pmatrix} \langle x_n^2 \rangle & -\langle x_n \rangle \\ -\langle x_n \rangle & 1 \end{pmatrix} \begin{pmatrix} \langle y_{0\,data} \rangle \\ \langle y_{n\,data} x_n \rangle \end{pmatrix} \tag{12.14}$$

For example, consider the set of data:

y_n	x_n
2.6481	20
4.3298	40
3.8263	60
4.6545	80
3.5495	100
4.2681	120
4.7791	140
4.4655	160
5.1356	180
6.7026	200
5.6331	220
6.9341	240
6.1637	260
6.7740	280
8.2683	300
7.2360	320

Here

$$x_n = 20n, \text{ for } n = 1, 2, 3, \ldots, 16$$

The line in Figure 12.2 has an intercept with the y_n axis, a_o is equal to 2.809235, and the slope of a_1 is equal to 0.01494152. The quantities used here have the following values:

$$\langle x_n \rangle = 170,\ \langle x_n^2 \rangle = 37{,}400,\ \langle y_n \rangle = 5.349294, \text{ and } \langle y_n x_n \rangle = 1{,}036.383.$$

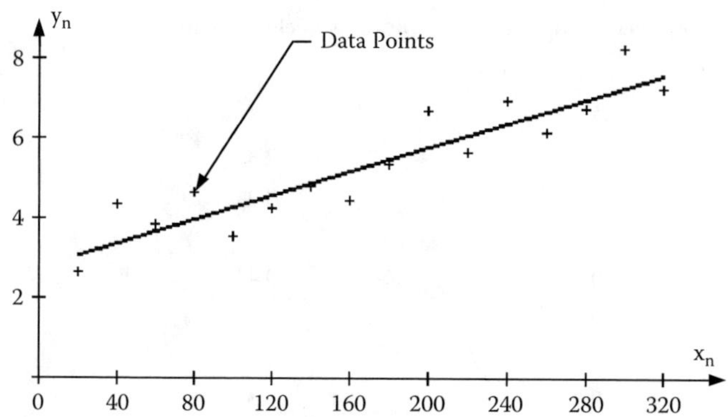

FIGURE 12.2 Data approximated by a straight line.

12.3 OTHER EXAMPLES

Another example is the interpretation of data obtained from measurements on a system in which a mathematical model exists. For example, the ionic conduction G of an organic compound depends on an activation energy E_A, as illustrated by Equation 12.15.

$$G_n = G_o \exp\left[-\frac{E_A}{kT_n}\right] \tag{12.15}$$

where k = 1.380662 ×10^{-23} J/°K is Boltzmann's constant, T is the absolute temperature, and G_o is a conductance constant. By taking the logarithm of Equation 12.15, it can be put into the following linear form:

$$\ln G_n = \left[-\frac{E_A}{k}\right]\frac{1}{T_n} + \ln G_o \tag{12.16}$$

The ionic conductance values G_n can be measured as a function of the values of the absolute temperature T_n. Here $y_n = \ln G_n$ and $x_n = \left[-1/T_n\right]$ can be calculated. These values can be used in a process similar to Equations 12.10 to 12.14 to calculate $\ln G_o = a_o$, and the activation energy divided by Boltzmann's constant $\left[-E_n/k\right] = a_1$.

PROBLEMS

PROBLEM 12.1

Given:

An electrical current is used for electroplating various components with chromium. Consider the following data:

Applied voltage (V)	Thickness of electrolytic film (μm)
–2.8	7.16
–2.4	7.61
–2	7.83
–1.6	8.94
–1.2	8.45
–0.8	9.44
–0.4	9.93
0	9.23
0.4	9.31
0.8	10.98
1.2	10.98
1.6	11.47
2	11.59
2.4	11.08
2.8	12.19
3.2	12.55
3.6	13
4	12.48
4.4	12.38
4.8	14.05
5.2	13.12
5.6	14.27
6	13.67
6.4	14.16
6.8	15.26
7.2	14.73

Problem:

a) Calculate the intercept and slope of a linear regression by linear approximation to this data.
b) Calculate the value χ of the least square differences.

Problem 12.2

Given:

In modern nanometer-scale digital circuits, the equation for the tunneling current I as a function of the applied voltage V across a potential barrier, such as the gate insulation of a field effect transistor, is used more often than Ohm's law.

$$I = I_o \sin h\left[AV^{\frac{1}{2}}\right]$$

Applied voltage (V)	Tunneling current (μA)
0.25	0.00011299866
0.268	0.00020314213
0.286	0.00031731618
0.304	0.00053697163
0.322	0.00086303084
0.34	0.00155783321
0.358	0.00246608472
0.376	0.00374336549
0.394	0.0060591567
0.412	0.00743222341
0.43	0.01289281787
0.448	0.01975244916
0.466	0.02505090338
0.484	0.04395447459
0.502	0.06027777387
0.52	0.09209867484
0.538	0.10990103892
0.556	0.1845988695
0.574	0.26222235254
0.592	0.35255983664
0.61	0.44312079083
0.628	0.61383609839
0.646	0.96256366475
0.664	1.329331388
0.682	1.6251536467
0.7	2.2398714557

Problem:

a) Using a least square difference technique, calculate I_o and A from the given data.

b) Calculate the value χ of the least square differences.

13 Chaotic Systems

13.1 FRACTALS

Complex patterns that resemble complicated terrain, clouds, turbulent liquid flows, etc., can be constructed by methods that are described in the following text. One such method is the use of *self-similar structures*. Self-similar structures are objects that when observed on any scale look the same. For example, consider the equilateral triangle of Figure 13.1a. Let us divide each side of the triangle into three parts and construct an equilateral triangle on the middle section of each side to obtain Figure 13.1b. The sides of the new equilateral triangles are ⅓ as long as the sides of the original triangle of Figure 13.1a. Therefore, each of the three sides of the resulting figure has four segments, each ⅓ as long as the original side. Thus, the boundary has grown in length by a factor of $4/3$. This process can be continued by dividing each resulting side again into three parts and erecting an equilateral triangle on the middle section to obtain Figure 13.1c. Each time, the boundary increases in length by a factor of $4/3$. The boundary of the object in Figure 13.1b is $(4/3)^2$ as long as the boundary of Figure 13.1a. After N iterations, one obtains an object having a boundary $(4/3)^N$ as long as the original object.

This iterative process can be continued, at least in theory, an infinite number of times. The boundary looks the same when viewed on any scale, as demonstrated in Figure 13.2. The boundary is going to become infinitely long. It appears that the boundary is acquiring some thickness. However, the boundary is only a line and thus has zero area. Therefore, there is an infinitely long object with zero area.

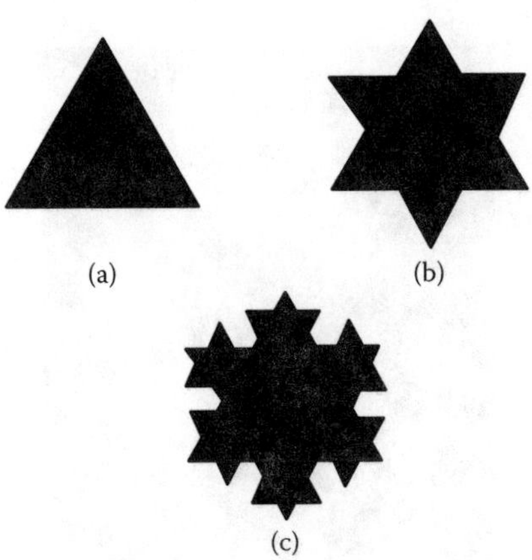

FIGURE 13.1 The Koch snowflake is an illustration of a self-similar object.

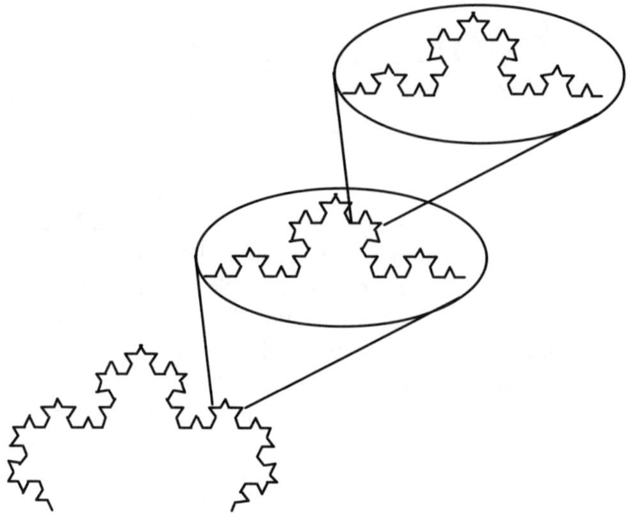

FIGURE 13.2 The boundary of the Koch snowflake of Figure 13.1 is self-similar. It looks the same no matter on what scale it is observed.

Most people are familiar with objects having integer dimensions. For example, lines are one dimensional, areas are two dimensional, and special relativity deals with four dimensions. However, the self-similar object curves have noninteger dimensions. They have fractional or fractal dimensions. Some aspect of fractals were discussed at the end of Chapter 10. Benoit Mandelbrot (1924–) developed the concept of fractals.

The fractal dimension d of a self-similar object is equal to the ratio of the logarithm of the multiplication factor m of the number of parts divided by the negative logarithm of the fractional change in linear dimension r of the part in going to the next smaller self-similar step.

Benoit Mandelbrot

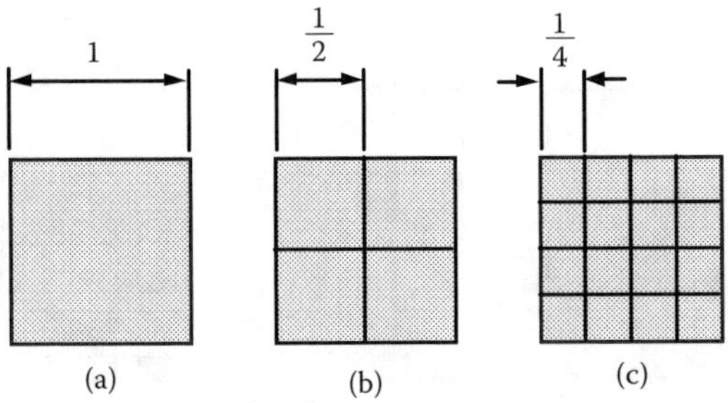

FIGURE 13.3 A square is subdivided successively into equal squares.

$$d = -\frac{\ln(m)}{\ln(r)} \tag{13.1}$$

The fractal dimension d can be used to characterize various types of terrain in aerial photographs. For example, aerial photographs of mountains, forests, and cities have different fractal dimensions. This concept is best illustrated by an example. Consider the square of Figure 13.3a. This square can be subdivided successively into equal squares, as shown in Figures 13.3b and 13.3c, etc.

In each step, the number of squares obtained increases by a factor 4 (m = 4). One obtains 1, 4, 16, etc., squares in each step. The squares in each step have linear dimensions one-half (r = ½) as large as the squares in the previous figure. The dimension of the square is

$$d = -\frac{\ln(4)}{\ln\left(\frac{1}{2}\right)}, \quad \text{or} \quad d = \frac{2\ln(2)}{\ln(2)}, \quad \text{or} \quad d = 2 \tag{13.2}$$

Indeed, as one would expect, the square is two dimensional, d = 2.

Another example is the triangle shown in the following figure. The triangle is successively subdivided, as was done with the square (see Figure 13.4). In each step, one generated four (m = 4) times as many triangles as in the previous step.

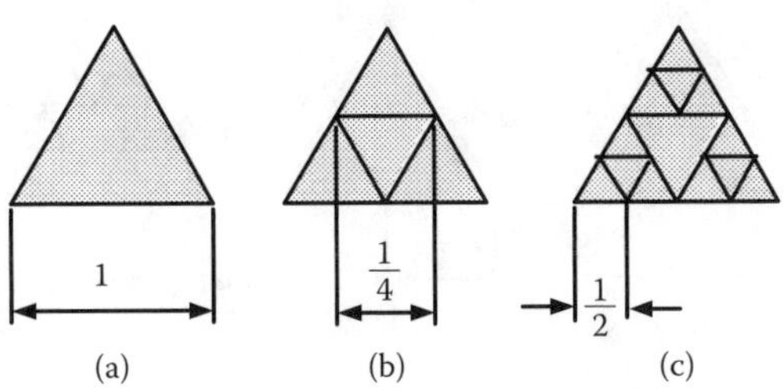

FIGURE 13.4 An equilateral triangle is subdivided successively into smaller equilateral triangles.

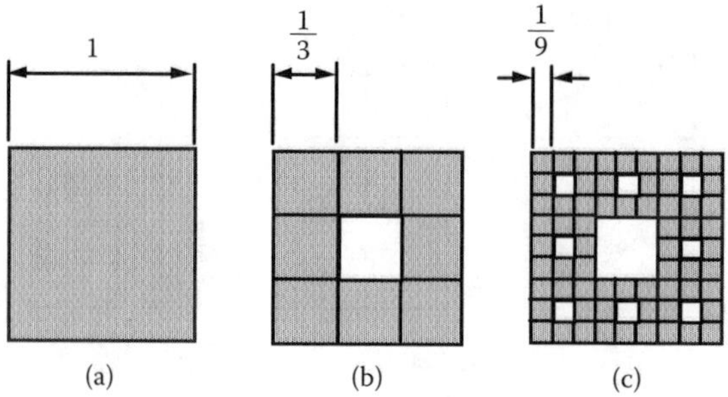

FIGURE 13.5 The shaded areas are solid ground, and the white areas are holes. One starts with a shaded square. In the next step, one obtains 8 shaded squares. In the third step, one obtains 64 shaded squares, etc.

The triangles in each step have linear dimensions one-half (r = ½) as large as the triangles in the previous figure. Because m and r are identical to the values of the square, the equilateral triangle, as expected, is also two dimensional.

Consider next the square with a square hole shown in Figure 13.5b. The next self-similar step is shown in Figure 12.5c, etc. In each step, one generates eight (m = 8) times as many squares as in the previous step. The squares in each step have linear dimensions one-third (r = ⅓) as large as the squares in the previous figure. The dimension d of the square with a square hole is

$$d = -\frac{\ln(8)}{\ln\left(\frac{1}{3}\right)}, \quad \text{or} \quad d = \frac{\ln(8)}{\ln(3)}, \quad \text{or} \quad d = 1.892789261 \tag{13.3}$$

Note that the square with a square hole has a dimension less than two. It is a fractal. A further example is the sparse triangles of Figure 13.6.

Consider the equilateral triangle with an equilateral triangle hole shown in Figure 13.6b. The next self-similar step is shown in Figure 13.5c. In each step, one generates three (m = 3) times as many shaded equilateral triangles as in the previous step. The equilateral triangles in each step have linear dimensions one-half (r = ½)

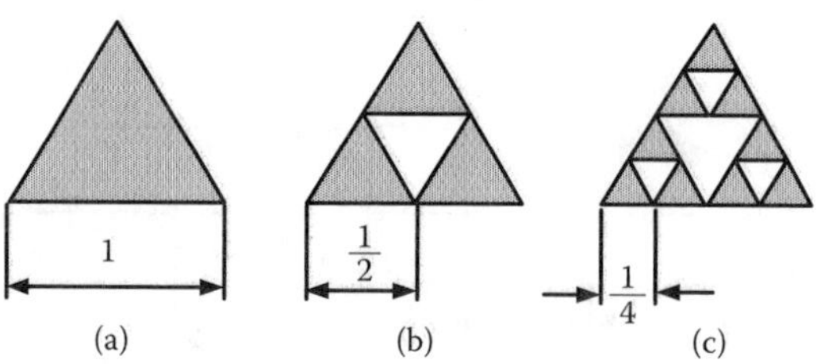

FIGURE 13.6 The shaded areas are solid ground, and the white areas are holes. One starts with a shaded triangle. In the next step, one obtains three shaded triangles. In the third step, one obtains 27 shaded triangles, etc.

as large as the equilateral triangles in the previous figure. The dimension d of the equilateral triangle with hole is

$$d = -\frac{\ln(3)}{\ln\left(\frac{1}{2}\right)}, \quad \text{or} \quad d = \frac{\ln(3)}{\ln(2)}, \quad \text{or} \quad d = 1.584962501 \tag{13.4}$$

The object in Figure 13.6 becomes sparser with each iteration. Therefore, it has a dimension less than two. It is a fractal.

Let us return to the Koch snowflake of Figure 13.1. The elements that change length in going to the next self-similar step are the sides of the triangle. The triangle sides increase by a factor of m = 4 in each self-similar step. The triangle sides of the succeeding self-similar step are r = ⅓ as long as the triangle sides of the previous self-similar figure. Therefore, the dimension of the boundary of the Koch snowflake is

$$d = -\frac{\ln(4)}{\ln\left(\frac{1}{3}\right)}, \quad \text{or} \quad d = \frac{\ln(4)}{\ln(3)}, \quad \text{or} \quad d = 1.261859507 \tag{13.5}$$

Indeed, the dimension d of the boundary of the Koch snowflake is larger than one.

For example, the line segments of the line of the Cantor set of Figure 10.18 of Chapter 10 doubles in each step (m = 2). The length of each line segment decreases in length after each iteration by a factor of r = ⅓. Therefore, its fractal dimensionality d is equal to 0.6309297536.

These fractal dimensions can be used to characterize classes of objects, such as aerial views of cities, forests, mountains, etc.

13.2 MANDELBROT SETS

Another method for constructing patterns that resemble complicated terrain, vortices in liquids, etc., is the use of the Mandelbrot set.[1,2] As will be shown, each of the images obtained with this method can be characterized by just two complex numbers. These two complex numbers are the macroscopic parameters of this stochastic system.

Consider the following complex difference equation:

$$z_{k+1} = z_k^2 + c \tag{13.6}$$

where z_k is a complex variable and c is a complex constant.

$$\text{a)}\, z_k = x_k + jy_k \quad \text{and} \quad \text{b)}\, c = a + jb \tag{13.7}$$

Here, $j = \sqrt{-1}$. The simplest way of interpreting a difference equation is that every value of the subscript k corresponds to an increment of time. One could say that k is the current time and k + 1 is the next instant of time, etc. Thus, one interpretation of Equation 13.6 is that the complex variable z_{k+1} at the next instant of time depends on the square of the current value of the complex variable z_k and a complex constant c. Difference equations will be discussed in more detail in the next section.

Equation 13.6 is solved iteratively; that is, one assumes a value of both the real and imaginary parts of the complex variable z_1 at some initial time k = 1. These values can be used to calculate the value of the complex variable z_2 at the next instant of time k = 2, etc.

There is a theorem that states that if the magnitude $|z_k|$ of the complex variable z_k starting from an initial value of $z_1 = 0 + j0$ after a sufficient number of iterations of Equation 13.6 reaches the quantity 2, the magnitude $|z_k|$ will tend to larger and larger numbers in the following iterations.

The Mandelbrot set consists of all solutions of the difference Equation 13.6 that remain finite for all iterations. Because any solutions that have a magnitude $|z_k|$ that reaches the value 2 will diverge in subsequent iterations, the Mandelbrot set consists of all solutions of the difference Equation 13.6 that have magnitudes that never reach the value 2.

The possibility that the magnitude $|z_k|$ will ever reach 2 depends on the constant c. For example, as required by the theorem, let $z_1 = 0 + j0$, and let $c = 0.1 + j0.2$, the first few iterations are

$$z_2 = 0.1 + j0.2 \qquad |z_2| = 0.2236$$

$$z_3 = 0.07 + j0.24 \qquad |z_3| = 0.25$$

$$z_4 = 0.0473 + j0.2336 \qquad |z_4| = 0.23834$$

etc.

Actually, the difference equation itself is only used in an indirect way in constructing patterns that resemble complicated terrain, vortices in liquids, etc. The procedure is as follows:[1]

1. One constructs a square matrix of picture elements. A square matrix has N rows and N columns. The rows are labeled by m and the columns by n. Iterations of the difference equation at every point of the matrix are going to be performed. This procedure will be described in the following text.
2. Because iterations of the difference equation are going to be performed at every point of the matrix, a value of the constant c_{nm} must be chosen at every matrix point nm. The value of the constant is chosen as follows:
   a. A value of the complex constant c_{11} at the lower left corner of the matrix is chosen.
   b. Next, a length L for the side of the square picture represented by the matrix is chosen. The length L is, of course, real.
   c. Then, the length d per picture element is calculated, d = L/N.
   d. The value of the constant at the nm matrix element is calculated as follows:

$$c_{nm} = c_{11} + nd + jmd$$

3. The procedure continues as follows:
   a. Using the values of the complex constants c_{nm} at each metric point and starting from $z_1 = 0 + j0$, one successively calculates the z_k's at each matrix point mn or pixel. The magnitude $|z_k|$ can be calculated for each value of z_k at each pixel.
   b. This calculation is continued at each matrix point until either the magnitude $|z_k|$ reaches 2 or the difference between two successive values of $|z_k|$ reaches some sufficiently small value, that is, $|z_{k+1}| - |z_k|$ becomes smaller than some small value such as 10^{-8}, or k reaches a large value such as 1000. One records at each matrix point mn, or pixel, the value of $k_{nm} = k$ when the magnitude $|z_k|$ reaches 2, or when the difference between two successive values of $|z_k|$ reaches some sufficiently small value, or when k reaches some large value.
4. The difference between the largest and smallest value of k_{nm} is divided into a number of regions. For example, suppose the largest value of k of the whole image is 1000, and the smallest value of k is 10. Then, the region between 10 and 1000 is divided into, say, five sections, say, 10–200, 209–406, 407–604, 605–802, and 803–1000. A color can be assigned to each numeric region. Thus, the color of each pixel or matrix point will be determined by the maximum value of k_{nm} the iteration reaches at the particular matrix point. The resulting picture can resemble mountains, clouds, complicated fluid flow patterns, etc.

Here are some suggested numerical values for c_{11} and d:

$$c_{11} = 0.26 + j0, \qquad d = 0.01$$

$$c_{11} = -0.76 + j0.01, \qquad d = 0.02$$

$$c_{11} = -1.26 + j0.01, \qquad d = 0.02$$

Of course, many other values of c_{11} and d will work. A Mandelbrot set image consisting of 50 × 50 pixels generated with $c_{11} = 0.16 + i0.2025$ and $L/N = 0.01$ is shown in Figure 13.7. The iteration in this case was continued until either $|z_k| > 2$, or $|z_{k+1}| - |z_k| < 10^{-4}$, or $k > 100$. The different shaded regions correspond to $0 < k \le 20$, $20 < k \le 40$, $40 < k \le 60$, $60 < k \le 80$, and $80 < k \le 100$. The last region, $80 < k \le 100$, is left white. The quantities c_{11} and $d = L/N$ are the macroscopic stochastic parameters of the Mandelbrot set; that is, the whole image obtained with this iterative process can be characterized by just two parameters, c_{11} and d.

13.3 DIFFERENCE EQUATIONS

Another method for constructing complex patterns that resemble complicated terrain, clouds, turbulent liquid flows, etc., is by the following iterative method. This iterative method is the consecutive solution of *difference equations*. A difference equation is an equation of the form:

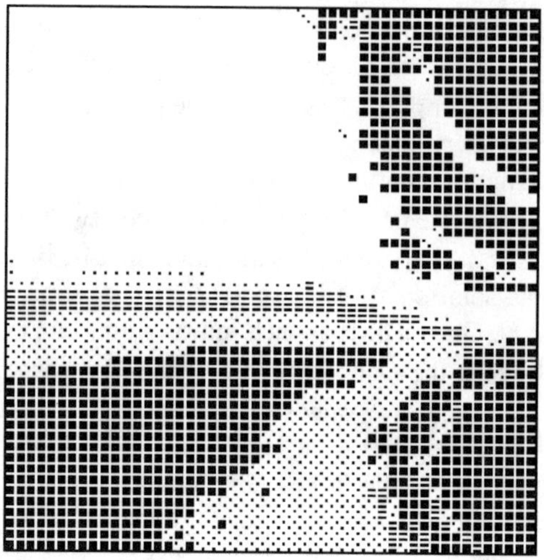

FIGURE 13.7 A Mandelbrot set image consisting of 50 × 50 pixels generated with $c_{11} = 0.16 + j0.2025$ and L/N = 0.01. The iteration was continued until either $|z_k| > 2$, or $|z_{k+1}| - |z_k| < 10^{-4}$, or k > 100. The different shaded regions correspond to $0 < k \le 20$, $20 < k \le 40$, $40 < k \le 60$, $60 < k \le 80$, and $80 < k \le 100$. The regions start at the bottom on the left side of the image.

$$x_{n+1} = Ax_{n-1} + Bx_n + Cx_n^2 + \ldots \tag{13.8}$$

The simplest way of interpreting a difference equation is that every value of the subscript n corresponds to an increment of time. One could say that n is the current time, n – 1 is the previous instant of time, and n + 1 is the next instant of time. Thus, one interpretation of Equation 13.8 is that the position x_{n+1} at the next instant of time depends on both the current position x_n and the position x_{n-1} at the previous instant of time. A difference equation that depends only on two instances of time is a first-order difference equation. A difference equation that depends on three instances of time, as Equation 13.8 does, is a second-order difference equation. The indices n could, of course, denote different positions or just increments of some abstract quantity. Indeed, in this section, the indices just label successive enumerations of some quantity.

Babbage used a variation of this method in his famous *difference engine* to calculate the values of various polynomials.

The consecutive solution of the difference equation proceeds as follows: One assumes *initial values* of the variable x_n. For a second-order difference equation, it is necessary to know the initial values x_o and x_1. These values can be used in Equation 13.8 to calculate x_2. Next, the quantities x_1 and x_2 are used to calculate x_3, etc. By consecutively solving the same difference equation starting from different initial values, exceedingly complicated patterns can be obtained.

Equation 13.8 is also a one-dimensional difference equation because it only contains one variable x_n. However, it is possible to construct multidimensional difference equations.

$$x_{n+1} = Ax_{n-1} + f(x_n, y_n) \tag{13.9}$$

and

$$y_{n+1} = By_{n-1} + g(y_n, x_n) \tag{13.10}$$

where $f(x_n,y_n)$ and $g(y_n,x_n)$ are, in general, nonlinear functions of the variables x_n and y_n. Equations 13.9 and 13.10 are a pair of two-dimensional, in general, nonlinear second-order difference equations. Indeed, one will deal with this type of equation here. The difference equation used in the Mandelbrot set, Equation 13.6, is a two-dimensional difference equation for the two variables x_k and y_k that constitute the complex variable z_k.

There are many ways of constructing difference equations. One way of constructing these equations is illustrated in the following section.

13.4 THE HÉNON DIFFERENCE EQUATION

Michel Hénon[3] constructed a set of difference equations that have attracted a lot of interest. These equations are

$$y_{t+1} = x_t \tag{13.11}$$

and

$$x_{t+1} = a - x_t^2 - by_t \tag{13.12}$$

where typical values for the constants a and b are a = −0.45 and b = 0.8.

A related problem will be investigated. Consider a particle moving subject to the potential f(x) shown in Figure 13.8:

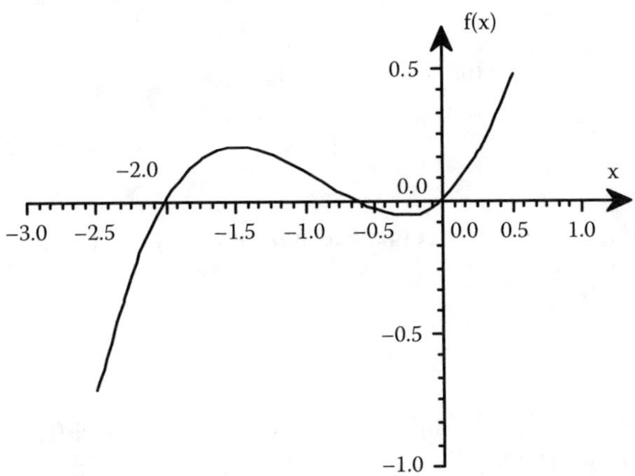

FIGURE 13.8 The potential f(x). The maximum is at x = −1.5. The minimum is at x = −0.3.

$$\phi(x) = ax - \frac{1}{3}x^3 - \frac{1+b}{2}x^2 + C \tag{13.13}$$

One can think of the one-dimensional potential as a hilly terrain on which the particle is sliding. The potential has a maximum at $x = -1.5$ and a minimum at $x = -0.3$. The particle experiences a friction force that is proportional to the particle velocity.

The equation of motion of a particle is

$$\frac{d^2x}{dt^2} + (1-b)\frac{dx}{dt} - \frac{d\phi}{dx} = 0 \tag{13.14}$$

The mass of the particle was taken as unity ($m = 1$). The first term is equal to the acceleration times the unit mass. The second term is equal to the friction force which is proportional to the particle velocity, and the last term is the force exerted on the particle by the potential.

Substituting the potential of Equation 13.13 into Equation 13.14:

$$\frac{d^2x}{dt^2} + (1-b)\frac{dx}{dt} - a + x^2 + (1+b)x = 0 \tag{13.15}$$

Owing to the complicated nature of the potential $\phi(x)$ of Equation 13.13, it is difficult to obtain a closed-form solution of Equation 13.15. One can attempt a numerical solution. To be able to solve this equation by an iterative numerical method, the second-time derivative can be approximated by the second difference and the first time derivative by the first difference.

$$\text{a)} \quad \frac{d^2x}{dt^2} \rightarrow \frac{x_{t+1} - 2x_t + x_{t-1}}{\Delta t^2} \quad \text{and} \quad \text{b)} \quad \frac{dx}{dt} \rightarrow \frac{x_t - x_{t-1}}{\Delta t} \tag{13.16}$$

Substituting Equation 13.6 into Equation 13.15:

$$(x_{t+1} - 2x_t + x_{t-1}) + (1-b)(x_t - x_{t-1}) - a + x_t^2 + (1+b)x_t = 0 \tag{13.17}$$

where the time increment Δt was taken as unity ($\Delta t = 1$). Collecting terms and solving for x_{t+1}:

$$x_{t+1} = a - x_t^2 - bx_{t-1} \tag{13.18}$$

Equation 13.18 can be put in the form of the Hénon difference equations, Equations 13.11 and 13.12, by replacing x_{t-1} by another variable y_t.

$$y_t \equiv x_{t-1} \tag{13.19}$$

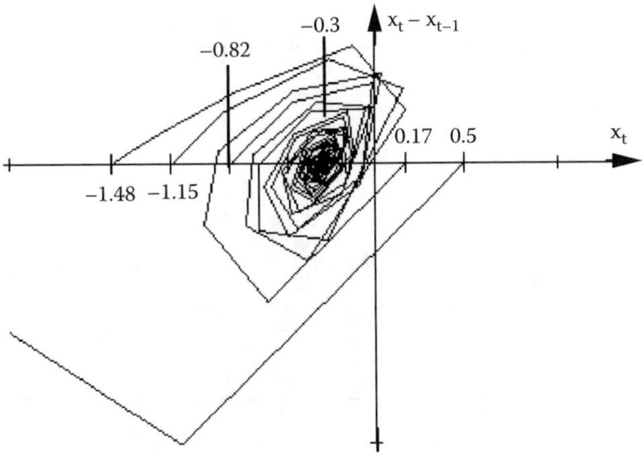

FIGURE 13.9 A phase space plot of the solutions of Equation 12.8 for different zero-velocity initial conditions are shown in the preceding figure. The particle starting at rest from most of the positions shown here comes to rest at the energy minimum at $x_t = -0.3$. However, a particle starting at $x_t = 0.5$ acquires a sufficiently large velocity to overcome the energy maximum at $x_t = -1.5$ and slide off to positions with very large negative values.

By substituting Equation 13.19 for x_{t-1} into Equation 13.18, Equations 13.18 and 13.19 can be put into the form of Equations 13.11 and 13.12.

Equation 13.18 is a nonlinear second-order difference equation. It can be solved by an iterative method as described in Section 13.3. Because Equation 13.18 is a second-order difference equation, two initial conditions are required. For example, one can specify the initial position x_o and the initial velocity $x_1 - x_o$. The solution for five different zero-velocity initial conditions are shown in Figure 13.9. Because of friction damping, the particle eventually comes to rest at the bottom of the potential at $x_t = -3$ for initial positions inside the potential valley of Figure 13.8. However, when the initial position x_t of the particle is more negative than −1.5, the particle will slide off to large negative values of x_t (see Figure 13.8). Likewise, when the particle starts too high up on the potential, the particle might be going fast enough to slide right over the potential maximum at $x_t = -1.5$. Indeed, for the initial condition at $x_t = 0.5$ and zero initial velocity in Figure 13.9, the particle eventually acquires sufficient speed to slide right over the potential maximum at $x_t = -1.5$. The behavior of the particle, of course, also depends on the magnitude and direction of the initial velocity. For example, a particle starting from a position with x_t more negative than −1.5, where it would slide off to positions with very large negative values if it started from rest, might come to rest at the energy minimum at $x_t = -0.3$ if it started with a sufficiently large positive velocity that it could overcome the potential maximum at $x_t = -15$.

A plot in phase space of the initial positions x_t and velocities $x_t - x_{t-1}$ at which the particle will spiral to the energy minimum at $x_t = -0.3$ is shown in Figure 13.10. The "stable region," the region of position and velocity initial conditions where the solution of Equation 13.18 converges to $x_t = -0.3$, is shown shaded. The phase space plot exhibits a general −45° slope because particles starting with initial positions

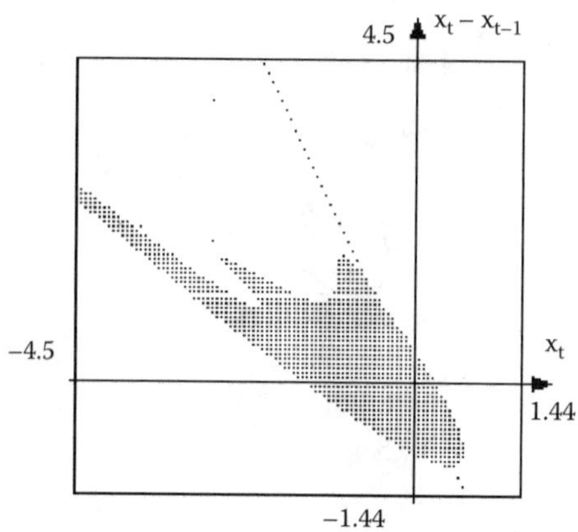

FIGURE 13.10 A phase space plot of the initial conditions. The region of initial conditions for which the solutions of the difference equation converge are shown shaded.

more negative than the potential maximum at $x_t = -1.5$ with, however, sufficiently large positive velocities, can overcome the potential maximum and come to rest at the potential minimum at $x_t = -0.3$.

13.5 SINGLE-PARTICLE SINGLE-WELL POTENTIAL

A difference equation can be constructed by analogy with the motion of real physical point particles. Consider, first, a single particle with mass M moving subject to a potential energy V(x,y). The particle is also subject to a damping force $\mathbf{F}_D$. The damping force $\mathbf{F}_D$ is equal to a damping constant D times the particle velocity vector **v**. Bold letters such as **v** denote vectors. We first construct a *Lagrangian* L_n as the difference between the kinetic and potential energies:

$$L_n = \text{Kinetic Energy} - \text{Potential Energy} \tag{13.20}$$

The *kinetic energy*, though quantized, has a form similar to the classical mechanical kinetic energy of a particle.

$$\text{Kinetic Energy} = \frac{1}{2}M\frac{\left(x_n - x_{n-1}\right)^2}{a^2} + \frac{1}{2}M\frac{\left(y_n - y_{n-1}\right)^2}{a^2} \tag{13.21}$$

where a is a time step, M is a mass, and $x_n - x_{n-1}/a$, the backward difference, is a discretized velocity component of the particle. However, because one is not interested in describing real physical systems, the time step a is taken to be equal to one.

The *potential energy*, though quantized, has also a form similar to the classical mechanical kinetic energy of a particle.

Next, one forms a sum I of Lagrangians:

$$I = \sum_{n=-\infty}^{n=\infty} L_n \tag{13.22}$$

The difference equation can be arrived at by differentiating the sum of Lagrangians with respect to each of the variables at the current time only, such as x_n. Setting the resulting equation equal to a loss force term of the form DM $x_n - x_{n-1}/a$:

$$\frac{\partial I}{\partial x_n} = DM\frac{(x_n - x_{n-1})}{a} \tag{13.23}$$

This term is called a loss term because it resembles a velocity-dependent friction force term a real physical particle might experience. Here, D is a loss constant. The terms in the resulting equation are terms analogous to forces our particle would experience.

A *simple example* is that of a particle moving in a harmonic potential of the form:

$$\text{Potential Energy} = \frac{1}{2}\omega_o^2 M\left(x_n^2 + y_n^2\right) \tag{13.24}$$

The potential energy is shown in Figure 13.11. Substituting Equations 13.21 and 13.24 into Equation 13.20 to form the Lagrangian:

$$L_n = \frac{1}{2}M\frac{(x_n - x_{n-1})^2}{a^2} + \frac{1}{2}M\frac{(y_n - y_{n-1})^2}{a^2} - \frac{1}{2}\omega_o^2 M\left(x_n^2 + y_n^2\right) \tag{13.25}$$

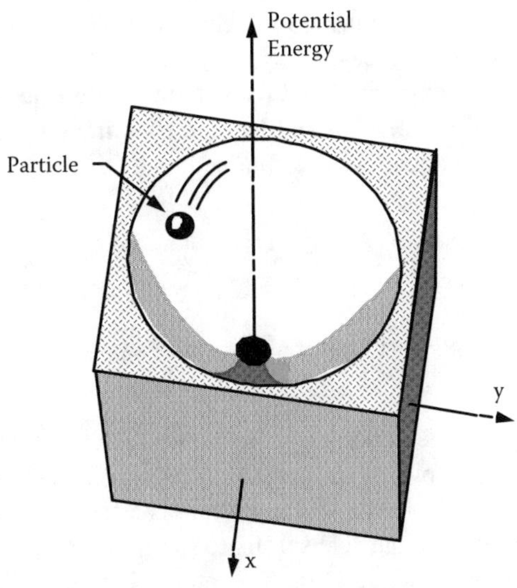

FIGURE 13.11 Simple potential energy proportional to $1/2\,\omega^2 M\left(x_n^2 + y_n^2\right)$.

Next, the sum I of Lagrangians is formed:

$$I = \ldots + \frac{1}{2}M\frac{\left(x_n - x_{n-1}\right)^2}{a^2} + \frac{1}{2}M\frac{\left(y_n - y_{n-1}\right)^2}{a^2} - \frac{1}{2}\omega_o^2 M\left(x_n^2 + y_n^2\right) + \frac{1}{2}M\frac{\left(x_{n+1} - x_n\right)^2}{a^2} + \frac{1}{2}M\frac{\left(y_{n+1} - y_n\right)^2}{a^2} - \frac{1}{2}\omega_o^2 M\left(x_{n+1}^2 + y_{n+1}^2\right) + .. \quad (13.26)$$

Only the two Lagrangians shown in the preceding sum I (Equation 13.26) contain x_n. Substituting Equation 13.26 into Equation 13.23, the following difference equations of motion are obtained:

$$\frac{M}{a^2}\left(-x_{n+1} + 2x_n - x_{n-1}\right) - \omega_o^2 M x_n = D\frac{M}{a}\left(x_n - x_{n-1}\right) \quad (13.27)$$

and

$$\frac{M}{a^2}\left(-y_{n+1} + 2y_n - y_{n-1}\right) - \omega_o^2 M y_n = D\frac{M}{a}\left(y_n - y_{n-1}\right) \quad (13.28)$$

These equations are called equations of motion because their solutions describe the path of the particle. The first terms on the left-hand side of Equations 13.27 and 12.38 are equal to a mass M times a second time difference $x_{n+1} - 2x_n - x_{n-1}/a^2$. This is analogous to a mass times the acceleration of a particle. The second time difference is analogous to the second time derivative. The second terms on the left-hand side of Equations 13.18 and 13.19 are springlike restoring forces. Here, $\omega_o^2 M$ is the spring constant. The terms on the right-hand side of Equations 13.27 and 13.28, as discussed before, are friction forces. These equations can be interpreted as describing the motion of a particle of mass M attached to a spring with a spring constant $\omega_o^2 M$ sliding on a friction surface.

Difference equations developed by this method are described here. One collects terms in Equations 13.27 and 13.28 and lets the time increment a be equal to one. The following difference equations are obtained:

$$x_{n+1} = \left(2 - \omega_o^2 - D\right)x_n - \left(1 - D\right)x_{n-1} \quad (13.29)$$

and

$$y_{n+1} = \left(2 - \omega_o^2 - D\right)y_n - \left(1 - D\right)y_{n-1} \quad (13.30)$$

A set of initial values for the variables x_n and y_n are selected; that is, one selects values for x_0, x_1, y_0, and y_1. A plot of the iterative solution of Equations 13.29 and 13.30 is shown in Figure 13.12. Note that because there is a friction-like damping term in the difference equations, the particle spirals to the bottom (see Figure 13.12) of the potential energy bowl of Figure 13.11 or the center of Figure 13.12. One can interpret

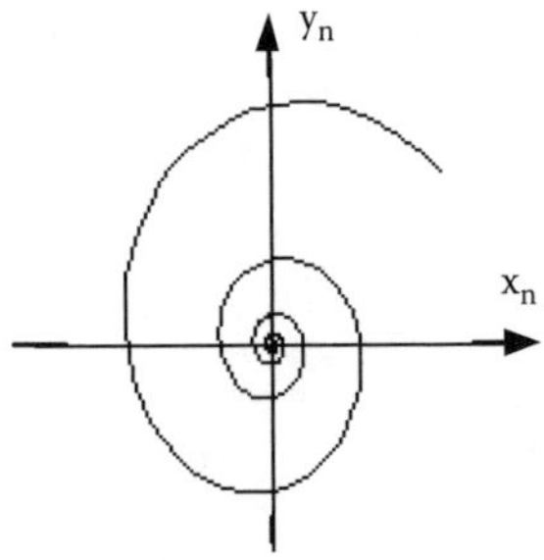

FIGURE 13.12 Particle subject to the potential energy of Figure 13.8 circling into the stable point at $x_n = 0$, $y_n = 0$.

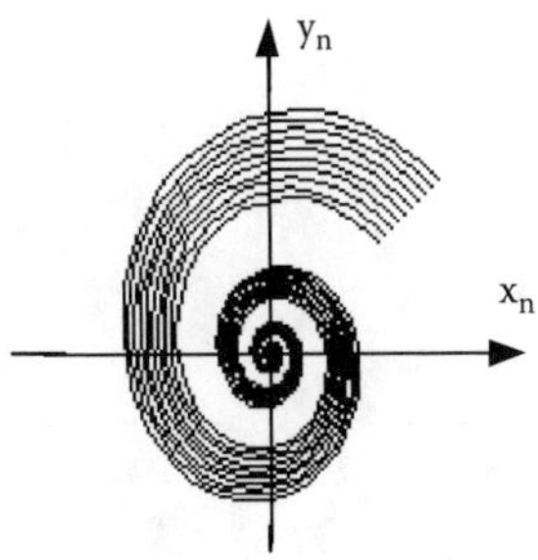

FIGURE 13.13 A particle subject to the potential energy of Figure 13.8 starting from 10 different initial conditions and circling into the stable point at $x_n = 0$, $y_n = 0$.

this as the particle losing energy by friction and spiraling down to a stable point, a so-called *strange attractor.*

Other initial values of the variables can be selected, and the curve can be recalculated. Plots of curves for ten different initial value sets are shown in Figure 13.13. Note that in every case, the curve spirals into the same strange attractor point located at $x_n = 0$ and $y_n = 0$.

REFERENCES

1. *Computer Recreations* by Dewdney, A.K., Scientific American, Vol. 253, No. 2, pp.16–24. August 1985.
2. *Chaos: Making a New Science*, James Gleick, Penguin Books, 1988.
3. *Basins of Attraction* by Helena E. Nusse and James A. Yorke, *Science,* Vol. 271, No. 5254, 8 March 1996, pp. 1376–1380.

Index

D

F

G

H

M

N

Q

R

T